Solar Power System Planning & Design

Solar Power System Planning & Design: Resource   Assessment, Site Evaluation, System Design, Production Forecasting and Feasibility Studies

Editor

Yosoon Choi

MDPI • Basel • Beijing • Wuhan • Barcelona • Belgrade • Manchester • Tokyo • Cluj • Tianjin

Editor
Yosoon Choi
Pukyong National University
Korea

Editorial Office
MDPI
St. Alban-Anlage 66
4052 Basel, Switzerland

This is a reprint of articles from the Special Issue published online in the open access journal *Applied Sciences* (ISSN 2076-3417) (available at: https://www.mdpi.com/journal/applsci/special_issues/Solar_Power_System).

For citation purposes, cite each article independently as indicated on the article page online and as indicated below:

LastName, A.A.; LastName, B.B.; LastName, C.C. Article Title. *Journal Name* **Year**, *Article Number*, Page Range.

ISBN 978-3-03943-637-8 (Hbk)
ISBN 978-3-03943-638-5 (PDF)

Cover image courtesy of Yosoon Choi.

Contents

About the Editor

Yosoon Choi received his BS degree at the School of Civil, Urban and Geosystem Engineering, Seoul National University, Korea, in 2004. He completed his PhD degree at the Department of Energy Systems Engineering, Seoul National University in 2009. He then served as a Postdoctoral Fellow at the Department of Energy and Mineral Engineering, Pennsylvania State University, USA, in 2010. He has served as Professor at the Department of Energy Resources Engineering, Pukyong National University, Korea, since his appointment in 2011. His research covers the areas of Smart Mining, Renewables in Mining, AICBM (AI, IoT, Cloud, Big Data, Mobile) Convergence, Mining Engineering, Geographic Information Systems (GIS), 3D Geomodeling, Operations Research, and Solar Energy Engineering.

Preface to "Solar Power System Planning & Design: Resource Assessment, Site Evaluation, System Design, Production Forecasting and Feasibility Studies"

With growing concerns about greenhouse gas emissions, the security of conventional energy supplies, and the environmental safety of conventional energy production techniques, renewable energy systems are becoming increasingly important and are receiving much political attention. Photovoltaic (PV) and concentrated solar power (CSP) systems for the conversion of solar energy into electricity are, in particular, technologically robust, scalable, and geographically dispersed, and they possess enormous potential as sustainable energy sources. Despite the advances in PV and CSP systems, inappropriate planning and design could impede the extensive penetration of solar energy. Systematic planning and design, considering various factors and constraints, are necessary for the successful deployment of PV and CSP systems.

This book on solar power system planning and design includes 14 publications from esteemed research groups worldwide. The research and review papers in this Special Issue fall within the following broad categories: resource assessments, site evaluations, system design, performance assessments, and feasibility studies.

Yosoon Choi
Editor

 *applied sciences*

Editorial

Solar Power System Planning and Design

Yosoon Choi

Department of Energy Resources Engineering, Pukyong National University, Busan 48513, Korea;
energy@pknu.ac.kr; Tel.: +82-33-570-6313

Received: 20 December 2019; Accepted: 2 January 2020; Published: 3 January 2020

1. Introduction

With growing concerns about greenhouse gas emissions, the security of conventional energy supplies, and the environmental safety of conventional energy production techniques, renewable energy systems are becoming increasingly important and are receiving much political attention [1]. Photovoltaic (PV) and concentrated solar power (CSP) systems for the conversion of solar energy into electricity are—in particular—technologically robust, scalable, and geographically dispersed, and they possess enormous potential as sustainable energy sources [2]. Despite the advances in PV and CSP systems, inappropriate planning and design could impede the extensive penetration of solar energy. Systematic planning and design considering various factors and constraints are necessary to deploy PV and CSP systems successfully [3].

This Special Issue on solar power system planning and design includes 14 publications from esteemed research groups worldwide. The research and review papers in this Special Issue fit in the following broad categories: resource assessment, site evaluation, system design, performance assessment, and feasibility study.

2. Resource Assessment

Solar radiation is the most important parameter to be considered when installing PV or CSP systems. Therefore, it is necessary to assess solar resources by analyzing and forecasting the spatiotemporal distribution of solar irradiance. Wang et al. [4] proposed an improved deep learning model based on discrete wavelet transformation (DWT), convolutional neural network (CNN), and long short-term memory (LSTM) for day-ahead solar irradiance forecasting. In the case study—which used two datasets from the Elizabeth City State University and Desert Rock Station in the United States—the performance of the proposed model, named DWT–CNN–LSTM, was compared with six other solar irradiance forecasting models. The results showed that DWT–CNN–LSTM is highly superior for solar irradiance forecasting, especially under extreme weather conditions.

Analyzing sky dynamics by processing a set of images of the sky dome is a new trend for solar resource assessment [5,6]. Valentín et al. [5] proposed a methodology based on implementing several image processing techniques to achieve a robust and automatic detection of the sun's position from a set of images acquired by a low-cost artificial vision system. The methodology could detect the position of the sun not only on clear but also on cloudy days, even if the sun was completely occluded. Richardson et al. [6] validated the all-sky imager technology using data obtained from three geographically diverse locations: in Golden, Colorado on the rooftop of the Energy Systems Integration Facility (ESIF) building at the National Renewable Energy Laboratory (NREL); in San Antonio, Texas at the CPS Energy microgrid facility of the Joint Base San Antonio (JBSA) and the Engineering Building of University of Texas at San Antonio (UTSA); and in the Canary Islands, Spain at Tenerife and Caleta de Sebo. The operations at the three locations provided several improvements to the UTSA SkyImager regarding weatherproofing techniques, environmental sensors, maintenance schedules, and optimal deployment locations.

Choi et al. [3] reviewed geographic information system (GIS)-based studies on solar resource assessments, especially for solar radiation mapping. GIS is beneficial for spatial and temporal analyses of solar resources while implementing location-specific technologies. The solar radiation analysis can be performed for individual points such as stations and for large areas represented by many pixels. The GIS analysis could also be conducted for specific administrative districts.

3. Site Evaluation

It is necessary to increase the effectiveness of installing solar power plants by prioritizing and selecting suitable locations to maximize electricity generation and minimize the damage that may occur. The results of such site evaluation can help solar utility companies, energy companies, and policymakers select potential sites for the construction of solar power plants [3].

Chen et al. [7] proposed an evaluation model of demand-side energy resources (DSER) for urban power grids based on geographic information. The commonality and individuality indices for five kinds of DSER, revolving wind power generation, photovoltaic power generation, electric vehicles, energy storage, and flexible load, were selected based on geographic information. Then, the weight of each sub-index of the commonality and individuality indices was determined by the analytic hierarchy process (AHP) and entropy weight method. Finally, the weighted overlay was generated according to the weights and quantized values of each index, and a comprehensive score was obtained from the commonality indices. The results depicted that the evaluation model is beneficial for the planning of the city and the power grid.

The installation of PV panels on the ground can cause some problems, especially in countries where there is not enough space for installation. As an alternative, floating PV, with advantages in efficiency and for the environment, attracted attention. Kim et al. [8] analyzed the water-level data from 3401 reservoirs in South Korea and selected suitable reservoirs for floating PV systems, with an average reservoir water depth greater than 5 m and minimum water depth greater than 1 m. The results were utilized to estimate priorities and potentiality prior to the actual floating PV installation and detailed analysis.

GIS is useful for site evaluations when installing solar power plants for PV or CSP on the regional scale. Choi et al. [3] reviewed the GIS-based methods for determining suitable locations for solar power plants. In most site evaluation studies, solar radiation is the primary consideration. However, it is also necessary to consider economic, environmental, technical, social, and risk factors. These factors can be used to exclude unsuitable regions through a Boolean overlay and can be employed in various multiple-criteria decision analysis (MCDA) methods to estimate suitability indices [3].

4. System Design

Before installing a solar power system, it is crucial to ensure that the system is not over- or undersized. Therefore, the designer should investigate the viability of the system carefully to operate in optimum conditions regarding produced unit costs and power reliability. Alsadi and Khatib [9] reviewed the sizing procedures of grid-connected and standalone PV systems, including system component modeling, available optimization software, optimization criteria, optimization methods, and sizing constraints. The study revealed that PV modeling and battery modeling are essential in system sizing optimization to predict the systems' performances.

The performance of a PV system depends significantly on the tilt angle of the PV panels. Chou et al. [10] conducted a wind-load analysis using wind tunnel experiments and numerical simulations for a stand-alone panel at high tilt angles. The effects of wind direction were also investigated. The findings of this study will be useful for the detailed structural design of offshore PV panels.

5. Performance Assessment

Chamkha and Selimefendigil [11] performed a numerical analysis of a photovoltaic–thermal (PV/T) unit with SiO_2–water nanofluid. The coupled heat conduction equations for the layers and convective heat transfer equations for the channel of the module were solved using the finite volume method. The effects of various particle shapes, solid volume fractions, water inlet temperature, solar irradiation, and wind speed on the thermal and PV efficiency of the unit were analyzed. The performance characteristics of the solar PV–thermal unit determined by the radial basis function artificial neural network were found to be in excellent agreement with the results obtained from computational fluid dynamics modeling.

Gulkowski et al. [12] carried out a comparative analysis of energy production by a grid-connected experimental PV system composed of various technology modules, which operates in the temperate-climate meteorological conditions of eastern Poland, for the year 2015. The study revealed that the copper indium gallium diselenide (CIGS) technology demonstrated the highest energy production and performance ratio, as well as the lowest observed temperature-related losses. These results can be useful for the prediction of electric energy production by different PV technologies at high latitudes under temperate climate conditions.

Rouibah et al. [13] determined the performance and viability of direct normal irradiation for three solar tower power plants to be installed in the Algerian highlands and the Sahara (Béchar, El Oued, and Djelfa regions). Each plant, with the annual production specification of 20 MW, is equipped with a supply of molten salt, an external receiver, and a field of heliostats. Results showed that there is a strong and direct relationship between the solar multiple, power generation, and storage capacity hours.

Machine learning methods were successfully applied in PV output prediction models. Xie et al. [14] proposed a hybrid short-term forecasting method based on the variational mode decomposition (VMD) technique, the deep belief network (DBN), and the auto-regressive moving average (ARMA) model to improve forecasting accuracy. The results showed that the hybrid forecasting method offers better accuracy and stability than the single prediction methods. Additionally, Mei et al. [15] developed an ultrashort-term forecasting model based on the phase space reconstruction and deep neural network (DNN) by considering the characteristics of the net load. The performance of this model was verified using real data, with superior accuracy in forecasting the net load under high PV penetration rates and different weather conditions.

Solar potential assessment using GIS can be placed in three different categories: (1) physical potential, which is the total amount of solar energy reaching a target surface or the total solar radiation on a surface or rooftop; (2) geographic potential, which is the spatial availability of a surface or building rooftop where solar energy can be obtained; and (3) technical potential, which represents the total amount of electricity considering the technical characteristics of the PV system. Choi et al. [3] reviewed 39 published articles on GIS-based solar potential assessment.

6. Feasibility Study

Within the agriculture sector, current solutions for groundwater pumping are primarily based on diesel technologies, with remarkable fossil-fuel dependence and emissions that must be reduced to fulfill both energy and environmental requirements. The integration of PV power plants into groundwater pumping installations was recently considered as a suitable alternative. Rubio-Aliaga et al. [16] presented a feasibility study with a multidimensional analysis of PV solar power systems connected to the grid as a groundwater pumping solution, including net-metering conditions and benefit estimations, in Castilla-La Mancha (Spain). Different surplus energy sale scenarios were analyzed based on crops typical in this location, the corresponding annual water requirements, and common grouping areas. The study found that PV power plants connected to the grid for the use of surplus energy could generate non-negligible global revenues: 10–18 million €/year with legislation promoting net-metering and 5–10 million €/year under the current legislation framework in Spain.

Appl. Sci. **2020**, *10*, 367

Funding: This work was supported by the Basic Science Research Program through the National Research Foundation of Korea (NRF) funded by the Ministry of Education (2018R1D1A1A09083947).

Acknowledgments: This Special Issue would not have been possible without the contributions of professional authors and reviewers and the excellent editorial team of Applied Sciences.

Conflicts of Interest: The author declares no conflicts of interest.

References

1. Choi, Y.; Rayl, J.; Tammineedi, C.; Brownson, J.R.S. PV Analyst: Coupling ArcGIS with TRNSYS to assess distributed photovoltaic potential in urban areas. *Sol. Energy* **2011**, *85*, 2924–2939. [CrossRef]
2. Zekai, S. Solar energy in progress and future research trends. *Prog. Energy Combust. Sci.* **2014**, *30*, 367–416.
3. Choi, Y.; Suh, J.; Kim, S.-M. GIS-Based Solar Radiation Mapping, Site Evaluation, and Potential Assessment: A Review. *Appl. Sci.* **2019**, *9*, 1960. [CrossRef]
4. Wang, F.; Yu, Y.; Zhang, Z.; Li, J.; Zhen, Z.; Li, K. Wavelet Decomposition and Convolutional LSTM Networks Based Improved Deep Learning Model for Solar Irradiance Forecasting. *Appl. Sci.* **2018**, *8*, 1286. [CrossRef]
5. Valentín, L.; Peña-Cruz, M.; Moctezuma, D.; Peña-Martínez, C.; Pineda-Arellano, C.; Díaz-Ponce, A. Towards the Development of a Low-Cost Irradiance Nowcasting Sky Imager. *Appl. Sci.* **2019**, *9*, 1131. [CrossRef]
6. Richardson, W.; Cañadillas, D.; Moncada, A.; Guerrero-Lemus, R.; Shephard, L.; Vega-Avila, R.; Krishnaswami, H. Validation of All-Sky Imager Technology and Solar Irradiance Forecasting at Three Locations: NREL, San Antonio, Texas, and the Canary Islands, Spain. *Appl. Sci.* **2019**, *9*, 684. [CrossRef]
7. Chen, F.; Guo, S.; Gao, Y.; Yang, W.; Yang, Y.; Zhao, Z.; Ehsan, A. Evaluation Model of Demand-Side Energy Resources in Urban Power Grid Based on Geographic Information. *Appl. Sci.* **2018**, *8*, 1491. [CrossRef]
8. Kim, S.; Oh, M.; Park, H. Analysis and Prioritization of the Floating Photovoltaic System Potential for Reservoirs in Korea. *Appl. Sci.* **2019**, *9*, 395. [CrossRef]
9. Alsadi, S.; Khatib, T. Photovoltaic Power Systems Optimization Research Status: A Review of Criteria, Constrains, Models, Techniques, and Software Tools. *Appl. Sci.* **2018**, *8*, 1761. [CrossRef]
10. Chou, C.; Chung, P.; Yang, R. Wind Loads on a Solar Panel at High Tilt Angles. *Appl. Sci.* **2019**, *9*, 1594. [CrossRef]
11. Chamkha, A.; Selimefendigil, F. Numerical Analysis for Thermal Performance of a Photovoltaic Thermal Solar Collector with SiO_2-Water Nanofluid. *Appl. Sci.* **2018**, *8*, 2223. [CrossRef]
12. Gulkowski, S.; Zdyb, A.; Dragan, P. Experimental Efficiency Analysis of a Photovoltaic System with Different Module Technologies under Temperate Climate Conditions. *Appl. Sci.* **2019**, *9*, 141. [CrossRef]
13. Rouibah, A.; Benazzouz, D.; Kouider, R.; Al-Kassir, A.; García-Sanz-Calcedo, J.; Maghzili, K. Solar Tower Power Plants of Molten Salt External Receivers in Algeria: Analysis of Direct Normal Irradiation on Performance. *Appl. Sci.* **2018**, *8*, 1221. [CrossRef]
14. Xie, T.; Zhang, G.; Liu, H.; Liu, F.; Du, P. A Hybrid Forecasting Method for Solar Output Power Based on Variational Mode Decomposition, Deep Belief Networks and Auto-Regressive Moving Average. *Appl. Sci.* **2018**, *8*, 1901. [CrossRef]
15. Mei, F.; Wu, Q.; Shi, T.; Lu, J.; Pan, Y.; Zheng, J. An Ultrashort-Term Net Load Forecasting Model Based on Phase Space Reconstruction and Deep Neural Network. *Appl. Sci.* **2019**, *9*, 1487. [CrossRef]
16. Rubio-Aliaga, A.; Molina-Garcia, A.; Garcia-Cascales, M.; Sanchez-Lozano, J. Net-Metering and Self-Consumption Analysis for Direct PV Groundwater Pumping in Agriculture: A Spanish Case Study. *Appl. Sci.* **2019**, *9*, 1646. [CrossRef]

Review

GIS-Based Solar Radiation Mapping, Site Evaluation, and Potential Assessment: A Review

Yosoon Choi [1], Jangwon Suh [2,*] and Sung-Min Kim [2]

[1]   Department of Energy Resources Engineering, Pukyong National University, Busan 48513, Korea; energy@pknu.ac.kr
[2]   Department of Energy Engineering, Kangwon National University, Samcheok 25913, Korea; smkim19@kangwon.ac.kr
*   Correspondence: jangwonsuh@kangwon.ac.kr; Tel.: +82-33-570-6313

Received: 11 February 2019; Accepted: 8 May 2019; Published: 13 May 2019

Abstract: In this study, geographic information system (GIS)-based methods and their applications in solar power system planning and design were reviewed. Three types of GIS-based studies, including those on solar radiation mapping, site evaluation, and potential assessment, were considered to elucidate the role of GISs as problem-solving tools in relation to photovoltaic and concentrated solar power systems for the conversion of solar energy into electricity. The review was performed by classifying previous GIS-based studies into several subtopics according to the complexity of the employed GIS-based methods, the type of solar power conversion technology, or the scale of the study area. Because GISs are appropriate for handling geospatial data related to solar resource and site suitability conditions on various scales, the applications of GIS-based methods in solar power system planning and design could be expanded further.

Keywords: geographic information system; solar energy; resource mapping; site evaluation; potential assessment; photovoltaic; concentrated solar power

---

1. Introduction

With growing concerns about greenhouse gas emissions, the security of conventional energy supplies, and the environmental safety of conventional energy production techniques, renewable energy systems are becoming increasingly important and are receiving a great deal of political attention [1]. In particular, photovoltaic (PV) and concentrated solar power (CSP) systems for the conversion of solar energy into electricity have been found to be technologically robust, scalable, and geographically dispersed and possess enormous potential as sustainable energy sources [2] (The full terms and their abbreviations referred to in the paper are summarized on the last page of this paper). Despite the advances in PV and CSP technology, inappropriate planning and design could impede the extensive penetration of solar energy. Systematic planning and design considering various factors and constraints are necessary to deploy PV and CSP systems successfully.

To achieve these objectives, geographic information system (GIS)-based methods have been applied effectively for rational solar power system planning and design. GISs have often been combined with analytical models and methods (e.g., probability/statistical, machine learning, and data mining methods) to complement the inherent capabilities of GISs in evaluating the spatial patterns or characteristics of events and their attributes.

The objective of this study was to review the GIS-based methods and applications currently used for solar radiation mapping, site evaluation, and potential assessment associated with PV and CSP systems. The scope of this review was confined to published literature related to GIS-based methods and applications in solar radiation mapping, site evaluation, and potential assessment of PV and CSP systems. There are numerous recent review papers dealing with solar potential [3,4]. The distinct difference between the aforementioned studies and this paper is the scope of GIS applications. In this study, we tried to investigate and summarize overall roles and applications of GIS to research associated with solar energy (not only focused to solar potential). Keywords (i.e., GIS, solar or PV or CSP, modeling or radiation or mapping or evaluation, or assessment) were inputted into the Google Scholar website to search the published literature, and 92 articles were selected for review. Specifically, three different keywords (A, B, and C) were combined and inputted to search papers (e.g., GIS, PV, and modeling). Then, 92 articles were selected for review by analyzing the title of the searched papers by authors. Even so, it should be noted that the number of papers included in this topic is not necessarily 92. Articles concerning remote sensing technologies for solar radiation mapping were excluded from this study if they did not involve any GIS-based methods.

This paper is organized into five sections. Section 1 introduced the main concept. Section 2 reviews the literature dealing with solar radiation mapping using GISs. Sections 3 and 4 describe the GIS-based methods and applications for solar site evaluation (geospatial suitability) and solar potential assessment (electricity production), respectively. Finally, the conclusions are summarized in Section 5.

2. Solar Radiation Mapping Using GISs

2.1. Overview

Solar radiation is the most important consideration when installing PV or CSP plants. Therefore, it is necessary to identify areas where solar radiation is abundant and to predict the spatial and temporal distribution of solar radiation for effective solar resource utilization. GISs are very useful in spatial and temporal analyses of solar resources while implementing location-specific technologies.

Solar radiation analysis can be performed for points such as stations as well as for many pixels over a large area. It may also be conducted for specific administrative districts. In addition to spatial mapping of solar radiation, analysis and visualization over time are performed. The spatial and temporal considerations are summarized in Table 1 based on the reviewed literature.

The solar radiation analysis methods using GISs can be classified according to the type of data employed. If solar radiation data are obtained at certain stations, it is useful to estimate the solar radiation through interpolation in the areas in which it is not measured. In addition, there have been some studies on solar radiation prediction at stations that acquire other weather information but do not obtain solar radiation information. If there are no data for stations at which observations are made, solar radiation models can be employed considering various geographical and terrain information. These methods are focused on predicting or mapping solar radiation. On the other hand, there have been many studies on analysis or simply visualization of the information about the solar resources in the area using the existing solar radiation map as a database (DB). The methods and types of data used for solar radiation analysis are summarized in Table 1 based on the reviewed literature.

2.2. Solar Radiation Map as a Spatial DB

There have been several studies that analyze the solar energy information in the area using the existing solar radiation map. These studies do not directly perform solar radiation mapping but use existing solar radiation maps to predict sunshine hours per year or to predict and compare regional solar radiation [5].

Some studies have compared the solar data extracted from the solar radiation map with the surface-measured data such as global horizontal irradiance (GHI), direct normal irradiance (DNI), diffuse horizontal irradiance (DHI), and the daily average clearness index to verify the accuracy and applicability of these solar radiation maps [6,7]. The solar radiation maps constructed by various sources such as Solargis (Slovakia), the National Renewable Energy Laboratory (NREL) in the US, and the Energy Sector Management Assistance Program (ESMAP) of the World Bank are used as the spatial database of solar energy information.

Table 1. Summary of previous solar radiation mapping studies using GISs.

References	Usage or Method	Scale	Location	Used Data Type	Spatial Unit	Temporal Unit
Bahrami and Abbaszadeh [5]	DB	National	Iran	Solar map	-	-
Tahir and Asim [6]	DB	National	Pakistan	Solar map	-	-
Zawilska and Brooks [7]	DB	Regional	Durban, South Africa	Solar map	-	-
Song et al. [8]	DB	Regional	Korea (mines)	Solar map	-	-
Burnett et al. [9]	DB	National	United Kingdom	Solar map	-	-
Šúri et al. [10]	DB	Continental	Europe	Solar map	-	-
Ramachandra and Shruthi [11]	Multivariate relationship	Regional	Karnataka, India	Station measurement	Administrative district	Seasonal
Nematollahi and Kim [12]	Interpolation	National	Korea	Station measurement	Pixel	Monthly, Annual
Alamdari et al. [13]	Interpolation	National	Iran	Station measurement	Pixel	Monthly, Annual
Stökler et al. [14]	Model validation	National	Pakistan	Satellite image, Station measurement	Pixel	Annual
Rumbayan et al. [15]	ANN [1]	National	Indonesia	Station measurement	Administrative district	Monthly
Koo et al. [16]	Advanced CBR [2]	National	Korea	Station measurement	Point	Monthly
Lee et al. [17]	Advanced CBR, Interpolation	National	Korea	Station measurement	Point, Pixel	Monthly
Koo et al. [18]	Advanced CBR, FEM [3], Interpolation	National	Korea	Station measurement	Point, Pixel	Monthly
Hofierka and Šúri [19]	r.sun	Continental, National	Central and Eastern Europe, Slovakia	DEM [4], Various parameters	Pixel	Daily, Monthly, Annual
Šúri and Hofierka [20]	r.sun	Continental	Central and Eastern Europe	DEM, Various parameters	Pixel	Monthly
Dubayah and Rich [21]	SOLARFLUX, ATM [5]	Regional	Tiefort Mountains and Grand Canyon, US	DEM, Plant canopies	Pixel	Monthly
Corripio [22]	Shadow analysis	Regional	Mont Blanc Massif, France	DEM	Pixel	15 min intervals
Redweik et al. [23]	Shadow analysis	Local	The University of Lisbon, Portugal (buildings)	DSM [6], Station measurement	Pixel	Hourly, Seasonal, Annual

[1] ANN: artificial neural network. [2] CBR: case based reasoning. [3] FEM: finite element method. [4] DEM: digital elevation model. [5] ATM: atmospheric and topographic model. [6] DSM: digital surface model.

Some studies were conducted for further analysis using solar radiation maps as DBs. The PV potential, power production, and economic effects can be estimated from a solar resource map using the software like RETScreen (Natural Resources Canada, Canada) shown in Figure 1 [8]. In addition, the impact of climate change can be analyzed using the solar radiation map [9]. In these studies, solar radiation maps constructed by the institution of each target country such as Korea Meteorological Administration (Seoul, Korea) and Met office (Exeter, UK) were used. Interactive web applications have also been developed that can assess the potential PV electricity generation in Europe, taking into account the solar energy data and climatic parameters from the solar radiation map [10].

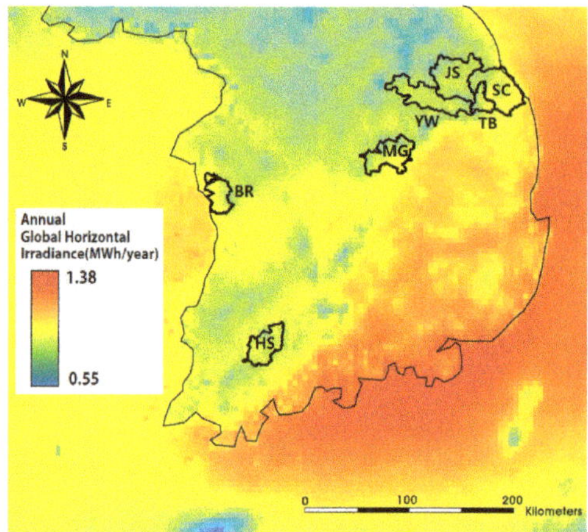

Figure 1. Locations of seven abandoned mine promotion districts in Korea (modified from Song et al. [8]).

2.3. Spatial Solar Radiation Mapping Using Interpolation Methods

GIS maps facilitate the assessment of radiation at various locations and times without measurement equipment. Because solar data are not available for all potential sites, there have been some attempts to estimate the spatial distributions of solar radiation or other parameters.

Probable relationships among the radiation-related parameters can be used to map the global solar radiation at sites where data were not available [11]. The interpolation method based on inverse distance weight or spatial autocorrelation is a very useful tool for solar radiation mapping. The interpolation methods have been widely used in many studies in various countries [12,13]. These methods have the advantage of predicting the entire area of the solar radiation in the form of grids based on the data of the stations where observations are made. In addition, solar radiation mapping based on satellite image can be performed using the certified method and it can be verified through observation data. For example, a solar atlas for Pakistan was made using Meteosat-7 satellite data by validating it with ground measurements based on the ESMAP approach. To adapt the radiation data to the measured datasets, the impacts of different aerosol models were evaluated [14].

2.4. Solar Radiation Estimation for Stations with No Solar Radiation Records

Because some nations have limitations on the number of stations recording solar radiation, due to the expense of acquiring data using precise sensors at all sites, the extent to which the accuracy of solar radiation mapping can be improved is limited. Therefore, attempts have been made to predict solar radiation at stations without solar radiation data by using the solar radiation measurements of other stations. This approach has the advantage that the accuracy of a specific point can be improved over that achievable using a mapping method in which spatial prediction is only performed via interpolation.

ANN is used to perform training in areas where data are present and to perform testing on areas without solar radiation records. It is important to know which input variables other than solar radiation are used for training. Variables such as average temperature, average relative humidity, average sunshine duration, average wind speed, average precipitation, longitude, latitude, and month of the year can be considered to estimate the monthly solar radiation [15]. To increase the accuracy of ANN, a multi-regression analysis and a genetic algorithm can be combined to develop a monthly average daily solar radiation (MADSR) estimation model for locations without measured data [16]. This method can also be combined with the Kriging interpolation to improve the prediction accuracy [17,18].

2.5. Solar Radiation Model

Solar radiation models integrated with GISs are also employed to generate spatial DBs. Solar radiation information over large territories can be provided by these models by considering surface inclination, aspect, and shadowing effects. In particular, these models can be useful for predicting solar radiation in areas without measured data.

The r.sun solar radiation model was developed for an open source environment of the geographic resources analysis support system GIS [19]. The model computes three components (i.e., the beam, diffuse, and reflected components) of global solar radiation for clear-sky or overcast conditions. The model can be applied to large regions, and the shadowing effects of the terrain can be modeled using a shadowing algorithm. In a case study, the model was applied to PV system planning in Central and Eastern Europe and for solar radiation modeling in mountainous terrain in Slovakia.

By integrating the clear-sky index based on a multivariate interpolation method with the r.sun model, the radiation estimation ability was improved [20]. This method can be especially helpful for data at higher resolutions and in regions lacking ground measurements. A study that models solar radiation considering the interaction over topographic and plant canopies was performed using both a GIS (SOLARFLUX) and an image processing system (ATM Model) [21]. The effects of topography and plant canopies on solar radiation were analyzed with various options for obtaining the data. Design issues, computational problems, and error propagation were considered to implement the model for mountainous areas.

An algorithm was presented for calculating the slope gradient, aspect, and cell surface area as a normal vector using the DEM [22]. Based on this algorithm, the sun position, the direct component of insolation, and the hillshade can be calculated. In addition, the horizon angles and sky view factor can be calculated more economically than they could with previous algorithms. A 3D urban solar model was developed for the calculation and visualization of building potential [23]. Light detection and ranging (LiDAR) data were used to build a DSM of an urban region (the University of Lisbon campus), and a shadow algorithm was developed to calculate shadow maps and sky view factors both for roofs and facades. In this study, climatic observations based on the typical meteorological year data were utilized for direct and diffuse solar radiation mapping at each point on the ground, roof, and facades.

3. Solar Site Evaluation Using GISs

3.1. Overview

As regulations for environmental pollution are becoming stricter, there is a growing need to install generating plants using solar energy. To increase the effects of installing such plants, it is necessary to prioritize and select suitable locations to maximize the electricity generation and to minimize the damage that may occur. The results of this site analysis can help solar utility companies, energy companies, and policy makers select potential sites for the construction of solar power plants.

Solar power generating plants are heavily influenced by solar radiation, and the distribution of solar resources varies considerably by location. In most studies in which GISs have been used for site selection, this solar radiation has been considered as an input layer.

Depending on the type of solar power plant, the type of solar radiation used for analysis can be distinguished. Solar power plants can be divided into PV plants and CSP plants. Unlike PV plants where diffuse radiation is also important, direct radiation plays an important role in CSP plants. Therefore, GHI is mainly used for studies on PV plants, and DNI is mainly used for studies on CSP plants. In some site evaluation studies, only general solar energy distributions have been analyzed, without classifying PV and CSP plants. The types of power plants are summarized in Table 2 based on the reviewed literature.

GIS-based methods of evaluating suitable locations for solar power plants can be largely classified depending on whether they are applied on a regional or local scale. Site selection studies using GISs have been widely conducted on the regional scale, and the results have been usefully applied for policy and installation planning. When performing site evaluation on the regional scale, it is necessary to consider various factors such as economic, environmental, technical, social, and risk factors related to the installation of solar power plants, as well as solar radiation. For example, it is difficult to install large-scale solar power plants in the center of cities, so land availability must be considered. In addition, solar power plants must be accessible for installation, so it is necessary to consider the distance to the road network. Solar power plants also should not be far from the power grid to transmit the electricity produced. Many other factors must be considered, and some of these factors serve as constraints to exclude areas where solar power plants should not be installed or as suitability criteria to quantify suitability. Even if the same GIS layer is used, it can be employed in different ways depending on the research.

GIS-based multi-criteria analysis basically relies on two main approaches: Boolean overlay operators and weighted summations procedures. Although there are some differences in the definition of the term multi-criteria decision analysis (MCDA) according to the literature, MCDA is specified as a method of quantifying suitability using weights (i.e., weighted summation procedures) in this paper. The term multi-criteria decision making (MCDM) is also used instead of MCDA in some literature.

The Boolean overlay simply determines whether any conditions are satisfied, and analyzes the suitable area that satisfies all conditions. These results can be used as constraints in weighted summation analysis methods. Weighted summation can be subdivided according to the method of weighting each factor. In this study, we use the term weighted sum when we use the same weight or a very simple weight. The Analytic Hierarchy Process (AHP) method is the most widely used method to quantify the weight according to the expert opinion. Fuzzy AHP (FAHP), which combines fuzzy theory with AHP, is used to mitigate the uncertainty that may arise in this process. A typical flow chart of these GIS-based MCDA methods is shown in Figure 2.

Table 2. Summary of previous solar site evaluation studies using GISs.

References	Methods	Scale	Location	Solar Map	Result	Energy Type
Fluri [24]	Boolean overlay	National	South Africa	Institution	Suitable area	CSP
Merrouni et al. [25]	Boolean overlay	Regional	Eastern Morocco	Analyst tool	Suitable area	CSP
Merrouni et al. [26]	Boolean overlay	Regional	Eastern Morocco	Interpolation	Suitable area	CSP
Wang et al. [27]	Boolean overlay	National	Tibet, China	Satellite image	Suitable area	PV
Hott et al. [28]	Boolean overlay	Regional	Wyoming, US	Institution	Suitable area	PV, CSP
Jahangiri et al. [29]	Boolean overlay	Continental	Middle-East	Interpolation	Suitable area	Solar, Wind
Anwarzai and Nagasaka [30]	Boolean overlay	National	Afghanistan	Institution	Suitable area	PV, CSP, Wind
Gherboudj and Ghedira [31]	Weighted sum	National	United Arab Emirates	Satellite image	Suitability map	PV, CSP
Castillo et al. [32]	Weighted sum	Continental	Europe	Institution	Suitability map	PV
Janke [33]	Weighted sum	Regional	Colorado	Institution	Suitability map	Solar, Wind
Cevallos-Sierra and Ramos-Martin [34]	Weighted sum	National	Ecuador	Institution	Suitability map	PV, CSP, Wind
Aydin et al. [35]	Weighted sum, Fuzzy	Regional	Western Turkey	Institution	Suitability map	PV
Brewer et al. [36]	Weighted sum, PVMapper	Regional	Southwestern US	Institution	Suitability map	PV
Vafaeipour et al. [37]	SWARA [1], WASPAS [2]	National	Iran	Institution	Suitability for cities	PV
Ziuku et al. [38]	AHP [3]	National	Zimbabwe	Interpolation	Suitability map	CSP
Al Garni and Awasthi [39]	AHP	National	Saudi Arabia	Analyst tool	Suitability map	PV
Uyan [40]	AHP	Regional	Karapinar, Turkey	Institution	Suitability map	Solar
Yushchenko et al. [41]	AHP	Continental	West Africa	Institution	Suitability map	PV, CSP
Merrouni et al. [42]	AHP	Regional	Eastern Morocco	Institution	Suitability map	PV

Table 2. *Cont.*

References	Methods	Scale	Location	Solar Map	Result	Energy Type
Aly et al. [43]	AHP	National	Tanzania	Institution	Suitability map	PV, CSP
Tahri et al. [44]	AHP	Regional	Southern Morocco	Analyst tool	Suitability map	PV
Watson and Hudson [45]	AHP	Regional	Southern England	Analyst tool	Suitability map	Solar, Wind
Asakereh et al. [46]	FAHP [4]	Regional	Shodirwan, Iran	Interpolation	Suitability map	PV
Noorollahi et al. [47]	FAHP	National	Iran	Institution	Suitability map	PV
Charabi and Gastli [48]	FLOWA [5]	National	Oman	Analyst tool	Suitability map	PV, CSP
Suh and Brownson [49]	FAHP	Regional	Ulleung Island, Korea	Analyst tool	Suitability map	PV
Sánchez-Lozano et al. [50]	AHP, TOPSIS [6]	Regional	Cartagena, Spain	Analyst tool	Suitability map	PV
Sánchez-Lozano et al. [51]	ELECTRE-TRI, IRIS [7]	Regional	Torre Pacheco, Spain	Analyst tool	Suitability map	PV
Mondino et al. [52]	ANN	Regional	Piedmont, Italy	Analyst tool	Suitability map	PV
Omitaomu et al. [53]	OR-SAGE [8]	Regional	Western US	Institution	Suitable area for each energy	CSP, Nuclear, Advanced coal, CAES [9]
Choi and Song [54]	Overlay	National	Korea (Mines)	Institution	Suitable mines	PV
Kim et al. [55]	Shadow analysis	National	Korea (Reservoirs)	Interpolation	Suitable reservoirs	Floating PV
Lukac et al. [56]	Shadow analysis	Local	Maribor, Slovenia (Buildings)	Measurement	Rooftop suitability	PV
Lee et al. [57]	Shadow analysis, Cluster analysis	Local	Gangnam, Korea (Buildings)	Institution	Rooftop suitability	PV

[1] SWARA: step-wise weight assessment ratio analysis. [2] WASPAS: weighted aggregates Sum Product Assessment. [3] AHP: analytic hierarchy process. [4] FAHP: fuzzy analytic hierarchy process. [5] FLOWA: fuzzy logic ordered weight averaging. [6] TOPSIS: technique for order preference by similarity to ideal solution. [7] IRIS: Interactive robustness analysis and parameters' inference for multi-criteria sorting problems. [8] OR-SAGE: Oak ridge siting analysis for power generation expansion. [9] CAES: compressed air energy storage.

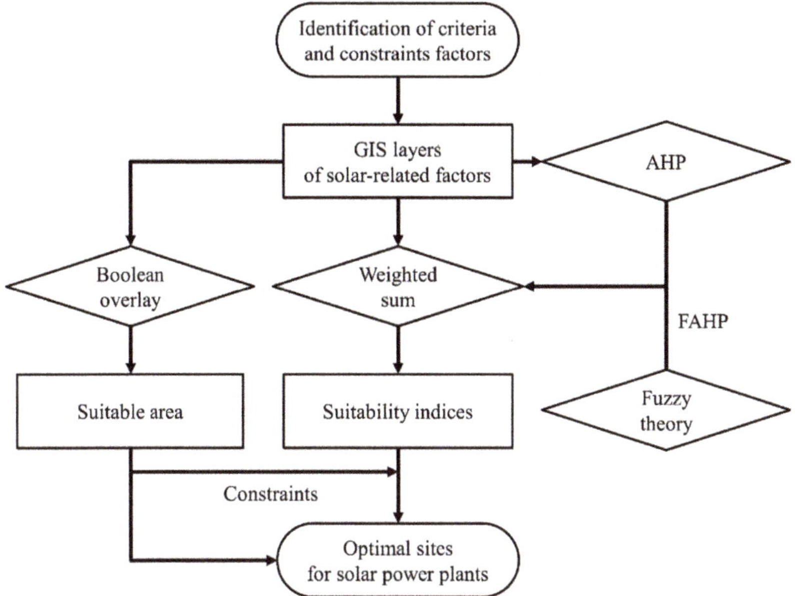

Figure 2. A representative flow chart of GIS-based MCDA methods for solar site evaluation. (Circle: goals, rectangle: GIS layer, and parallelogram: method).

3.2. Boolean Overlay

Although various MCDA techniques in which GISs are employed are useful for site selection for solar power plants, several studies have also been performed to find suitable sites by considering constraints using Boolean overlay operators.

The potential of large-scale CSP has been mainly studied in Africa, which has strong solar energy [24,25]. Conditions such as solar radiation, proximity to transmission lines, terrain, land cover, and hydrology can be considered to select a suitable area for CSP plant installation. Although the detailed constraints are different for each study, the suitable area for large-scale PV plant installation can be analyzed using similar approaches [26,27]. The constraints such as population areas and infrastructures are also important for evaluating suitability as well as the geographic factors such as slope, aspect, and land cover. Some studies have used GHI and DNI to analyze both PV and CSP [28]. After selecting a suitable area, the installation capacity or power generation of the area can be analyzed. This Boolean overlay method has the advantage that it is the simplest method for solar site evaluation, but it has a disadvantage that it cannot quantify the relative suitability for the appropriate region.

3.3. Weighted Sum

These methods, based on the Boolean overlay, can elucidate suitable areas for solar power plant installation, but are limited in that they cannot prioritize suitable areas. To quantify suitability, various MCDA approaches have been developed, and the results have been presented mostly as suitability indices.

The simplest way to quantify the suitability indices for a solar power plant installation is to normalize and add each factor linearly. At the method, the same or a very simple weight is given. To produce the suitability indices, scaled solar radiation maps can be multiplied times a land capability map or the environmental risk map [31]. There are studies that have analyzed solar power plant suitability indices for various regions using a similar method. The geographical factors such as slope, land use, urban extent, population distribution, and proximity to the power grid, as well as solar radiation are combined to generate a suitability map [32,33]. In addition, by considering not only several spatial factors, but also social acceptance data collected through surveys regarding the potential public resistance to development, solar energy suitability can be assessed [36]. However, these studies are logically limited in determining scaling and weighting criteria.

3.4. Analytic Hierarchy Process (AHP)

Evaluating the relative importance between different factors and calculating the overall suitability is a controversial and difficult task. Some techniques have been applied to reflect the opinions of experts in order to reasonably calculate and quantify the weight of each of these factors.

As a method of the expert system, SWARA and WASPAS were integrated with the MCDA method to determine the relative significance of every effective criterion and to evaluate specified alternatives [37]. The AHP is the most widely used and useful systematic expert tool for handling MCDA. In the AHP, hierarchical structures are used to represent a problem and make judgments based on experts to derive priority scales. The overall weights and importance of each input parameter are obtained using a pair-wise comparison matrix.

There have been a number of studies evaluating appropriate sites for CSP or PV power plant installations using AHP [38–44]. In the process of AHP, solar radiation, transmission lines, water bodies, slope, land use, the possible electricity generation, and various economic and technical factors can be considered as assessment factors. In most MCDA studies using the AHP, the unsuitable area is first excluded considering the constraints and then the suitability for each analysis unit (pixel or administrative district) is calculated considering the weights determined via the AHP.

3.5. Fuzzy AHP (FAHP)

Although the AHP is broadly utilized to deal with the complexity of many problems by prioritizing the alternatives, the AHP does not consider the uncertainty associated with the process. To alleviate this issue, fuzzy set theory has been combined with the AHP. The FAHP considers the vagueness, imprecision, and uncertainty associated with the process.

The FAHP has been widely applied in Asia to locate the most appropriate sites for PV or CSP power plants [46–48]. In these studies, various criteria for climatology, location, environment, and meteorology related to solar power plant installation were considered. In addition, the annual electricity production for solar energy plant was estimated in some studies. The FAHP has also been applied to the PV plant installation suitability analysis for Ulleung Island, Korea [49]. PV solar farm criteria were evaluated for an island-based case region having complex topographic and regulatory criteria (Figure 3a), along with high demand for low-carbon local electricity production. Six factor variables (solar radiation, sunshine hours, average temperature in summer, proximity to transmission lines, proximity to roads, and slope) (Figure 4) were normalized via a fuzzy theory to calculate an on-site suitability index (Figure 3b).

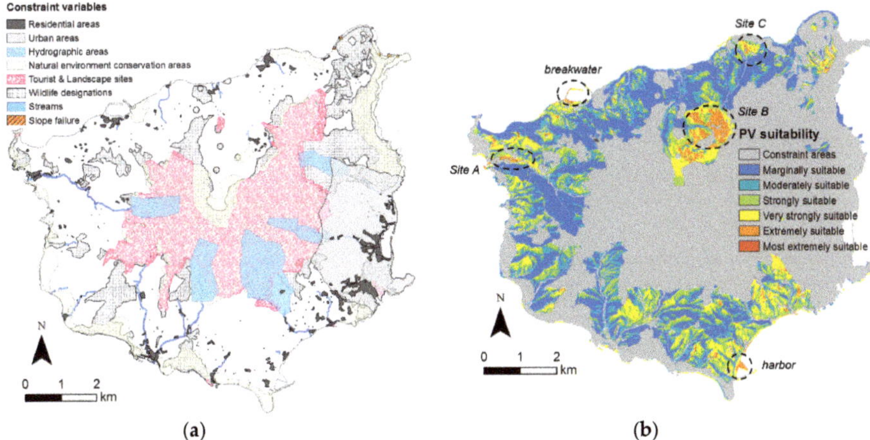

(**a**) (**b**)

Figure 3. PV solar farm evaluation for Ulleung Island, Korea using the FAHP method: (**a**) Distribution of constraint areas; (**b**) Suitability index for a PV solar farm (Suh and Brownson [49]).

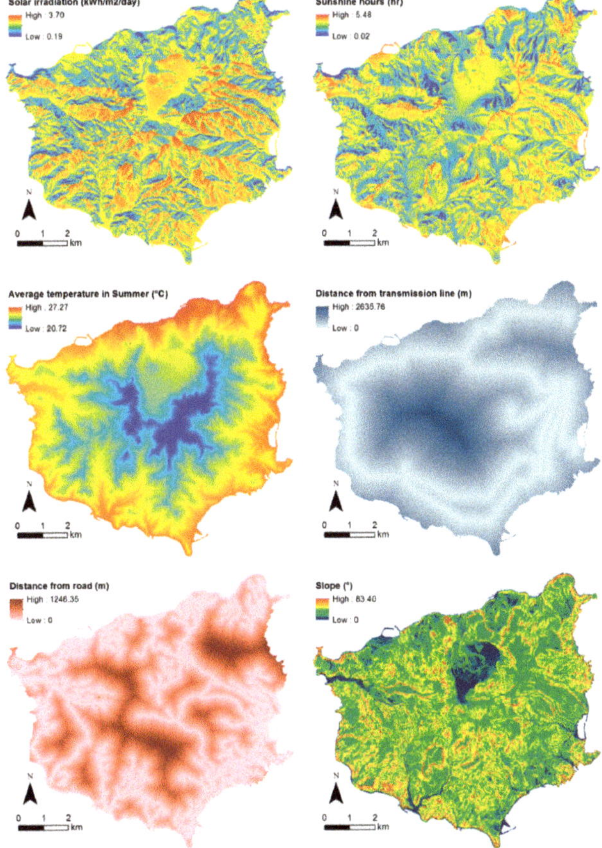

Figure 4. Factor layer inputs to identify a suitable area for PV installation on Ulleung Island (Suh and Brownson [49]).

3.6. Other Approaches

The AHP and specific models have also been used together, and new attempts have been made to support reasonable decision making.

To evaluate PV plant location alternatives, the alternatives were assessed through the AHP and TOPSIS method based on the concept that the chosen alternative should have the shortest distance from the positive ideal solution and the farthest from the negative ideal solution [50]. The final ranking is obtained using a closeness index. An MCDA model was developed by applying the ELECTRE-TRI method and the decision support system IRIS to classify the suitable areas for PV solar farms, into ordered categories of merit according to multiple evaluation criteria [51]. ELECTRE-TRI was used to classify the alternatives by utilizing IRIS, which implements the most common variant of the ELECTRE-TRI method (pessimistic variant). This approach involves classification of each location based on its absolute merits and drawbacks and does not require setting a precise numerical value to express the importance of each criterion. In addition, ANN was applied to identify suitable areas for the installation of PV systems [52]. The final index was determined by combining the quantitative criteria using an ANN trained with values corresponding to the sites of existing PV plants in the region.

3.7. Solar Site Evaluation for Specific Objects

As mentioned above, most solar site evaluation studies involving GISs have been focused on selecting and prioritizing suitable areas from a macroscopic view. These studies were conducted mainly from the regional scale to the national scale and rarely on the continental scale, because solar site evaluation studies using GISs can play an important role in decision making. However, some studies have been performed on more specific objects, rather than land for solar power plant installation.

Choi and Song [54] assessed the PV potential at abandoned mine reclamation sites in Korea. A spatial DB was constructed for 218 abandoned mine reclamation sites (Figure 5) according to reclamation type. By combining the solar energy resource and mine reclamation maps using overlay analysis, mine sites with high annual GHI were selected and energy simulations were conducted based on the system advisor model (SAM) by NREL in the US. Kim et al. [55] estimated the priorities and potential of floating PV for 3,401 reservoirs in Korea. To select a suitable reservoir for floating PV installation, a water depth DB of reservoirs was constructed using OpenAPI (Figure 6a). The annual power production for all possible reservoirs was calculated by considering solar radiance, topographical information (Figure 6b), and solar panel parameters.

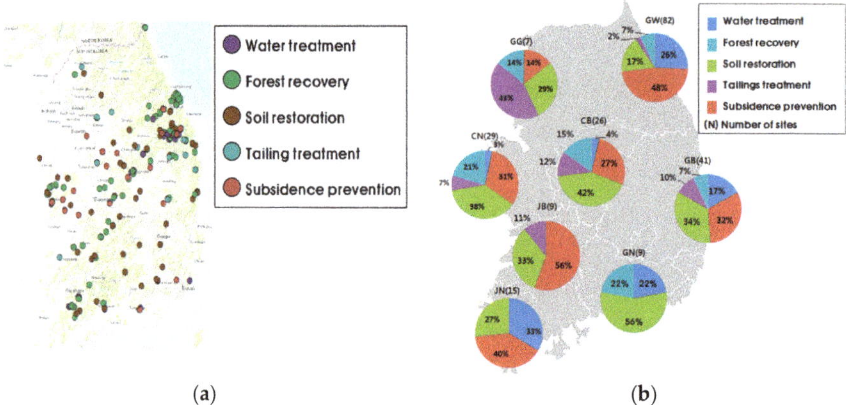

(a)   (b)

Figure 5. Spatial DB for abandoned mine reclamation sites in Korea: (**a**) Locations of abandoned mine reclamation sites; (**b**) Composition according to reclamation type in each administrative district (Choi and Song [54]).

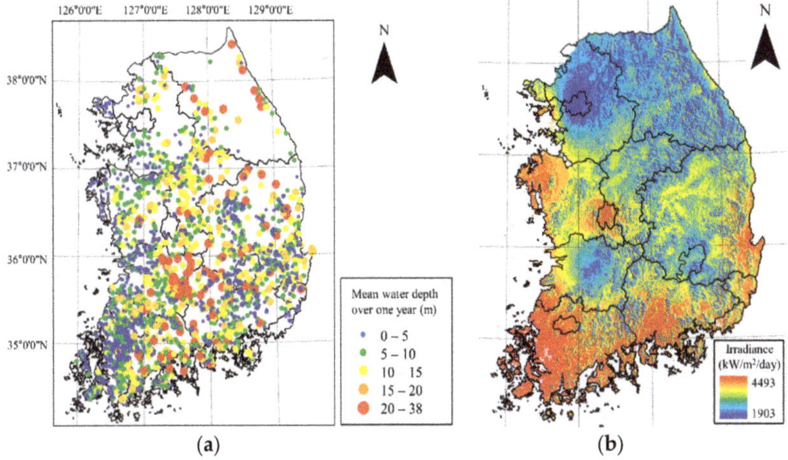

Figure 6. Floating PV analysis for reservoirs in Korea using a GIS: (**a**) Map of average water depth for each reservoir; (**b**) Solar radiance map considering terrain (Kim et al. [55]).

Lukac et al. [56] rated roof surfaces in terms of solar potential and suitability for PV systems. The solar potential was determined by combining the urban topography extracted from LiDAR data with pyranometer measurements and analyzing the shadowing effect. After the roofs were split into segments, a filtering process was conducted to identify buildings unsuitable for PV installation. Then, the solar potential rating was employed for the suitable buildings. Lee et al. [57] evaluated the rooftop solar PV suitability of a building from a microscopic view. To consider not only its technical performance, but also its economic performance, hillshade analysis and life cycle cost analysis were conducted. To develop a rooftop solar PV rating system, cluster analysis based on the technical and economic suitability criteria was performed. The rating system was applied to 21,681 buildings in the Gangnam district in Seoul, South Korea by dividing them into four grades according to their rooftop solar PV potentials, investment returns, and payback periods.

Because only one type of renewable energy system cannot provide continuous power generation, there have been site selection studies for two or more renewable energy systems including solar PV or CSP energy. When multiple renewable energy sources are used in combination, they can compensate for each other when one of them is not available.

In some studies, a Boolean overlay method or weighted sum method was applied to find suitable locations for hybrid solar-wind power station construction by considering factors such as the resources, topography, and environmental and economic viewpoints [29,30,34,45]. In addition, an MCDA framework incorporating an FAHP was applied for hybrid solar-wind renewable energy system site selection [35]. The OR-SAGE tool was developed to analyze the impacts of future energy technology while balancing competing resource use [36]. The tool considers population growth, water availability, environmental indicators, and tectonic and geological hazards and was applied to the western US. The final map categorized and showed the most suitable region for each energy source (solar, nuclear, advanced coal, and CAES).

4. Solar Potential Assessment Using GISs

4.1. Overview

Solar potential can essentially be categorized into three different types [30]. First, physical potential is the total amount of solar energy reaching the target surface, which can be referred to as the total solar radiation on the surface or rooftop. Second, geographic potential is the spatial availability of the surface or building rooftop where solar energy can be obtained, which can be referred to as the available area for solar PV installation. Third, technical potential is the total amount of electricity considering the technical characteristics of the solar PV system (e.g., module efficiency, inverter capacity, and system design), which can be referred to as electricity generation. A GIS can be used for physical potential and geographic potential.

In this study, 39 published articles on GIS-based solar potential assessment were reviewed. Table 3 summarizes the classification and number of articles according to the role of the GIS, solar type, or scale. GISs were utilized for various purposes in solar potential assessment, such as DB and visualization, rooftop extraction, radiation modeling, shading analysis, and spatial analysis tools. Regarding solar type, most studies were focused on PV or both PV and CSP, while few dealt with CSP. In terms of analysis scale, most of the solar potential assessment studies were conducted on national or regional scales, while few were performed on global, continental, or object (building) scales.

This section consists of five sub-sections corresponding to the roles of GISs in research as DB and visualization, rooftop extraction, radiation modeling, shading analysis, and spatial analysis tools.

Table 3. Classification and number of articles according to the role of the GIS, solar type, or scale.

Role of GIS	No.	Solar Type	No.	Analysis Scale	No.
DB & Visualization tools	9	PV	23	Global	1
Rooftop extraction tool	2	CSP	6	National	16
Radiation modeling tool	6	General	1	Region	20
Shading analysis tool	7	PV & CSP	9	Object	2
Spatial analysis tool	15				
Sum	39	Sum	39	Sum	39

4.2. DB and Visualization Tools

This sub-section discusses the use of GISs as solar potential DB (map) or solar potential visualization (map production) tools.

A few researchers have used GISs as only solar DB maps in solar potential assessment. Song et al. [58] employed a georeferenced raster-formatted solar resource map (i.e., annual mean daily global horizontal radiation at the surface) in Kangwon province, created by the Korea Meteorological Administration, to assess the PV potential at three mines (Figure 7). By overlapping the solar resource map with a mine map, the annual mean daily global horizontal radiation at the three mines was extracted. Then, these values were entered into RETScreen software to analyze the electricity production and greenhouse gas emission reduction.

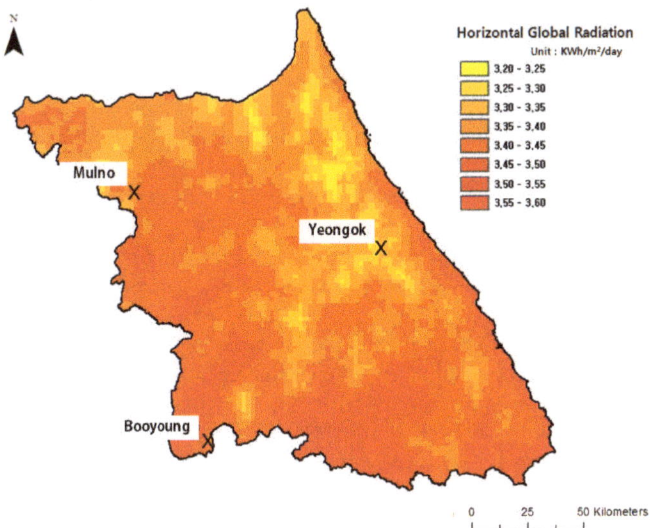

Figure 7. GIS DB map showing annual mean daily global horizontal radiation (kWh/m^2/day) at the surface (Song et al. [58]).

Many researchers have used GISs as both solar DB maps and visualization tools in solar potential assessment. Tarigan et al. [59] performed a SolarGIS-pvPlanner simulation to assess PV power generation for household in Surabaya, Indonesia. This simulator provides assessment results at any selected site online by integrating numerical simulation models generated from the latest climate DB. Specifically, the SolarGIS method is based on using statistically aggregated solar and temperature data stored in the DB with a time step of 15 min. The simulator provides meteorological and geographical data as inputs to assess power generation from PV systems.

A GIS plays a role in the web-based simulator as both a DB system and a visualization system. Besarati et al. [60] generated solar radiation maps for five different tracking modes to compare the applicability of PV and CSP power plants. Then, a 5 MW PV power plant was considered for 50 cities in Iran to investigate the viability of PV power plants for each city. The capacity factors, electricity generated, and annual greenhouse gases emission reductions were compared. Fichter et al. [61] considered a DNI map sourced from the German Aerospace Center to extract CSP hotspots in Northeast Brazil. Using these data, the optimal CSP plant configuration was derived and the potential of CSP for the northeast power system of Brazil was assessed. Lukač et al. [62] utilized rooftop LiDAR point cloud data to generate a grid-type map with 1 m resolution to assess the PV potential of building roofs considering the nonlinear efficiency characteristics of a given PV module type and the solar inverter.

Bergamasco and Asinari [63] generated a solar resource map showing the yearly sum of global irradiation in the Piedmont region by performing interpolation among all of the cell values of the solar radiation map within the municipality. Using this map, the roof surface area available for installation was calculated, and the PV energy potentials of the rooftop integrated PV systems were assessed. Malagueta et al. [64] explored a DNI map and meteorological data for the northeastern region of Brazil to assess incentive policies for integrating CSP generation into the Brazilian electric power system. As a result, subsystems and interchanges in the model in the national grid were proposed. Martín-Pomares et al. [65] generated GHI and DNI average monthly sum maps (2003–2013) in Qatar by improving satellite-derived data using ground measurement data. By considering the characteristics of reference solar power plants and the aforementioned solar resource map using the SAM software, various solar potential electricity generations in Qatar were calculated according to plant type.

Milbrandt et al. [66] assessed and mapped renewable energy potential, including PV, CSP, wind, geothermal, and biomass energy and landfill gas-to-energy on marginal lands (e.g., abandoned, disturbed, under-utilized, wasteland, limbo-land, degraded, and idle) in the US, representing about 11% of the US mainland. Solar technologies exhibited the highest potentials. It was estimated that about 4.5 PWh and 4 PWh of electricity could be produced from PV and CSP energy, respectively, on marginal lands in the contiguous US.

4.3. Rooftop Extraction Tool

A rooftop solar PV system is a PV system in which electricity-generating solar panels are mounted on the rooftop of a residential or commercial building or structure. With the ever-increasing population and unavailability of large-scale solar power plant installation in urban areas, interest in rooftop solar systems that can be installed on buildings in urban area is increasing. GISs have been utilized to extract rooftops to assess their solar potentials on the national and regional scales.

Khan and Arsalan [67] extracted rooftops from Google Earth satellite imagery of the Karachi region of Pakistan using a feature extraction tool of ENVI EX software employing object-based image recognition. From the extracted rooftop data and their characteristics (i.e., building orientation, shading effect, and other roof uses), the PV energy and power output were assessed. The characteristics of previous GIS-based rooftop PV potential studies were also summarized.

Izquierdo et al. [68] calculated roof surface area using vector-type roof data compiled throughout Spain. The installation of solar hot water systems (SHWSs) and PV systems was considered. With assumptions for SHWS demand coverage and payback evaluation, the results showed that SHWSs could contribute up to 1662 ktoe/year of primary energy. In addition, the PV potential of each rooftop was calculated by multiplying the area of the roof times simple constants derived from technical parameters, and it was found that PV systems would provide 10 TWh/year of electricity.

4.4. Radiation Modeling Tool

Solar radiation modeling enables prediction of the average daily and hourly global horizontal radiation, beam radiation, and diffuse radiation. GISs were utilized to map and analyze the effects of the sun over a geographical area for specific time periods. In general, GIS-based radiation modeling was implemented, on the national or regional scales to calculate the insolation across an entire landscape (area) or to calculate the amount of radiant energy for a given location (point).

Charabi and Gastli [69] discussed solar power prospects to assess a large CSP plant in Wilayat Duqum, Oman in a geospatial context. A solar radiation map of Wilayat Duqum was generated by employing solar radiation tools in ArcGIS software and DEM data. Subsequently, the yearly electric power generation potential was calculated according to the type of CSP technology. In a similar manner, Gastli et al. [70] mapped the solar radiation over Wilayat Duqum by employing the aforementioned identical tools and data to investigate the potential of implementing a combined CSP electric power and seawater desalination plant. This study dealt with two options (i.e., the combination of a CSP plant with a thermal desalination unit and exploitation of only the electricity output of a CSP plant with a reverse osmosis desalination unit) and showed where each concept is preferable considering local conditions. Hofierka and Kaňuk [71] estimated the annual PV electricity production in kilowatt-hours per building in the urban areas of Bardejov in eastern Slovakia. In this study, 3D city model input data, the r.sun solar radiation model, and PVGIS estimation utility, an open-source solar radiation tool, were utilized to assess the PV potential of each building. The analysis revealed a high PV potential that could cover about two-thirds of the current electricity consumption of the city.

Meanwhile, Catita et al. [72] modeled the vertical facades of buildings as well as solar radiation on roofs to inspect the potential of building an integrated PV system in the region of the University of Lisbon, Portugal. To achieve this objective, a 3D building model and DSM data were employed as input data for the self-developed software. The DSM data indicated the digital surface height, including objects (e.g., buildings, facilities, and trees), while the DEM data excluded the built (e.g., power lines, buildings, and towers) and natural (trees and other types of vegetation) elements. Izquierdo et al. [73] listed hierarchical methodologies for potentials such as the physical, geographic, technical, economic, and social potentials. The horizontal irradiance on roofs was derived by using radiation computed based on the geometry of the sun–earth system, land use maps, and building maps. Subsequently, considering the PV arrangement and model, the technical potential of a roof-integrated PV system in Spain was estimated. Polo et al. [74] presented maps of the solar resources in Vietnam and the solar potential for CSP and grid-connected PV technology. GHI and DNI maps were derived from the Meteosat Indian Ocean Data Coverage satellite imagery based on the Heliosat method. Subsequently, the solar potential was assessed by performing simple simulations. GISs were used to combine the solar potential with the land availability determined based on the slope conditions and to map the technical solar potential.

4.5. Shading Analysis Tool

Surfaces and rooftops are shaded if the direct path of the light from the sun is obstructed. Shading of PV modules is a common phenomenon that can affect the performances of PV systems. As such, shading analysis is essential during solar PV project design or analysis since it is associated with solar access or unobstructed solar gains. In PV systems, it is important to analyze shading caused by surrounding objects (buildings) and/or vegetation.

Several authors have employed the hillshade algorithm and analysis to compute shadow areas and sunshine hours. Ko et al. [75] determined the shadow areas on rooftops to obtain the hourly sun and shade grayscale values to evaluate rooftop solar PV potential in Taiwan. A national-scale hillshade map was generated as a sun–shadow model showing the illuminance of each grid of the surface by setting the location of the assumed light source and calculating the related illuminance value between each grid and its neighboring grids. The preset conditions ranged from 0 (black; more shadowing) to 255 (white; less shadowing). The grayscale values were then integrated into binary images to calculate the shadow areas on rooftops. Lee et al. [76] and Hong et al. [77] analyzed shadows based on the location of the sun via hillshade analysis to examine the available building rooftop areas for solar PV installation. The altitude and azimuth of the sun were calculated every hour from 6 a.m. until 7 p.m. (when the sun is over the horizon) on the 15th (when the sun is at the average position for the present month) of each month from January until December in Seoul. Using these input data, hillshade analysis was conducted on 12 days (on the 15th of each month from January until December) in hourly intervals (from 6 a.m. to 7 p.m.), for a total of 156 times. Thereby, the unshaded rooftop area in which the solar PV system could perform at the optimal level without any disturbances by building shadows was extracted.

Many researchers have used DSMs (or 3D building models) to examine regional shadow characteristics. Song and Choi [78] conducted a simple shadow analysis to evaluate rooftop PV electricity generation systems to establish a green campus in a GIS environment. Prior to shadow analysis, 3D modeling of the study area was performed using a DEM (0.1 m grid spacing) easily accessed and building height data. As mentioned above, since the DEM data only describe the relief of the natural terrain, the DEM and building height data were integrated to create 3D building data. Subsequently, shadow analysis was conducted using the solar radiation tools in ArcGIS to analyze the yearly shadow patterns based on the changes in the angle of the sun (Figure 8). Locations, that were unaffected by shadows for more than 9 h/day, were regarded as usable areas for a rooftop PV system at the university. Choi et al. [1] developed a PV analyst that couples ArcGIS with a transient systems simulation (TRNSYS) to assess the distributed PV potential in urban areas. In this study, a DSM was input into the area solar radiation model of the solar radiation tools in ArcGIS to calculate the sun hours, which is the duration (in hours) of direct solar irradiation at each grid cell in the DSM. The duration of direct incoming solar irradiance from 9 a.m. until 3 p.m. (solar time) at the winter solstice was analyzed to fulfill the requirement that the PV systems to be installed in the study area would work solar time throughout the year without shading effects. The areas satisfying this requirement were regarded as usable areas for rooftop PV systems on each building in the study.

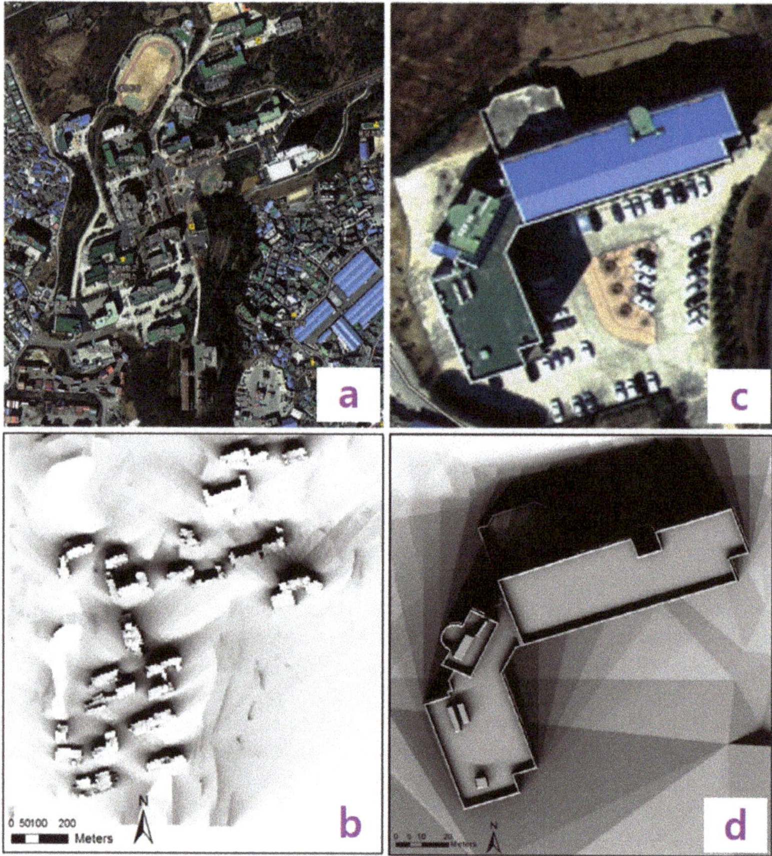

Figure 8. Shadow analysis. (**a**) Satellite image of the study area; (**b**) Shadow analysis of the overall study area; (**c**) Satellite image of a building; (**d**) Shadow analysis of the rooftop area of a building (modified from Song and Choi [78]).

In a few studies, detailed field investigations using fish-eye lens cameras as well as regional shadow analysis using DSMs were performed. Choi and Song [79] analyzed the usable area for installing a PV system, considering the surrounding topography, for PV potential assessment at the tailing dam of an abandoned mine. For this purpose, two-step shading analysis, including local shading analysis and detailed field investigation, were conducted. In the first step, the study area was represented in 3D using a DEM with 10 m grid spacing in ArcGIS. To assess the shading effects on the surface of the tailing embankment for a conservative perspective, the daily hours of sunshine on the surface were analyzed at the winter solstice using the DEM and solar radiation analysis tool in ArcGIS. The number of sunshine hours was found to range from 6.1 h/day to 7.9 h/day without interference from shadows (Figure 9). In the second step, the detailed shade effects associated with nearby light barriers such as trees and plants were evaluated by performing a field investigation. By using a fish-eye lens camera, a skyline image at the solar site was captured, and the image was used to analyze on-site barriers to light reception (Figure 10). From the results generated by checking the distribution of light obstructions from on-site barriers, a shading matrix was generated (Table 4). The shading matrix stored month-by-hour shading data for the shading elements surrounding the solar site, expressed by values ranging from 0.0 (complete shielding of direct radiation from the PV system) to 1.0 (no shading effects).

Figure 9. Daily sunshine hours on the surface of the tailing embankment on the winter solstice (Choi and Song [79]).

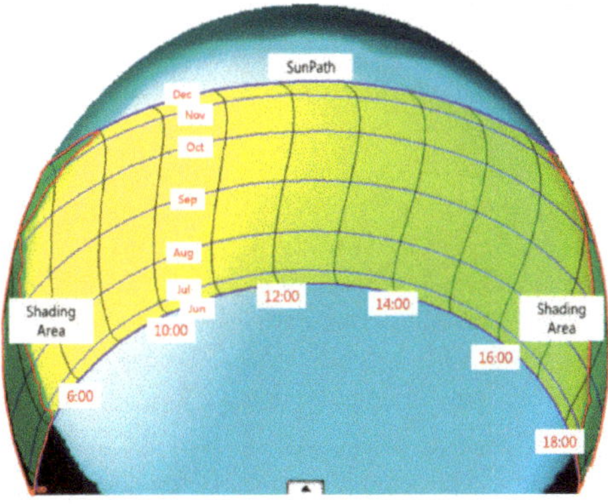

Figure 10. Results of shading analysis using a fish-eye lens camera (Sun Eye 210) (Choi and Song [79]).

Table 4. Shading matrix generated from the onsite solar assessment (Choi and Song [79]).

Month	Time														
	5	6	7	8	9	10	11	12	13	14	15	16	17	18	19
January	0	0	0	0.02	0.63	1	1	1	1	1	0.96	0.72	0.02	0	0
February	0	0	0.04	0.64	0.99	1	1	1	1	1	1	0.99	0.44	0.06	0
March	0	0.10	0.56	1	1	1	1	1	1	1	1	1	0.84	0	0
April	0	0.09	0.83	1	1	1	1	1	1	1	1	1	0.92	0.03	0
May	0	0.24	0.92	1	1	1	1	1	1	1	1	1	1	0.33	0
June	0	0.25	0.92	1	1	1	1	1	1	1	1	1	1	0.65	0
July	0	0.15	0.89	1	1	1	1	1	1	1	1	1	1	0.69	0
August	0	0.10	0.84	1	1	1	1	1	1	1	1	1	0.99	0.21	0
September	0	0.17	0.84	1	1	1	1	1	1	1	1	1	0.73	0	0
October	0	0.04	0.64	1	1	1	1	1	1	1	1	1	0.26	0	0
November	0	0	0.02	0.54	0.98	1	1	1	1	1	0.97	0.60.	0	0	0
December	0	0	0	0.01	0.61	1	1	1	1	1	0.89	0.42	0	0	0

Song and Choi [80] performed shading analysis to identify the area suitable for installing a floating PV system on a mine pit lake in Korea. A four-step procedure was suggested to assess the potential of a floating PV system, including solar site assessment, design of the PV system, simulation of the PV system, and evaluation of economic feasibility. In the first step, shading analysis was performed in a GIS environment using a method similar to that described above. The maximum amount of sunshine was 6.5 h/day. The area of the pit lake surface with more than 6 h/day of sunshine was found to be 87,650 m^2, accounting for 38.9% of the total water surface area. Moreover, small obstructions around the pot lake were captured from the east and west of the skyline images. Based on these results, a month-by-hour shading matrix was generated.

4.6. Spatial Analysis Tool

Spatial analysis in a GIS environment is a concept in contrast to general analysis, which does not take into account spatial characteristics. Spatial analysis includes any of the formal techniques used to investigate entities based on their topological, geometric, or geographic properties. These techniques include various approaches such as rooftop extraction and solar radiation modeling. However, the roles of GISs for spatial analysis discussed in this sub-section are confined to simple geospatial analysis tools such as map algebra, overlaying, reclassification, slope, orientation, distance, and so on.

In some studies, map algebra analysis (including filtering) or overlaying of two or more maps was employed to generate new maps. Clifton and Boruff [81] identified the potential of using CSP to generate electricity in a rural region of Western Australia. Potential CSP sites were defined by overlaying environmental variables and electricity infrastructure on a high-resolution grid using widely available datasets and ArcGIS. The CSP potential map regions were classified as high potential, medium potential, and low potential, and the statistics of each class were calculated. Wong et al. [82] generated transmissivity and diffuse proportion maps by applying map algebra analysis to maps showing the proportions of clear days, partly cloudy days, and cloudy days in Hong Kong. Moreover, the slope of the terrain was analyzed and rooftops with angles of more than 40°, which is regarded as maximum for installing PV modules, were excluded. Subsequently, building footprint and solar potential maps were spatially joined, within a GIS environment, to calculate the rooftop PV potential in Hong Kong. Massimo et al. [83] overlapped solar radiation in the areas exploitable for solar energy production. Subsequently, the electricity consumption and potential production were investigated for each district in Central Italy. Sun et al. [84] generated a map showing the spatial distributions of geographical constraint areas such as built-up, non-built-up, and unsuitable areas. Then, technical potential and unit generation cost maps for each area were created by using a few equations. Furthermore, the simple payback period value, net present value, and carbon reduction potential for each area were calculated and visualized via map algebra analysis based on several equations.

Bayrakci et al. [85] generated two TRNSYS-based solar power output maps, one in which temperature was ignored and one in which it was considered, to examine the effects of temperature on PV systems. The power output differences for 236 cities across the US were used to generate contour maps indicating a continuous surface of differences between these two approaches via map algebra analysis. Mahtta et al. [86] included wasteland areas for solar potential calculation using a filtering approach in a GIS environment. Conversely, areas with average slopes greater than 2.1% and those with less than minimum thresholds of 4.0 kWh/m^2/day for GHI and 5.47 kWh/m^2/day for DNI, were excluded to calculate the solar potential for suitable areas. Lopez et al. [87] proposed GIS-based approaches to analyze various renewable energy technical potentials in the US. Various spatial analysis tools were utilized to estimate the system performance and for mapping, including various solar PV and CSP systems.

Several researchers applied spatial joining of two maps (of different types) or spatial comparison of maps. Monforti et al. [88] spatially compared solar and wind potential maps to calculate hourly, daily, and monthly wind and solar electricity correlation coefficients in Italy for 2005. Based on the results, the complementarity of wind and solar resources for energy production in Italy was assessed using a Monte Carlo approach. Niblick and Landis [89] assessed five different renewable energy potentials on US marginal and contaminated sites that could be re-used. For this purpose, point data for three different types of marginal land (i.e., Brownfield, Landfill, and abandoned mine lands) and grid cell data for five different renewable energy sources were spatially joined to map energy resources to regions of interest in a GIS environment.

Many researchers examined surface characteristics, including slope, aspect, orientation, and curvature. Choi et al. [90] analyzed the slope of terrain using the surface toolset and DEM data in ArcGIS to identify regions with slopes of less than 10° for a fixed-type PV system. The purpose of setting this terrain condition was to minimize the environmental damage caused by constructing a solar power plant in an abandoned mine area. Buffat et al. [91] computed the slope, orientation, and horizontal angles of each building, using a DSM coupled with building footprints, to assess solar irradiation on building rooftops, because shadows occur when the elevation angle of the sun is less than the angle of the local horizon of the observer. Gastli and Charabi [92] analyzed the slope in Oman to calculate the yearly electricity generation potential for different CSP technologies, such as parabolic trough, parabolic dish, tower, and concentrated PV. It was determined that if only the land of Oman with a slope less than 1%, which constitutes 10% of the total land area, is considered to be exploitable for parabolic trough CSP technology, then the total potential yearly electricity generation would be about 7.6 million GWh. He and Kammen [93] calculated the available land for solar development in each province by applying the following filters in the GIS modeling: DEM with an elevation less than 3000 m and a slope less than 1% and land use categorized as barren land, as defined in the land use data available for solar development. Forests, cropland, wetlands, water, woody savannas, shrub lands, savannas, grasslands, snowy and glacial areas, and protected land were excluded from this study for stationary PV.

In a few studies, distance or buffer analysis was performed to derive specific regions or values related to distance. Köberle et al. [94] generated two worldwide maps showing the cost of electricity by PV systems with and without transmission lines to analyze the effects of the cost of building new transmission lines to the nearest load center on the cost of PV electricity. At this point, distance analysis and map algebra were employed to estimate the cost derived from the length of the transmission line (distance to the nearest load center). Peterseim et al. [95] identified regions with DNI of more than 18 MJ/m^2/day to determine the most promising regions for CSP systems. In addition, transmission lines with more than 66 kV were considered as they can technically absorb the output of 5–60 MWe CSP-biomass hybrid plants. In addition, 50 km buffer analysis was conducted around transmission lines as biomass transport for 50 km is possible. These two conditions were integrated to generate a map showing the areas within 50 km of the existing transmission lines, overlapped with different DNI values.

5. Discussion and Conclusions

In this paper, numerous published articles on GIS-based methods and applications for the planning and design of solar power systems were reviewed. For solar radiation mapping, site evaluation, and potential assessment of PV and CSP systems, GISs have been used for purposes ranging from simple to complex, such as (i) DB and visualization, (ii) rooftop extraction, (iii) shading analysis, (iv) solar radiation modeling, (v) spatial analysis, and (vi) MCDA. In addition, GIS-based methods can be applied differently according to the scale of study area, as well as the type of solar energy conversion technology. Based on the detailed literature review, the following conclusions were drawn.

(1) Solar radiation maps can be useful spatial DBs in spatial and temporal analyses of solar resources. Interpolation methods can be employed for solar radiation mapping in areas with sufficient measured data. The results are presented as solar maps over large areas in pixels. In the absence of sufficient measured solar radiation data, other weather information can be used to predict solar radiation. This method is performed for a specific point, such as a station, by predicting the solar radiation relatively accurately based on various parameters. In recent years, machine learning

techniques such as ANNs have been used. After accurately estimating the solar radiation for additional stations, more accurate solar radiation mapping can be achieved via interpolation. Solar radiation models using topography were also developed to predict solar radiation where there is no measured data. Various algorithms and shadow analysis based on GISs are mainly utilized in these models. The solar radiation maps produced using these methods can facilitate the identification of solar resources in specific areas and the determination of suitable areas for solar power plants. Validation using measured data is necessary to verify the accuracy of such maps.

(2) GISs are useful for site evaluation when installing solar power plants for PV or CSP. While diffuse radiation is also an important factor in PV suitability analysis, only direct radiation is considered in CSP suitability analysis. In most site evaluation studies, solar radiation is the primary consideration and is obtained in various ways depending on the presence or absence of data. In particular, site evaluation by employing a GIS is useful for supporting decision making on the regional scale, and it is necessary to consider economic, environmental, technical, social, and risk factors in addition to solar radiation. These factors can be used to exclude unsuitable regions through Boolean overlay. They can also be employed in various MCDA methods to estimate suitability indices. Some researchers have performed suitability analysis for buildings, mines, and reservoirs.

(3) The assessment of solar PV potential is critical in the development of planning policies and financing schemes for successful PV system deployment. Most of the reviewed literature focused on assessing the technical potential in the region of interest among the three types of solar potential. GISs are effective for assessing physical potential (total solar radiation on the surface/rooftop) and geographic potential (available surface/rooftop area considering the shadow), serving various purposes such as DB and visualization, rooftop extraction, radiation modeling, shading analysis, and spatial analysis tools. In addition, GISs can be utilized to visualize and interpret solar energy-based power output or economic values in the geospatial context in technical potential assessment.

Despite the advances of PV and CSP technology, a lack of information regarding the feasibility of solar power systems among installers and consumers, financial groups that broker large installations, policymakers who enable the deployment of technology, and even scientists and engineers from other complementary disciplines, has become a formidable barrier to their extensive penetration. The widespread use of GISs has provided information to various social groups with the modeling and assessment capabilities of solar power systems associated with resource mapping, site evaluation, and potential assessment. The roles of GISs have extended beyond data inventory and visualization to sophisticated modeling, evaluation, assessment, and interdisciplinary studies of solar energy.

Nonetheless, a gap is still present between the solar energy related maps generated by researchers and their practical use in solar energy system design and management works by engineers, planners, and designers. Consequently, greater efforts are required to minimize this gap by maximizing the applicability and practicality of future modeling and assessment results. All the solar energy's potential, economic, environmental aspects are associated with geospatial context and thus GIS can be used as a basis platform. Therefore, it is necessary to conduct convergence research or to develop to couple GIS technology and solar energy related potential/economic/environmental analysis software.

In the future, related research will continue to improve and develop via the use of high-resolution geospatial data, advancement of spatial data analysis techniques, and coupling of GIS technology with various (empirical, theoretical, and analytical) models and methods. Furthermore, the future of solar PV energy is in going to a distributed power system or a smart grid that reflects a spatial perspective. Therefore, it is necessary to conduct research on the optimization of solar energy system design using GIS in the future.

Author Contributions: Y.C. gathered published literature; all the authors reviewed and analyzed the published literature; all the authors wrote the paper. J.S. implemented the final work for the paper submission.

Acknowledgments: This work was supported by the Basic Science Research Program through the National Research Foundation of Korea (NRF) funded by the Ministry of Education (2018R1D1A1A09083947).

Conflicts of Interest: The authors declare no conflict of interest.

Abbreviations and Symbols

The following abbreviations are used in this manuscript:

AHP	analytic hierarchy process
ANN	artificial neural network
ATM	atmospheric and topographic model
CAES	compressed air energy storage
CBR	case based reasoning
CSP	concentrated solar power
DB	database
DEM	digital elevation model
DHI	diffuse horizontal irradiance
DNI	direct normal irradiance
DSM	digital surface model
ESMAP	energy sector management assistance program
FAHP	fuzzy analytic hierarchy process
FEM	finite element method
FLOWA	fuzzy logic ordered weight averaging
GHI	global horizontal irradiance
GIS	geographic information system
IODC	Indian Ocean Data Coverage
IRIS	Interactive robustness analysis and parameters' inference for multicriteria sorting problems
LiDAR	Light Detection and Ranging
MADSR	monthly average daily solar radiation
MCDA	multi-criteria decision analysis
MCDM	multi-criteria decision making
NREL	National Renewable Energy Laboratory
OR-SAGE	Oak ridge siting analysis for power generation expansion
PV	photovoltaics
SAM	system advisor model
SHWS	solar hot water systems
SWARA	step-wise weight assessment ratio analysis
TOPSIS	technique for order preference by similarity to ideal solution
TRNSYS	transient systems simulation
US	United States
WASPAS	weighted aggregates Sum Product Assessment

References

1. Choi, Y.; Rayl, J.; Tammineedi, C.; Brownson, J.R.S. PV Analyst: Coupling ArcGIS with TRNSYS to assess distributed photovoltaic potential in urban areas. *Sol. Energy* **2011**, *85*, 2924–2939. [CrossRef]
2. Zekai, S. Solar energy in progress and future research trends. *Prog. Energy Combust. Sci.* **2014**, *30*, 367–416.
3. Yang, D.; Kleissl, J.; Gueymard, C.A.; Pedro, H.T.C.; Coimbra, C.F.M. History and trends in solar irradiance and PV power forecasting: A preliminary assessment and review using text mining. *Sol. Energy* **2018**, *168*, 60–101. [CrossRef]
4. Freitas, S.; Catita, C.; Redweik, P.; Brito, M.C. Modelling solar potential in the urban environment: State-of-the-art review. *Renew. Sustain. Energy Rev.* **2015**, *41*, 915–931. [CrossRef]
5. Bahrami, M.; Abbaszadeh, P. An overview of renewable energies in Iran. *Renew. Sustain. Energy Rev.* **2013**, *24*, 198–208. [CrossRef]
6. Tahir, Z.R.; Asim, M. Surface measured solar radiation data and solar energy resource assessment of Pakistan: A review. *Renew. Sustain. Energy Rev.* **2018**, *81*, 2839–2861. [CrossRef]

7.   Zawilska, E.; Brooks, M.J. An assessment of the solar resource for Durban, South Africa. *Renew. Energy* **2011**, *36*, 3433–3438. [CrossRef]

8.   Song, J.; Choi, Y.; Yoon, S. Analysis of photovoltaic potential at abandoned mine promotion districts in Korea. *Geosyst. Eng.* **2015**, *18*, 168–172. [CrossRef]

9.   Burnett, D.; Barbour, E.; Harrison, G.P. The UK solar energy resource and the impact of climate change. *Renew. Energy* **2014**, *71*, 333–343. [CrossRef]

10. Šúri, M.; Huld, T.A.; Dunlop, E.D. PV-GIS: A web-based solar radiation database for the calculation of PV potential in Europe. *Int. J. Sust. Energy* **2005**, *24*, 55–67. [CrossRef]

11. Ramachandra, T.; Shruthi, B. Spatial mapping of renewable energy potential. *Renew. Sustain. Energy Rev.* **2007**, *11*, 1460–1480. [CrossRef]

12. Nematollahi, O.; Kim, K.C. A feasibility study of solar energy in South Korea. *Renew. Sustain. Energy Rev.* **2017**, *77*, 566–579. [CrossRef]

13. Alamdari, P.; Nematollahi, O.; Alemrajabi, A.A. Solar energy potentials in Iran: A review. *Renew. Sustain. Energy Rev.* **2013**, *21*, 778–788. [CrossRef]

14. Stökler, S.; Schillings, C.; Kraas, B. Solar resource assessment study for Pakistan. *Renew. Sustain. Energy Rev.* **2016**, *58*, 1184–1188. [CrossRef]

15. Rumbayan, M.; Abudureyimu, A.; Nagasaka, K. Mapping of solar energy potential in Indonesia using artificial neural network and geographical information system. *Renew. Sustain. Energy Rev.* **2012**, *16*, 1437–1449. [CrossRef]

16. Koo, C.; Hong, T.; Lee, M.; Park, H.S. Estimation of the Monthly Average Daily Solar Radiation using Geographic Information System and Advanced Case-Based Reasoning. *Environ. Sci. Technol.* **2013**, *47*, 4829–4839. [CrossRef]

17. Lee, M.; Koo, C.; Hong, T.; Park, H.S. Framework for the Mapping of the Monthly Average Daily Solar Radiation Using an Advanced Case-Based Reasoning and a Geostatistical Technique. *Environ. Sci. Technol.* **2014**, *48*, 4604–4612. [CrossRef]

18. Koo, C.; Hong, T.; Jeong, K.; Kim, J. Development of the monthly average daily solar radiation map using a-CBR, FEM, and Kriging method. *Technol. Econ. Dev. Econ.* **2018**, *24*, 489–512. [CrossRef]

19. Hofierka, J.; Šúri, M. The solar radiation model for Open source GIS: Implementation and applications. In Proceedings of the Open Source GIS—GRASS Users Conference 2002, Trento, Italy, 11–13 September 2002; pp. 51–70.

20. Šúri, M.; Hofierka, J. A New GIS-based Solar Radiation Model and Its Application to Photovoltaic Assessments. *Trans. GIS* **2004**, *8*, 175–190. [CrossRef]

21. Dubayah, R.; Rich, P.M. Topographic solar radiation models for GIS. *Int. J. Geogr. Inf. Syst.* **1995**, *9*, 405–419. [CrossRef]

22. Corripio, J.G. Vectorial algebra algorithms for calculating terrain parameters from dems and solar radiation modelling in mountainous terrain. *Int. J. Geogr. Inf. Sci.* **2003**, *17*, 1–23. [CrossRef]

23. Redweik, P.; Catita, C.; Brito, M. Solar energy potential on roofs and facades in an urban landscape. *Sol. Energy* **2013**, *97*, 332–341. [CrossRef]

24. Fluri, T.P. The potential of concentrating solar power in South Africa. *Energy Policy* **2009**, *37*, 5075–5080. [CrossRef]

25. Merrouni, A.A.; Mezrhab, A.; Mezrhab, A. CSP sites suitability analysis in the Eastern region of Morocco. *Energy Procedia* **2014**, *49*, 2270–2279. [CrossRef]

26. Merrouni, A.A.; Mezrhab, A.; Mezrhab, A. PV sites suitability analysis in the Eastern region of Morocco. *Sustain. Energy Technol. Assess.* **2016**, *18*, 6–15. [CrossRef]

27. Wang, S.; Zhang, L.; Fu, D.; Lu, X.; Wua, T.; Tong, Q. Selecting photovoltaic generation sites in Tibet using remote sensing and geographic analysis. *Sol. Energy* **2016**, *133*, 85–93. [CrossRef]

28. Hott, R.; Santini, R.; Brownson, J.R.S. GIS-based Spatial Analysis for Large-Scale Solar Power and Transmission Line Issues: Case Study of Wyoming, U.S. In Proceedings of the 41st American Solar Energy Society Meeting, Denver, CO, USA, 13–17 May 2012; pp. 1–6.

29. Jahangiri, M.; Ghaderi, R.; Haghani, A.; Nematollahi, O. Finding the best locations for establishment of solar-wind power stations in Middle-East using GIS: A review. *Renew. Sustain. Energy Rev.* **2016**, *66*, 38–52. [CrossRef]

30. Anwarzai, M.A.; Nagasaka, K. Utility-scale implementable potential of wind and solar energies for Afghanistan using GIS multi-criteria decision analysis. *Renew. Sustain. Energy Rev.* **2017**, *71*, 150–160. [CrossRef]

31. Gherboudj, I.; Ghedira, H. Assessment of solar energy potential over the United Arab Emirates using remote sensing and weather forecast data. *Renew. Sustain. Energy Rev.* **2016**, *55*, 1210–1224. [CrossRef]

32. Castillo, C.P.; Silva, F.B.E.; Lavalle, C. An assessment of the regional potential for solar power generation in EU-28. *Energy Policy* **2016**, *88*, 86–99. [CrossRef]

33. Janke, J.R. Multicriteria GIS modeling of wind and solar farms in Colorado. *Renew. Energy* **2010**, *35*, 2228–2234. [CrossRef]

34. Cevallos-Sierra, J.; Ramos-Martin, J. Spatial assessment of the potential of renewable energy: The case of Ecuador. *Renew. Sustain. Energy Rev.* **2018**, *81*, 1154–1165. [CrossRef]

35. Aydin, Y.N.; Kentel, E.; Duzgun, H.S. GIS-based site selection methodology for hybrid renewable energy systems: A case study from western Turkey. *Energy Convers. Manag.* **2013**, *70*, 90–106. [CrossRef]

36. Brewer, J.; Ames, D.P.; Solan, D.; Lee, R.; Carlisle, J. Using GIS analytics and social preference data to evaluate utility-scale solar power site suitability. *Renew. Energy* **2015**, *81*, 825–836. [CrossRef]

37. Vafaeipour, M.; Zolfani, H.S.; Mohammad, H.M.V.; Derakhti, A.; Eshkalag, M.K. Assessment of regions priority for implementation of solar projects in Iran: New application of a hybrid multi-criteria decision-making approach. *Energy Convers. Manag.* **2014**, *86*, 653–663. [CrossRef]

38. Ziuku, S.; Seyitini, L.; Mapurisa, B.; Chikodzi, D.; Kuijk, K. Potential of Concentrated Solar Power (CSP) in Zimbabwe. *Energy Sustain. Dev.* **2014**, *23*, 220–227. [CrossRef]

39. Al Garni, H.Z.; Awasthi, A. Solar PV power plant site selection using a GIS-AHP based approach with application in Saudi Arabia. *Appl. Energy* **2017**, *206*, 1225–1240. [CrossRef]

40. Uyan, M. GIS-based solar farms site selection using analytic hierarchy process (AHP) in Karapinar region, Konya/Turkey. *Renew. Sustain. Energy Rev.* **2013**, *28*, 11–17. [CrossRef]

41. Yushchenko, A.; de Bono, A.; Chatenoux, B.; Patel, M.K.; Ray, N. GIS-based assessment of photovoltaic (PV) and concentrated solar power (CSP) generation potential in West Africa. *Renew. Sustain. Energy Rev.* **2018**, *81*, 2088–2103. [CrossRef]

42. Merrouni, A.A.; Elalaoui, F.E.; Mezrhab, A.; Mezrhab, A.; Ghennioui, A. Large scale PV sites selection by combining GIS and Analytical Hierarchy Process. Case study: Eastern Morocco. *Renew. Energy* **2018**, *119*, 863–873. [CrossRef]

43. Aly, A.; Jensen, S.S.; Pedersen, A.B. Solar power potential of Tanzania: Identifying CSP and PV hot spots through a GIS multicriteria decision making analysis. *Renew. Energy* **2017**, *113*, 159–175. [CrossRef]

44. Tahri, M.; Hakdaoui, M.; Maanan, M. The evaluation of solar farm locations applying Geographic Information System and Multi-Criteria Decision-Making methods: Case study in southern Morocco. *Renew. Sustain. Energy Rev.* **2015**, *51*, 1354–1362. [CrossRef]

45. Watson, J.J.W.; Hudson, M.D. Regional Scale wind farm and solar farm suitability assessment using GIS-assisted multi-criteria evaluation. *Landsc. Urban Plan.* **2015**, *138*, 20–31. [CrossRef]

46. Asakereh, A.; Omid, M.; Alimardani, R.; Sarmadian, F. Developing a GIS-based Fuzzy AHP Model for Selecting Solar Energy Sites in Shodirwan Region in Iran. *Int. J. Adv. Sci. Technol.* **2014**, *68*, 37–48. [CrossRef]

47. Noorollahi, E.; Fadai, D.; Shirazi, M.A.; Ghodsipour, S.H. Land Suitability Analysis for Solar Farms Exploitation Using GIS and Fuzzy Analytic Hierarchy Process (FAHP)-A Case Study of Iran. *Energies* **2016**, *9*, 643. [CrossRef]

48. Charabi, Y.; Gastli, A. PV site suitability analysis using GIS-based spatial fuzzy multi-criteria evaluation. *Renew. Energy* **2011**, *36*, 2554–2561. [CrossRef]

49. Suh, J.; Brownson, J.R.S. Solar farm suitability using geographic information system fuzzy sets and analytic hierarchy processes: Case study of Ulleung Island, Korea. *Energies* **2016**, *9*, 648. [CrossRef]

50. Sánchez-Lozano, J.M.; Teruel-Solano, J.; Soto-Elvira, P.L.; García-Cascales, M.S. Geographical Information Systems (GIS) and Multi-Criteria Decision Making (MCDM) methods for the evaluation of solar farms locations: Case study in south-eastern Spain. *Renew. Sustain. Energy Rev.* **2013**, *24*, 544–556. [CrossRef]

51. Sánchez-Lozano, J.M.; Antunes, H.C.; García-cascales, M.S.; Dias, L.C. GIS-based photovoltaic solar farms site selection using ELECTRE-TRI: Evaluating the case for Torre Pacheco, Murcia, Southeast of Spain. *Renew. Energy* **2014**, *66*, 478–494. [CrossRef]

52. Mondino, E.B.; Fabrizio, E.; Chiabrando, R. Site Selection of Large Ground-Mounted Photovoltaic Plants: A GIS Decision Support System and an Application to Italy. *Int. J. Green Energy* **2015**, *12*, 515–525. [CrossRef]
53. Omitaomu, O.A.; Blevins, B.R.; Jochem, W.C.; Mays, G.T.; Belles, R.; Hadley, S.W.; Harrison, T.J.; Bhaduri, B.L.; Neish, B.S.; Rose, A.N. Adapting a GIS-based multicriteria decision analysis approach for evaluating new power generating sites. *Appl. Energy* **2012**, *96*, 292–301. [CrossRef]
54. Choi, Y.; Song, J. Assessment of Photovoltaic Potentials at Abandoned Mine Reclamation Sites in Korea using Renewable Energy Resource Maps. *New Renew. Energy* **2016**, *12*, 44–52. [CrossRef]
55. Kim, S.M.; Oh, M.; Park, H.D. Analysis and Prioritization of the Floating Photovoltaic System Potential for Reservoirs in Korea. *Appl. Sci.* **2019**, *9*, 395. [CrossRef]
56. Lukac, N.; Žlaus, D.; Seme, S.; Žalik, B.; Štumberger, G. Rating of roofs' surfaces regarding their solar potential and suitability for PV systems, based on LiDAR data. *Appl. Energy* **2013**, *102*, 803–812. [CrossRef]
57. Lee, M.; Hong, T.; Jeong, J.; Jeong, K. Development of a rooftop solar photovoltaic rating system considering the technical and economic suitability criteria at the building level. *Energy* **2018**, *160*, 213–224. [CrossRef]
58. Song, J.; Choi, Y.; Jang, M.; Yoon, S. A Comparison of Wind Power and Photovoltaic Potentials at Yeongok, Mulno and Booyoung Abandoned Mines in Kangwon Province, Korea. *J. Korean Soc. Miner. Energy Resour. Eng.* **2014**, *51*, 525–536. [CrossRef]
59. Tarigan, E.; Djuwari; Purba, L. Assessment of PV Power Generation for Household in Surabaya Using SolarGIS—PvPlanner Simulation. *Energy Procedia* **2014**, *47*, 85–93. [CrossRef]
60. Besarati, S.M.; Padilla, R.V.; Goswami, D.Y.; Stefanakos, E. The potential of harnessing solar radiation in Iran: Generating solar maps and viability study of PV power plants. *Renew. Energy* **2013**, *53*, 193–199. [CrossRef]
61. Fichter, T.; Soria, R.; Szklo, A.; Schaeffer, R.; Lucena, A.F.P. Assessing the potential role of concentrated solar power (CSP) for the northeast power system of Brazil using a detailed power system model. *Energy* **2017**, *121*, 695–715. [CrossRef]
62. Lukač, N.; Seme, S.; Žlaus, D.; Štumberger, G.; Žalik, B. Buildings roofs photovoltaic potential assessment based on LiDAR (Light Detection and Ranging) data. *Energy* **2014**, *66*, 598–609. [CrossRef]
63. Bergamasco, L.; Asinari, P. Scalable methodology for the photovoltaic solar energy potential assessment based on available roof surface area: Application to Piedmont Region (Italy). *Sol. Energy* **2011**, *85*, 1041–1055. [CrossRef]
64. Malagueta, D.; Szklo, A.; Borba, B.S.M.C.; Soria, R.; Aragão, R.; Schaeffer, R.; Dutra, R. Assessing incentive policies for integrating centralized solar power generation in the Brazilian electric power system. *Energy Policy* **2013**, *59*, 198–212. [CrossRef]
65. Martín-pomares, L.; Martínez, D.; Polo, J.; Perez-astudillo, D.; Bachour, D.; San, A. Analysis of the long-term solar potential for electricity generation in Qatar. *Renew. Sustain. Energy Rev.* **2017**, *73*, 1231–1246. [CrossRef]
66. Milbrandt, A.R.; Heimiller, D.M.; Perry, A.D.; Field, C.B. Renewable energy potential on marginal lands in the United States. *Renew. Sustain. Energy Rev.* **2014**, *29*, 473–481. [CrossRef]
67. Khan, J.; Arsalan, M.H. Estimation of rooftop solar photovoltaic potential using geo-spatial techniques: A perspective from planned neighborhood of Karachi Pakistan. *Renew. Energy* **2016**, *90*, 188–203. [CrossRef]
68. Izquierdo, S.; Montanés, C.; Dopazo, C.; Fueyo, N. Roof-top solar energy potential under performance-based building energy codes: The case of Spain. *Sol. Energy* **2011**, *85*, 208–213. [CrossRef]
69. Charabi, Y.; Gastli, A. GIS assessment of large CSP plant in Duqum, Oman. *Renew. Sustain. Energy Rev.* **2010**, *14*, 835–841. [CrossRef]
70. Gastli, A.; Charabi, Y.; Zekri, S. GIS-based assessment of combined CSP electric power and seawater desalination plant for Duqum—Oman. *Renew. Sustain. Energy Rev.* **2010**, *14*, 821–827. [CrossRef]
71. Hofierka, J.; Kaňuk, J. Assessment of photovoltaic potential in urban areas using open-source solar radiation tools. *Renew. Energy* **2009**, *34*, 2206–2214. [CrossRef]
72. Catita, C.; Redweik, P.; Pereira, J.; Brito, M.C. Extending solar potential analysis in buildings to vertical facades. *Comput. Geosci.* **2014**, *66*, 1–12. [CrossRef]
73. Izquierdo, S.; Rodrigues, M.; Fueyo, N. A method for estimating the geographical distribution of the available roof surface area for large-scale photovoltaic energy-potential evaluations. *Sol. Energy* **2008**, *82*, 929–939. [CrossRef]
74. Polo, J.; Bernardos, A.; Navarro, A.A.; Fernandez-Peruchena, C.M.; Ramírez, L.; Guisado, M.V.; Martínez, S. Solar resources and power potential mapping in Vietnam using satellite-derived and GIS-based information. *Energy Convers. Manag.* **2015**, *98*, 348–358. [CrossRef]

75. Ko, L.; Wang, J.; Chen, C.; Tsai, H. Evaluation of the development potential of rooftop solar photovoltaic in Taiwan. *Renew. Energy* **2015**, *76*, 582–595. [CrossRef]

76. Lee, M.; Hong, T.; Jeong, K.; Kim, J. A bottom-up approach for estimating the economic potential of the rooftop solar photovoltaic system considering the spatial and temporal diversity. *Appl. Energy* **2018**, *232*, 640–656. [CrossRef]

77. Hong, T.; Lee, M.; Koo, C.; Jeong, K.; Kim, J. Development of a method for estimating the rooftop solar photovoltaic (PV) potential by analyzing the available rooftop area using Hillshade analysis. *Appl. Energy* **2017**, *194*, 320–332. [CrossRef]

78. Song, J.; Choi, Y. Evaluation of rooftop photovoltaic electricity generation systems for establishing a green campus. *Geosyst. Eng.* **2015**, *18*, 51–60. [CrossRef]

79. Choi, Y.; Song, J. Sustainable development of abandoned mine areas using renewable energy systems: A case study of the photovoltaic potential assessment at the tailings dam of abandoned Sangdong mine, Korea. *Sustainability* **2016**, *8*, 1320. [CrossRef]

80. Song, J.; Choi, Y. Analysis of the Potential for Use of Floating Photovoltaic Systems on Mine Pit Lakes: Case Study at the Ssangyong Open-Pit Limestone Mine in Korea. *Energies* **2016**, *9*, 102. [CrossRef]

81. Clifton, J.; Boruff, B.J. Assessing the potential for concentrated solar power development in rural Australia. *Energy Policy* **2010**, *38*, 5272–5280. [CrossRef]

82. Wong, M.S.; Zhu, R.; Liu, Z.; Lu, L.; Peng, J.; Tang, Z.; Lo, C.H.; Chan, W.K. Estimation of Hong Kong' s solar energy potential using GIS and remote sensing technologies. *Renew. Energy* **2016**, *99*, 325–335. [CrossRef]

83. Massimo, A.; Dell'Isola, M.; Frattolillo, A.; Ficco, G. Development of a Geographical Information System (GIS) for the Integration of Solar Energy in the Energy Planning of a Wide Area. *Sustainability* **2014**, *6*, 5730–5744. [CrossRef]

84. Sun, Y.; Hof, A.; Wang, R.; Liu, J.; Lin, Y.; Yang, D. GIS-based approach for potential analysis of solar PV generation at the regional scale: A case study of Fujian Province. *Energy Policy* **2013**, *58*, 248–259. [CrossRef]

85. Bayrakci, M.; Choi, Y.; Brownson, J.R.S. Temperature dependent power modeling of photovoltaics. *Energy Procedia* **2014**, *57*, 745–754. [CrossRef]

86. Mahtta, R.; Joshi, P.K.; Jindal, A.K. Solar power potential mapping in India using remote sensing inputs and environmental parameters. *Renew. Energy* **2014**, *71*, 255–262. [CrossRef]

87. Lopez, A.; Roberts, B.; Heimiller, D.; Blair, N.; Porro, G.; Lopez, A.; Roberts, B.; Blair, N.; Porro, G.U.S. Renewable Energy Technical Potentials A GIS-Based Analysis. Available online: https://www.nrel.gov/docs/fy12osti/51946.pdf (accessed on 24 December 2018).

88. Monforti, F.; Huld, T.; Bódis, K.; Vitali, L.; D'Isidoro, M.; Lacal-arántegui, R. Assessing complementarity of wind and solar resources for energy production in Italy. A Monte Carlo approach. *Renew. Energy* **2014**, *63*, 576–586. [CrossRef]

89. Niblick, B.; Landis, A.E. Assessing renewable energy potential on United States marginal and contaminated sites. *Renew. Sustain. Energy Rev.* **2016**, *60*, 489–497. [CrossRef]

90. Choi, Y.; Choi, Y.; Suh, J.; Park, H.; Jang, M.; Go, W.R. Assessment of Photovoltaic Potentials at Buguk, Sungsan and Younggwang Abandoned Mines in Jeollanam-do, Korea. *J. Korean Soc. Miner. Energy Resour. Eng.* **2013**, *50*, 827–837. [CrossRef]

91. Buffat, R.; Grassi, S.; Raubal, M. A scalable method for estimating rooftop solar irradiation potential over large regions. *Appl. Energy* **2018**, *216*, 389–401. [CrossRef]

92. Gastli, A.; Charabi, Y. Solar electricity prospects in Oman using GIS-based solar radiation maps. *Renew. Sustain. Energy Rev.* **2010**, *14*, 790–797. [CrossRef]

93. He, G.; Kammen, D.M. Where, when and how much solar is available? A provincial-scale solar resource assessment for China. *Renew. Energy* **2016**, *85*, 74–82. [CrossRef]

94. Köberle, A.C.; Gernaat, D.E.H.J.; van Vuuren, D.P. Assessing current and future techno-economic potential of concentrated solar power and photovoltaic electricity generation. *Energy* **2015**, *89*, 739–756. [CrossRef]

95. Peterseim, J.H.; Herr, A.; Miller, S.; White, S.; O'Connell, D.A. Concentrating solar power/alternative fuel hybrid plants: Annual electricity potential and ideal areas in Australia. *Energy* **2014**, *68*, 698–711. [CrossRef]

 *applied sciences*

Article

Wavelet Decomposition and Convolutional LSTM Networks Based Improved Deep Learning Model for Solar Irradiance Forecasting

Fei Wang [1,2,3,*], Yili Yu [2], Zhanyao Zhang [2], Jie Li [2], Zhao Zhen [2,*] and Kangping Li [2]

[1] State Key Laboratory of Alternate Electrical Power System with Renewable Energy Sources, North China Electric Power University, Baoding 071003, China

[2] Department of Electrical Engineering, North China Electric Power University, Baoding 071003, China; ylfisher@sina.com (Y.Y.); zzyaoo@sina.com (Z.Z.); ncepu_lijie@sina.com (J.L.); kangpingli@ncepu.edu.cn (K.L.)

[3] Hebei Key Laboratory of Distributed Energy Storage and Micro-grid, North China Electric Power University, Baoding 071003, China

* Correspondence: feiwang@ncepu.edu.cn (F.W.); zhenzhao@ncepu.edu.cn (Z.Z.); Tel.: +86-139-0312-5055 (F.W.); +86-136-6339-9230 (Z.Z.)

Received: 8 July 2018; Accepted: 24 July 2018; Published: 1 August 2018

Abstract: Solar photovoltaic (PV) power forecasting has become an important issue with regard to the power grid in terms of the effective integration of large-scale PV plants. As the main influence factor of PV power generation, solar irradiance and its accurate forecasting are the prerequisite for solar PV power forecasting. However, previous forecasting approaches using manual feature extraction (MFE), traditional modeling and single deep learning (DL) models could not satisfy the performance requirements in partial scenarios with complex fluctuations. Therefore, an improved DL model based on wavelet decomposition (WD), the Convolutional Neural Network (CNN), and Long Short-Term Memory (LSTM) is proposed for day-ahead solar irradiance forecasting. Given the high dependency of solar irradiance on weather status, the proposed model is individually established under four general weather type (i.e., sunny, cloudy, rainy and heavy rainy). For certain weather types, the raw solar irradiance sequence is decomposed into several subsequences via discrete wavelet transformation. Then each subsequence is fed into the CNN based local feature extractor to automatically learn the abstract feature representation from the raw subsequence data. Since the extracted features of each subsequence are also time series data, they are individually transported to LSTM to construct the subsequence forecasting model. In the end, the final solar irradiance forecasting results under certain weather types are obtained via the wavelet reconstruction of these forecasted subsequences. This case study further verifies the enhanced forecasting accuracy of our proposed method via a comparison with traditional and single DL models.

Keywords: solar irradiance forecasting; wavelet decomposition; convolutional neural network; recurrent neural network; long short term memory

---

1. Introduction

1.1. Background and Motivation

With the global attention to environmental issues, the solar photovoltaic (PV) power has been increasingly regarded as an important kind of renewable energy used to supply clean energy for the power grid [1]. Nearly 60% of power generated in 2040 is projected to come from renewables, which wind and solar PV accounts for more than 50%. Additionally, International Energy Agency (IEA) reported that the installed solar PV capacity has already reached more than 300 GW by the end of 2016 [2]. The annual market of solar PV power has increased by nearly 50%. The top five countries, led by China, accounted for 85% of additions [3]. The above phenomena verified that solar PV power was the world's leading source of renewables in 2016.

However, the high dependence of solar PV power on geographical locations and weather conditions can lead to the dynamic volatility and randomness characteristics of solar PV output power. This unavoidable phenomenon makes PV power forecasting become an important challenge for the power grid in terms of the effective integration of large-scale PV plants, because accurate solar PV power forecasting can provide expected future PV output power, which provides good guidance for the system operator to design a rational dispatching scheme and maintain the balance between supply and demand sides. At the same time, scheduling PV power and other power reasonably may be helpful for effectively addressing the problems, such as system stability and electric power balance [4]. Therefore, accurate solar PV forecasting is essential for the sustainable and stable operation of the whole power system.

In the actual PV stations, its final PV output is affected by a variety of meteorological factors, such as solar irradiance [5], moisture, ambient temperature, wind velocity and barometric pressure. There are two categories of the existing PV forecasting approaches: direct forecast and step-wise forecast. Direct forecast creates a map between historical power data and power forecast values [6,7]. Differently, the step-wise forecast is comprised of two steps. In the first step, each meteorological factor is predicted at the target time. In the next step, these predicted meteorological factors are then utilized to create a map that can reflect the relationship between these meteorological factors and PV power forecast value. In sum, the reliable information of the relevant meteorological factors is the key to PV power forecasting. Therefore, as the main influence factor of PV power generation, the solar irradiance and its accurate forecasting are the prerequisite for solar PV power forecasting.

1.2. Literature Review

With the fast advancement of forecasting theories [8,9], solar physics [10], stochastic learning [11], and machine learning [12], the relevant technology of the solar irradiance forecasting research area has also developed rapidly. In general, the existing various forecasting models are correspondingly designed for solar PV prediction with different time horizon. For example, the forecasting horizon of Numerical Weather Prediction (NWP) forecasting models is from several hours to several days [13]. Time series forecasting models generate forecast outputs with a time scale that ranges from 5 min to 6 h [14]. Statistical forecasting models based on cloud motion images and satellite information can generate PV forecast value with a time sclerosis of 6 h [15]. In this paper, we focus on day-ahead solar irradiance forecasting which the forecasting horizon is 24 h.

Among the previous studies, solar irradiance forecasting approaches can be generally divided into several categories: statistical approaches, physical approaches and machine learning approaches and ensemble approaches. In physical approaches, three kinds of basic methods are NWP forecasting model [16], Total Sky Imagery (TSI) [17] and cloud moving based satellite imagery models, which can also help to estimate the output power of distributed PV system [18]. These kinds of physical based forecasting models require additional information about the sky image.

As for the statistical approaches, persistence forecasting, time series, and Model Output Statistics (MOS) models [19] are involved. In this model, it is supposed that the forecasting data at time $t + 1$ is equal to the historical data at time t.

Time series approaches primarily aim at the modeling of long-term solar irradiance forecast, which includes Moving Average (MA), Autoregressive (AR) [20], Autoregressive Moving Average (ARMA) [21], and Autoregressive Integrated Moving Average (ARIMA) [22] models. The time series forecasting model only requires historical irradiance data, in which the relevant meteorological factors are not involved. In addition, time series approaches can merely capture linear relationships and require stationary input data or stationary differencing data.

In recent years, machine learning based forecasting methods have also been successfully applied in many fields [23–26]. Machine learning models that have been done widely applied in solar forecasting field are non-linear regression models such as Artificial Neural Network (ANNs) [27,28], the Support Vector Machine (SVM) [29], and the Markov chain [30]. These nonlinear regression models are also frequently used together with the classification models [31].

Regarding the ensemble approach, this kind of integrated model consists of multiple trained forecasting sub-models. Additionally, all the outputs of these forecasting sub-models are taken into consideration to determine the best output of the ensemble model. This method can well leverage the advantages of different forecasting sub-models to achieve the performance optimization of the ensemble model to provide better forecasting results for application [32,33].

Based on the abovementioned forecasting theories, many researchers have carried out important research work in the field of solar irradiance forecasting and PV power forecasting (both referred to as "solar forecasting" in what follows). Considering this abundant literature on solar forecasting, Yang et al. [34] have conducted an adequate literature review work on the history and trends in solar irradiance and PV power forecasting through text mining. Furthermore, Wan et al. [35] have also reviewed the state-of-the-art of PV and solar forecasting methodologies developed over the past decade. Regarding the forecasting of grid-connected photovoltaic plant production, Ferlito et al. [36] implemented a comparative analysis of eleven forecasting data-driven models online and offline. The above eleven models include: (1) simple linear models, such as Multiple Linear Regression; (2) nonlinear models, such as Extreme Learning Machines and weighted k-Nearest Neighbors; and (3) ensemble methods, such as Random Forests and Extreme Gradient Boosting. To improve real-time control performance and reduce possible negative impacts of PV systems, Yang et al. [37] proposed a weather-based hybrid method for 1-day ahead hourly forecasting of PV power output with the application of Self-organizing Map (SOM), Learning Vector Quantization (LVQ) and Support Vector Regression (SVR). Gensler et al. [38] used auto-encoder to reduce the dimension of historical data, and employed LSTM to forecast solar power.

In the field of solar forecasting, a few researchers have also paid attention to the prediction of solar irradiance due to its important influence on PV power output. For example, Hussain et al. [39] applied a simple and linear statistical forecasting technique named ARIMA to day ahead hourly forecast of solar irradiance for Abu Dhabi, UAE. In another relevant study, five novel semi-empiric models for hourly solar radiation forecasting are developed and then compared with the Angstrom-Prescott (A-P) type models [40]. Differently, a multi-level wavelet decomposition is applied by Zhen et al. [41] to preprocess the solar irradiance data in order to further improve the day-ahead solar irradiance forecasting accuracy. In Zhen's another paper, a new day-ahead solar irradiance ensemble forecasting model was developed based on time-section fusion pattern classification and mutual iterative optimization [42]. With the emergence of deep learning (DL) models, Qing et al. [43] turned to Long Short Term Memory (LSTM) to catch the dependence between consecutive hours of daily solar irradiance data.

In general, the DL algorithm is more promising compared to the abovementioned traditional machine learning. Recently, DL approaches have been not only successfully applied in image processing [44], but also utilized to address the classification and regression issues of one-dimensional data [45]. In the DL system, there are various branches, including LSTM, Convolutional Neural Networks (CNN), and Recurrent Neural Network (RNN) and so on. In spite of the superior performance of DL algorithms, few studies have applied the DL methods in the day-ahead solar irradiance forecasting. Researchers need to validate whether the introduction of DL can improve the solar irradiance forecasting accuracy. Moreover, there are various versions of DL models just like those mentioned above. Different DL models have their own advantages and disadvantages. Therefore, in the practice of solar irradiance forecasting, three important issues should be taken into consideration, namely how to select the rational DL models, how to well combine them, and how to further improve the performance of the hybrid DL model.

1.3. The Content and Contribution of the Paper

According to the literature review work, we have found that the previous forecasting approaches using manual feature extraction (MFE), traditional modeling and single DL models could not satisfy the performance requirements in partial solar irradiance forecasting scenarios with complex fluctuations. In this paper, we proposed an improved DL model to achieve the performance improvement of day-ahead solar irradiance forecasting. This proposed model is named the DWT-CNN-LSTM model. It should be noted that the historical daily solar irradiance curve always presents high variability and fluctuation since the solar irradiance is influenced by the non-stationary weather conditions. Therefore, the forecasting accuracy of day-ahead solar irradiance strongly depends on the weather statuses no matter what kinds of forecasting models we choose. Given this fact, the DWT-CNN-LSTM models are independently constructed for four general weather types (i.e., sunny, cloudy, rainy, and heavy rainy days). This is because a single forecasting model cannot well reflect the temporal relationships between historical and future solar irradiance under different weather conditions. In other words, classification modeling could reduce the complexity and difficulty of intro-class data fitting to improve the corresponding forecasting accuracy [1,28].

The basic pipeline framework behind data-driven DWT-CNN-LSTM models consists of three major parts: (1) Discrete Wavelet Transformation (DWT) based solar irradiance sequence decomposition, (2) a CNN-based local feature extractor, and (3) an LSTM based sequence forecasting model. In solar irradiance forecasting under certain weather types, the raw solar irradiance sequence is decomposed into several subsequences via discrete wavelet transformation. Then, each subsequence is fed into the CNN-based local feature extractor, which leverages the advantage of CNN to automatically learn the abstract feature representation from the raw subsequence data. Since the extracted features are also time series data, they are individually transported to LSTM to construct the subsequence forecasting model. In the end, the final solar irradiance forecasting results under certain weather types are obtained via the wavelet reconstruction of these forecasted subsequences. Compared to the existing studies for solar irradiance forecasting, the contributions of this paper can be summarized as follows:

(1) Discrete wavelet transformation is applied in our proposed DWT-CNN-LSTM model to decompose the raw solar irradiance sequence data of certain weather types into several stable parts (i.e., low-frequency signals) and fluctuant parts (i.e., high-frequency signals). These decomposed subsequences have better behaviors (e.g., more stable variances and fewer outliers) in terms of regularity than the raw solar irradiance sequence data. Such wavelet decomposition (WD) is helpful for precision improvement of the solar irradiance forecasting model.

(2) The CNN and LSTM are perfectly combined in our proposed DWT-CNN-LSTM model, in which the abstract feature representation from the raw subsequence data is effectively extracted by CNN and then these features are fed into LSTM. CNN is good at automatically extracting abstract features from its input, and LSTM is able to find the long dependencies of the time series input.

(3) The validity of the proposed DWT-CNN-LSTM model is verified based on the two measured dataset, namely the dataset of Elizabeth City State University and Desert Rock Station.

The rest of paper is constructed as follows. Section 2 illustrates the three main parts of the proposed DWT-CNN-LSTM model, including DWT based solar irradiance sequence decomposition, the CNN-based local feature extractor, and the LSTM based sequence forecasting model. In Section 3, the details of the experimental simulation are introduced and the relevant analysis results are discussed. Finally, conclusions are drawn in Section 4.

2. Improved Deep Learning Model for Day-Ahead Solar Irradiance Forecasting

The historical daily solar irradiance curve always presents high variability and fluctuation since solar irradiance is influenced by non-stationary weather conditions. This makes the forecasting accuracy of day-ahead solar irradiance strongly depend on the weather statuses no matter what kinds of forecasting models we choose.

Therefore, as shown in Figure 1, the solar irradiance forecasting models are independently constructed for four general weather types, because according to different weather types, classification modeling could reduce the complexity and difficulty of intro-class data fitting so as to improve the corresponding forecasting accuracy.

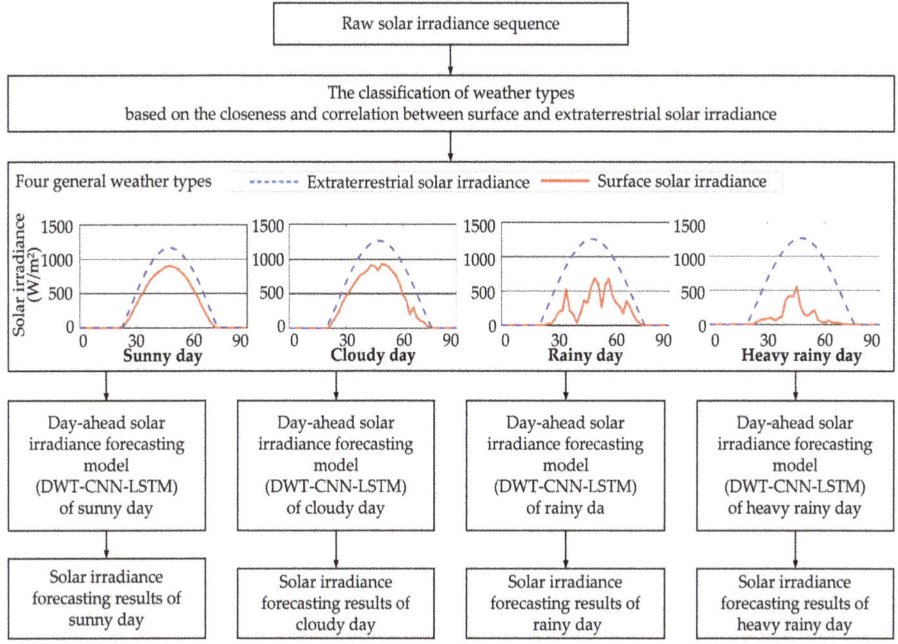

Figure 1. The flowchart of the day–ahead solar irradiance forecasting for four general weather types. The DWT-CNN-LSTM forecasting model is based on discrete wavelet transformation (DWT), convolutional neural network (CNN) and long short term memory (LSTM) network.

In terms of the proposed model (i.e., DWT-CNN-LSTM model) for day-ahead solar irradiance forecasting, its integrated framework is illustrated in Figure 2. The basic pipeline framework behind data-driven DWT-CNN-LSTM models consists of three major parts: (1) DWT based solar irradiance sequence decomposition; (2) CNN based local feature extractor; and (3) LSTM based sequence forecasting model. As for certain weather types, the raw historical solar irradiance sequence is decomposed into approximate subsequence and several detailed subsequences. Then each subsequence is fed to the CNN based local feature extractor, which leverages the advantage of CNN to automatically learn the abstract feature representation from the raw subsequence data. Since the features extracted by the CNN are also time series data that have rich temporal dynamics, then they are input to LSTM to construct the subsequence forecasting model. In the end, the final solar irradiance forecasting results under certain weather types are obtained through the wavelet reconstruction of these forecasted subsequences. More details about three major parts above are respectively illustrated in Sections 2.1–2.3.

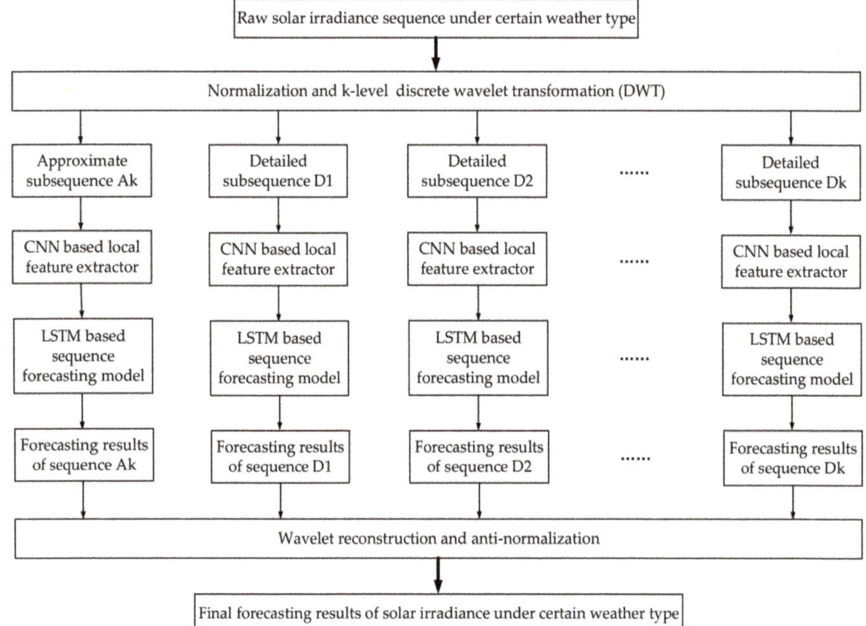

Figure 2. The detailed framework of DWT-CNN-LSTM day-ahead forecasting model for solar irradiance under certain weather type. The DWT-CNN-LSTM forecasting model is based on discrete wavelet transformation (DWT), convolutional neural network (CNN) and long short term memory (LSTM) network.

2.1. Discrete Wavelet Transformation Based Solar Irradiance Sequence Decomposition

In general, solar irradiance sequence data always presents high volatility, variability and randomness due to its correlation to non-stationary weather conditions. Therefore, the raw solar irradiance sequence probably includes nonlinear and dynamic components in the form of spikes and fluctuations. The existence of these components will undoubtedly deteriorate the precision of the solar irradiance forecasting models. In practice, high-frequency signals and low-frequency signals are contained in solar irradiance sequence data. The former primarily results from the chaotic nature of the weather system. The latter is caused by the daily rotation of the earth. As for each signal with certain frequency, it is easier for a specific sequence forecasting model to predict the corresponding outliners and behaviors of that signal. Given the above considerations, DWT is employed here to decompose the raw solar irradiance sequence data into several stable parts (i.e., low-frequency signals) and fluctuant parts (i.e., high-frequency signals). These decomposed subsequences have better behaviors (e.g., more stable variances and fewer outliers) in terms of regularity than the raw solar irradiance sequence data, which is helpful for the precision improvement of the solar irradiance forecasting model [46].

In numerical analysis, DWT is a kind of wavelet transform for which the wavelets are discretely sampled. The key advantage of DWT over Fourier transforms is that DWT is able to capture both frequency and location information (location in time). In addition, DWT is good at the processing of multi-scale information processing [47]. These superiorities make DWT an efficient tool for complex data sequence analysis. In wavelet theory, the original sequence data are generally decomposed into two parts called approximate subsequence and detailed subsequence via DWT. The approximate subsequence captures the low-frequency features of the original sequence, while the detailed subsequence contains the high-frequency features. This process is regarded as wavelet decomposition (WD), and the approximate subsequences obtained from the original sequence can also be further decomposed by WD process. Then the high-frequency noise in the forms of the fluctuation and randomness in original sequence can be extracted and filtered through WD process.

Given a certain mother wavelet function $\psi(t)$ and its corresponding scaling function $\varphi(t)$, a sequence of wavelet $\psi_{j,k}(t)$ and binary scale-functions $\varphi_{j,k}(t)$ can be calculated as follows:

$$\psi_{j,k}(t) = 2^{\frac{j}{2}}\psi\left(2^j t - k\right) \tag{1}$$

$$\varphi_{j,k}(t) = 2^{\frac{j}{2}}\varphi\left(2^j t - k\right) \tag{2}$$

in which t, j and k respectively denote the time index, scaling variable and translation variable. Then the original sequence $os(t)$ can be expressed as follows:

$$os(t) = \sum_{k=1}^{n} c_{j,k}\varphi_{j,k}(t) + \sum_{j=1}^{J}\sum_{k=1}^{n} d_{j,k}\psi_{j,k}(t) \tag{3}$$

in which $c_{j,k}$ is the approximation coefficient at scale j and location k, $d_{j,k}$ denotes the detailed coefficient at scale j and location k, n is the size of the original sequence, and J is the decomposition level. Based on the fast DWT proposed by Mallat [48], the approximate sequence and detailed sequence under a certain WD level can be obtained via multiple low-pass filters (LPF) and high-pass filters (HPF).

Figure 3 exhibits the specific WD process in our practical work. During a certain k-level WD process, the raw solar irradiance sequence of certain weather types is first decomposed into two parts: approximate subsequence A1 and detailed subsequence D1. Next, the approximate subsequence A1 is further decomposed into another two parts namely A2 and D2 at WD level 2, and continues to A3 and Ds at WD level 3, etc. Therefore, as shown in Figure 2, the approximate subsequence Ak and detailed subsequences D1 to Dk can be individually forecasted by various time sequence forecasting models (i.e., our proposed CNN-LSTM model, autoregressive integrated moving average model, support vector regression, *etc*). Then the final forecasting results of solar irradiance sequence can be obtained through the wavelet reconstruction on the forecasting results of Ak and D1 to Dk.

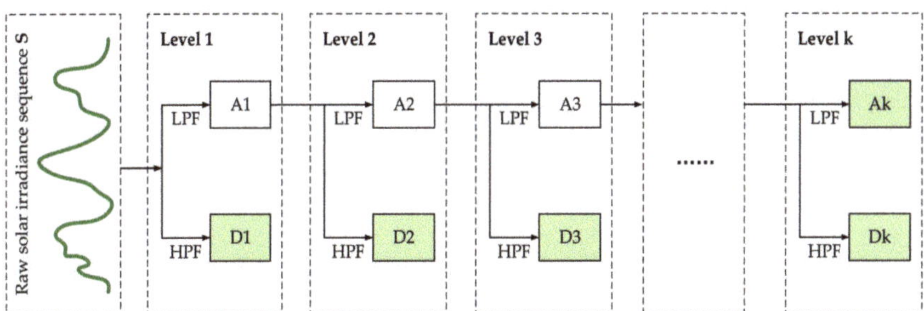

Figure 3. The detailed process of k-level wavelet decomposition. A1 to Ak are the approximate subsequences, and D1 to Dk are the detailed subsequences. All of these subsequences can be forecasted individually using some kind of time sequence forecasting models.

2.2. Convolutional Neural Networks Based Local Feature Extractor

Generally speaking, the historical solar irradiance sequence data is the most important input that contains abundant information for forecasting the day-ahead solar irradiance. In our proposed DWT-CNN-LSTM model, the original solar irradiance sequence under certain weather type is decomposed through DWT into several subsequences. These subsequences also include relevant and significant information that is useful for the later forecasting of subsequences. Therefore, the effective extraction of local features that are robust and informative from the sequential input is very important for enhancing the forecasting precision. Traditionally, many previous works primarily focused on multi-domain feature extractions [49], including statistical (variance, skewness, and kurtosis) features, frequency (spectral skewness) features, time frequency (wavelet coefficients) features, etc. However, these hand-engineered features require intensive expert knowledge of the sequence characteristics and cannot necessarily capture the intrinsic sequential characteristic behind the input data. Moreover, knowing how to select these manually extracted features is another big challenge. Unlike manual feature extraction, CNN is an emerging branch of DL that is used for automatically generating useful and discriminative features from raw data, which has already been broadly applied in image recognition, speech recognition, and natural language processing [50].

As for application, the subsequences decomposed from solar irradiance sequence can be regarded as 1-dimensional sequences. Thus 1-dimensional CNN is adopted here to work as a local feature extractor. The key idea of CNN lies in the fact that abstract features can be extracted by convolutional kernels and the pooling operation. In practice, to address the sequences, the convolutional layers (convolutional kernels) firstly convolve multiple local filters with the sequential input. Each feature map corresponding to each local filter can be generated by sliding the filter over the whole sequential input. Subsequently, the pooling layer is utilized to extract the most significant and fixed-length features from each feature map. In addition, the convolution and pooling layers can be combined in a stacked way.

First of all, the most simply constructed CNN with only one convolutional layer and one pooling layer is introduced to briefly show how the CNN directly process the raw sequential input. It is assumed that K filters with a window size of m are used in the convolutional layer. The details of the relevant mathematical operation in these two layers are presented in the following two subsections.

(1)   Convolutional Layer

Convolution operation is regarded as a specific linear process that aims to extract local patterns in the time dimension and to find local dependencies in the raw sequences. The raw sequential input S and filter sequence FS is defined as follows. Here vectors are expressed in bold according to the convention.

$$S = [s_1, s_2, s_3, \cdots, s_L] \tag{4}$$

$$FS = [\boldsymbol{w}_1, \boldsymbol{w}_2, \boldsymbol{w}_3, \cdots, \boldsymbol{w}_K] \tag{5}$$

in which $s_i \in R$ is the single sequential data point that is arrayed according to time, and $\boldsymbol{w}_j \in R^{m \times 1}$ is one of the filter vectors. L is the length of the raw sequential input S, and K is the number of total filters in the convolutional layer. Then the convolution operation is defined as a multiplication operation between a filter vector $\boldsymbol{w}_j$ and a concatenation vector representation $s_{i:i+m-1}$.

$$s_{i:i+m-1} = s_i \oplus s_{i+1} \oplus s_{i+2} \oplus \cdots \oplus s_{i+m-1} \tag{6}$$

in which $\oplus$ is the concatenation operator, and $s_{i:i+m-1}$ denotes a window of m continuous time steps starting from the i-th time step. Moreover, the bias term $b \in R$ should also be considered into the convolution operation. Thus, the final calculation equation is written as follows.

$$c_i = f\left(\boldsymbol{w}_j^{\mathrm{T}} s_{i:i+m-1} + b\right) \tag{7}$$

in which $\boldsymbol{w}_j^{\mathrm{T}}$ represents the transpose of a filter matrix $\boldsymbol{w}_j$, and f is a nonlinear activation function. In addition, index i denotes the i-th time step, and index j is the j-th filter.

The application of activation function aims to enhance the ability of models to learn more complex functions, which can further improve forecasting performance. Applying suitable activation function can not only accelerate the convergence rate but also improve the expression ability of model. Here, Rectified Linear Units (ReLu) are adopted in our model due to their superiority over other kinds of activation functions [51].

(2) Pooling layer

In the above subsection, the given example only introduces the detailed convolution operation process between one filter and the input sequence. In actual application, one filter can only generate one feature map. Generally, multiple filters are set in the convolution layer in order to better excavate the key features of input data. Just as assumed above, there are K filters with a window size of m in the convolutional layer. In Equations (5) and (7), each vector w_j represents a filter, and the sing value c_i denotes the activation of the window.

The convolution operation over the whole sequential input is implemented via sliding a filtering window from the beginning time step to the ending time step. So the feature map corresponding to that filter can be denoted in the form of a vector as follows.

$$F_j = [c_1, c_2, c_3, \cdots, c_{L-m+1}] \tag{8}$$

in which index j is the j-th filter, and the elements in F_j corresponds to the multi-windows as $\{s_{1:m}, s_{2:m}, \cdots, s_{l-m+1:L}\}$.

The function of pooling is equal to subsampling as it subsamples the output of convolutional layer based on the definite pooling size p. That means the pooling layer can effectively compress the length of feature map so as to further reduce the number of model parameters. Based on the max-pooling applied in our model, the compressed feature vector $F_{j-compress}$ can be obtained as follows. In addition, the max operation takes a max function over the p consecutive values in feature map F_j.

$$F_{j-compress} = [h_1, h_2, h_3, \cdots, h_{\frac{L-m}{p}+1}] \tag{9}$$

in which $h_j = \max\left(c_{(j-1)p}, c_{(j-1)p+1}, \cdots, c_{jp-1}\right)$.

In the application in our solar irradiance forecasting, the solar irradiance sequence input is a vector with only one dimension. The subsequences that are decomposed from the solar irradiance sequence are also a vector with only one dimension. Therefore, the size of the input subsequences in the convolution layer is $n \times L \times 1$. n is the number of data samples and L is the length of the subsequences. The size of the corresponding outputs after the pooling layer is $n \times ((L - m)/p + 1) \times K$. It can be obviously noted that the length of the input sequence is compressed from L to $((L - m)/p + 1)$.

In sum, the CNN based feature extractor can provide more representative and relevant information than the raw sequential input. Moreover, the compression of the input sequence's length also increases the capability of the subsequent LSTM models to capture temporal information.

To give a brief illustration, the framework for the CNN-based local feature extractor is shown in Figure 4. Additionally, in the actual application, some important parameters need to be set according to the specific circumstances. These parameters include the number of the convolutional and pooling layers, the number of filters in each convolution layer, the sliding steps, the size of sliding window, the pooling size, etc.

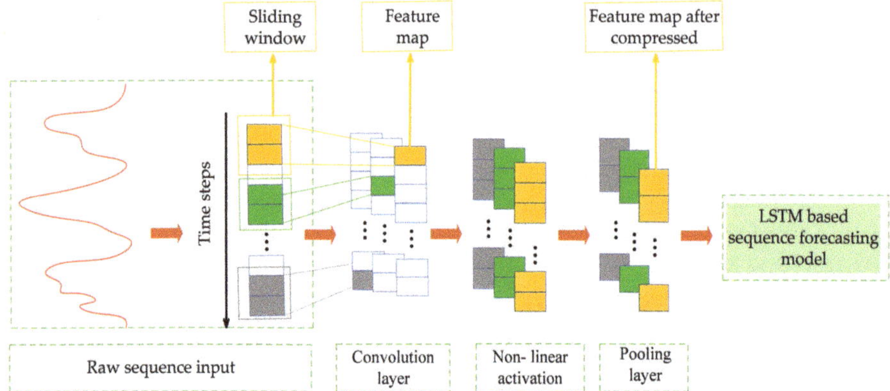

Figure 4. The picture shows the framework of the CNN based local feature extractor. The convolution layer consists of different filters marked by yellow, green and grey colors. Each filter can generate a specific feature map to extract the key information of the raw sequence input through sliding the corresponding windows. The activation function is used to enhance the ability of models to learn more complex functions. The function of pooling is equal to subsampling as it subsamples the output of convolutional layer based on the definite pooling size.

2.3. Long Short Term Memory Based Sequence Forecasting Model (from RNN to LSTM)

In the previous works, some sequence models (e.g., Markov models, Kalman filters and conditional random fields) are commonly used tools to address the raw sequential input data. However, the biggest drawback of these traditional sequential models is that they are unable to adequately capture long-range dependencies. In the application of day-ahead solar irradiance, many indiscriminative or even noisy signals that exist in the sequential input during a long time period may bury informative and discriminative signals. This can lead to the failure of these above sequences models. Recently, RNN has emerged as one effective model for sequence learning, which has already been successfully applied in the various fields, including image captioning, speech recognition, genomic analysis and natural language processing [52].

In our proposed DWT-CNN-LSTM model, LSTM that overcomes the problems of gradient exploding or vanishing in RNN, is adopted to take the output of CNN based local feature extractor to further predict the targeted subsequences. As mentioned in Section 2.1, these subsequences are decomposed from solar irradiance data. In the following two subsections, the principle of RNN is simply introduced and the construction of its improved variant (i.e., LSTM) is then illustrated in detail.

2.3.1. Recurrent Neural Network

The traditional neural network structure is characterized by the full connections between neighboring layers, which can only map from current input to target vectors. However, RNN has the ability to map target vectors from the whole history of the previous inputs. Thus RNN is more effective at modeling dynamics in sequential data when compared to traditional neural networks. In general, RNN builds connections between units from a directed cycle and memorizes the previous inputs via its internal state. Specifically speaking, the output of RNN at time step $t-1$ could influence the output of RNN at time step t. This makes RNN able to establish the temporal correlations between present sequence and previous sequences. The structure of RNN is shown in Figure 5.

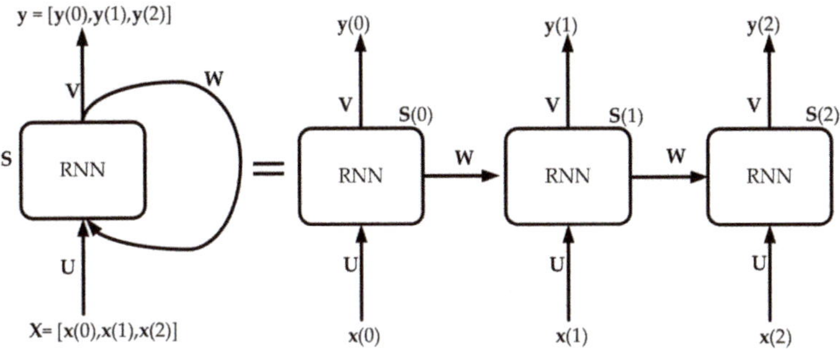

Figure 5. The structure of Recurrent Neural Network.

In Figure 5, the sequential vectors $X = [x(0), x(1), x(2)]$ are passed into RNN one by one according to the set time step. This is obviously different from the traditional feed-forward network in which all the sequential vectors are fed into the model at one time. The relevant mathematical equation can be described as follows.

$$S(t) = \sigma(U{\cdot}x(t) + W{\cdot}S(t-1) + b) \tag{10}$$

$$y(t) = \sigma(V{\cdot}s(t) + c) \tag{11}$$

in which $x(t)$ is the input variable at t time step, W, U and V are weight matrixes, b and c are the biases vectors, σ is activation functions, and $y(t)$ is the expected output at t time step.

Although RNN is very effective at modeling dynamics in sequential data, it can suffer from the gradient vanishing and explosion problem in its backpropagation based model training when modeling long sequences [53]. Considering the inherent disadvantages of typical RNN, its improved variant named LSTM is adopted in our work, which is illustrated in the following subsection.

2.3.2. Long-Short-Term Memory

LSTM network proposed by Hochreiter et al. [53] in 1997 is a variant type of RNN, which combines representation learning with model training without requiring additional domain knowledge. The improved construction of LSTM is helpful for the achievement of avoiding gradient vanishing and explosion problems in typical RNN. This means that LSTM is superior at capturing long-term dependencies and modeling nonlinear dynamics when addressing the sequential data with a longer length. The structure of LSTM cell is shown in Figure 6.

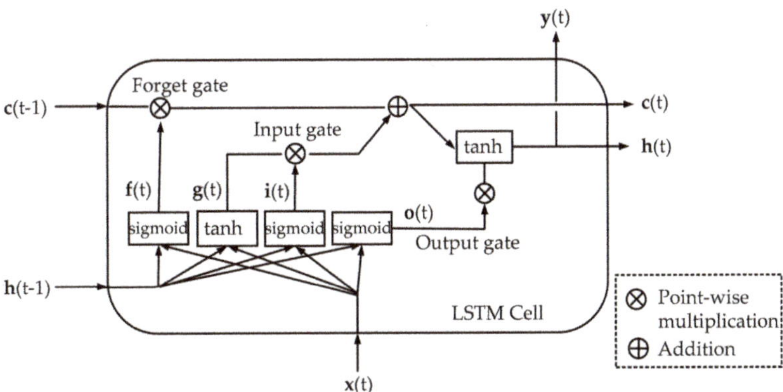

Figure 6. The structure of Long Short-Term Memory Cell.

LSTM is explicitly designed to overcome the problem of gradient vanishing, by which the correlation between vectors in both short and long-term can be easily remembered. In LSTM cell, $h(t)$ can be considered as a short-term state, and $c(t)$ can be considered as a long-term state. The significant characteristic of LSTM is that it can learn what needs to be stored in the long-term, what needs to be thrown away and what needs to be read. When $c(t-1)$ point enters into cell, it first goes through a forget gate to drop some memory; then, some new memories are added to it via an input gate; finally, a new output $y(t)$ that is filtered by the output gate is obtained. The process of where the new memories come from and how these gates work is shown below.

(1)   Forget

This part reveals how LSTM controls what kinds of information can enter into the memory cell. After $h(t-1)$ and $x(t)$ has passed through sigmoid function, a value $f(t)$ between 0 and 1 is generated. The value of 1 means that $h(t-1)$ will be completely absorbed in the cell state $c(t-1)$. On the contrary, if the value is 0, $h(t-1)$ will be abandoned by cell state $c(t-1)$. The formula of this process is shown below.

$$f(t) = \sigma(w_f \cdot [h(t-1), x(t)] + b_f) \tag{12}$$

in which W_f weight matrix, b_f is biases vectors, and σ is activation function.

(2)   Store

This part shows how LSTM decides what kinds of information can be stored in the cell state. First, $h(t-1)$ passes through sigmoid function, and a value $i(t)$ between 0 and 1 is then obtained. Next, $h(t-1)$ passes through tanh function and then a new candidate value $g(t)$ is obtained. In the end, the above two steps can be integrated to update the previous state.

$$i(t) = \sigma(W_i \cdot [h(t-1), x(t)] + b_i) \tag{13}$$

$$g(t) = tanh(W_g \cdot [h(t-1), x(t)] + b_g) \tag{14}$$

Then the previous cell state $c(t-1)$ considers what information should be abandoned and stored and then creates a new cell state $c(t)$. This process can be formulated as follows.

$$c(t) = f(t) \cdot c(t-1) + i_t \cdot g_t \tag{15}$$

(3)   Output

The output of LSTM is based on the updated cell state $c(t)$. First of all, we employ the sigmoid function to generate a value $o(t)$ to control the output. Then tanh and the output of sigmoid function $o(t)$ are further utilized to generate the cell state $h(t)$. Thus we can output $y(t)$ after the above process as shown in the following two steps.

$$o(t) = \sigma(W_o \cdot [h(t-1), x(t)] + b_o) \tag{16}$$

$$y(t) = h(t) = o(t) * tanh(C(t)) \tag{17}$$

The training process of LSTM is called BPTT (backpropagation through time) [54].

3. Case Study

3.1. Data Source and Experimental Setup

The historical irradiance data applied in the above proposed solar irradiance forecasting models is based on the dataset of Elizabeth City State University and Desert Rock Station. The first irradiance dataset in our simulation is downloaded from the National Renewable Energy Laboratory (NREL), which is measured by the Elizabeth City State University at Elizabeth City from 2008 to 2012 [55]. There are 1817 days of solar irradiance data available with 5 min time resolution. The second irradiance dataset in our simulation is downloaded from the National Oceanic & Atmospheric Administration (NOAA) Earth System Research Laboratory website, which is measured by the Surface Radiation station at Desert Rock from 2014 to 2017 [56]. There are 1196 days of solar irradiance data available with 1min time resolution.

To meet the international standard of short-period solar irradiance forecasting, the irradiance data should be further transformed to be the data with 15 min time resolution by taking the average of irradiance points data in the span of every 15 min. Therefore, there are total 96 irradiance data points in one day. Considering the earliest sunrise time and the latest sunset time in three years, we only use daily data points that range from 18th to 78th. As for the forecast periodicity, we use the historical irradiance data from the previous three days to predict the irradiance value for the next day. Therefore, in the solar irradiance forecasting model, the input variable is the historical irradiance data from the previous three days and the output variable is the predicted irradiance value for the next day.

All experimental platforms are built on high-performance Lenovo desktop computer equipped with the Win10 operating system, Intel(R) Core(TM) i5-6300HQ CPU@2.30GHz, 8.00 GB RAM, and NVIDIA GeForce GTX 960M GPU. We use Python 3.6.1 with Keras [57] and Scikit-learn [58] to establish the DWT-CNN-LSTM forecasting models for day-ahead solar irradiance.

3.2. Model Training and Hyperparameters Selection

In the DL based forecasting models, the mean square error (MSE) is chosen as loss function, and Adam Optimization is selected as an optimizer. During the deep learning training process, weight initialization and bias initialization play a vital role. Therefore, we choose the data from truncated normal distribution with 0 mean and 0.05 standard deviation as weight initialization method of CNN and fully connected layer. This method is the recommended initializer for neural network weights and filters. Orthogonal method, a popular initialization way, is selected as weight initializer for LSTM block. The bias for all hidden layers is set as 0.1. The learning rate is 0.001, the batch size is 24 and the epoch is 200.

In addition, for two dataset, the numbers of training set and the testing set are different under four general weather types. The training set is used for training forecasting model, the testing set for evaluating forecasting result. All the above mentioned details of the division of training and testing sets, as well as parameter setting of DWT-CNN-LSTM model, are listed in Tables 1 and 2.

Table 1. The division detail of samples sets under four general weather types.

Weather Types	Elizabeth City	Desert Rock Station
Sunny type	The number of training set: {288} The number of validation set: {32} The number of testing set: {80}	The number of training set: {412} The number of validation set {46} The number of testing set: {115}
Cloudy type	The number of training set: {504} The number of validation set {56} The number of testing set: {140}	The number of training set: {230} The number of validation set{25} The number of testing set:{65}
Rainy type	The number of training set: {366} The number of validation set {40} The number of testing set: {100}	The number of training set: {147} The number of validation set {14} The number of testing set: {40}
Heavy Rainy type	The number of training set: {153} The number of validation set {16} The number of testing set: {42}	The number of training set: {72} The number of validation set {10} The number of testing set: {20}

Table 2. The parameter setting detail of DWT-CNN-LSTM model.

Option	Parameter Setting
Training Method	Adam Optimizer
Learning rate	{0.001}
Batch size	{24}
Epoch	{200}
Training stop strategy	{early stopping}
Loss Function	MSE

We set the split proportion of training set, validation set and testing set as 0.7:0.1:0.2. The training set is used to train the solar irradiance forecasting models. The validation set is used to adjust the hyper-parameters of these DL forecasting models. The testing set is used to verify the model performance.

For the proposed model, we first design two CNN layers with 64 filters, and the filter size and pooling size are both set to 3. Then, two LSTM layers are connected to CNN output with 100 neurons. The outputs of LSTM are fed into two fully connected layers with linear activation function. The Relu activation function is applied to CNN and LSTM layers. To overcome the overfitting problems in models, dropout method with 0.2 parameter is applied after CNN and LSTM layers. In addition, early stopping method is also applied. In addition, the output data format of the input layer, each intermediate layer, and the output layer are accordingly shown in Table 3. Additionally, Table 4 illustrates the structure of the other forecasting models used as benchmarks.

Table 3. The output data format of the input layer, each intermediate layer, and the output layer in DWT-CNN-LSTM model.

Layer	Output Shape
Input layer	(180,1)
Conv-1D layer	(180,64)
Max-Pooling layer	(60,64)
Conv-1D layer	(60,64)
Max-Pooling layer	(20,64)
LSTM layer	(20,100)
LSTM layer	(100)
Fully connected Layer	(100)
Output layer	(60,1)

Table 4. The structure of the other forecasting models used as benchmarks.

Forecasting Models	Structure
CNN	Convolutional layer (64 filters + 3 filter size) + maxpooling (3 pooling size) + convolutional layer (64 filters + 3 filter size) MaxPooling (3 pooling size) + Fully connected layer (100 neurons)
LSTM	2 LSTM layers (100 neurons)
ANN	2 fully connected layers (100 neurons)
ARIMA	Determined by the minimum AIC of each input sample

3.3. Performance Criterion

To evaluate the performance of solar irradiance forecasting models, we employ three effective error indexes that are Root Mean Squared Error (RMSE), Mean Absolute Error (MAE), and Correlation Coefficient (R). The smaller RMSE and MAE, together with the higher R denote the good performance of a forecasting model. The mathematical calculation methods of these three error indexes are shown in the following equations in turn.

$$RMSE = \sqrt{\frac{\sum_{t=1}^{N}(y_t - \hat{y}_t)^2}{N}} \tag{18}$$

$$MAE = \frac{\sum_{t=1}^{N}|y_t - \hat{y}_t|}{N} \tag{19}$$

$$R = \frac{Cov(y, \hat{y})}{\sqrt{V(y)}\sqrt{V(\hat{y})}} \tag{20}$$

in which $\hat{y}_t$, y_t are, respectively, the forecasting value and actual value at time t. $\bar{y}$ refers to the mean value of the whole y_t, and N is the sample size of the test set.

3.4. Model Performance Analysis for DWT-CNN-LSTM Model with Different WD Level

In the proposed DWT-CNN-LSTM model, the first step is to decompose the raw solar irradiance sequence of certain weather type into several approximate subsequences and detailed subsequences. The key of this step is the determination of decomposition level. As for the solar irradiance forecasting based on certain dataset, both the higher and lower WD level are not conducive to the performance improvement of subsequent forecasting models. Therefore, in this part, the performance comparison of DWT-CNN-LSTM model with different WD level is conducted using two different datasets, namely the dataset of Elizabeth City State University and Desert Rock Station. The detailed results are respectively shown in Tables 5 and 6. As shown in Table 5, under the sunny weather type, the DWT-CNN-LSTM model without WD performs better than that with WD level 1 to 4. This is mainly because the solar irradiance curve of sunny days is smooth and less fluctuating. Therefore, the application of WD will not bring very obvious improvement of the forecasting performance.

Table 5. The performance comparison of DWT-CNN-LSTM model at different WD levels using the dataset of Elizabeth City State University.

Weather Types	Error Index	Wavelet Decomposition (WD) Level				
		Level 1	Level 2	Level 3	Level 4	without WD
Sunny	MAE	23.174	23.474	24.213	24.848	22.560
	RMSE	36.548	36.363	40.323	41.244	36.226
	R	0.991	0.991	0.989	0.989	0.992
Cloudy	MAE	86.313	81.466	83.547	88.731	86.754
	RMSE	121.506	118.645	124.364	126.149	121.922
	R	0.926	0.928	0.925	0.919	0.925
Rainy	MAE	95.1758	89.503	93.126	93.695	93.694
	RMSE	145.219	139.133	143.919	142.998	142.194
	R	0.748	0.757	0.741	0.741	0.743
Heavy rainy	MAE	41.234	38.642	39.981	42.774	43.435
	RMSE	68.742	67.574	68.981	70.885	70.410
	R	0.628	0.641	0.634	0.611	0.615

Table 6. The performance comparison of DWT-CNN-LSTM model at different WD levels using the dataset of Desert Rock Station.

Weather Types	Error Index	Wavelet Decomposition (WD) Level				
		Level 1	Level 2	Level 3	Level 4	without WD
Sunny	MAE	17.131	17.379	18.249	18.498	16.573
	RMSE	34.299	34.429	35.844	36.477	33.101
	R	0.992	0.991	0.989	0.987	0.993
Cloudy	MAE	62.144	66.499	67.425	68.552	66.661
	RMSE	91.099	95.377	96.374	98.551	96.641
	R	0.965	0.963	0.958	0.957	0.959
Rainy	MAE	131.384	130.194	136.847	138.257	132.83
	RMSE	181.392	180.079	184.963	187.241	182.97
	R	0.865	0.866	0.847	0.832	0.857
Heavy rainy	MAE	68.212	62.160	64.161	65.840	63.448
	RMSE	96.490	94.977	97.203	103.880	96.373
	R	0.657	0.663	0.651	0.619	0.647

Nevertheless, for other three weather types (i.e. cloudy, rainy and heavy rainy) shown in Table 5, DWT based solar irradiance sequence decomposition does enhance the corresponding forecasting performance to a different extent. This can be explained by the fact that the solar irradiance curve of cloudy, rainy and heavy rainy days presents higher volatility, variability and randomness than that of sunny days. Therefore, the raw solar irradiance sequence of cloudy, rainy and heavy rainy days probably includes nonlinear and dynamic components in the form of spikes and fluctuations. The existence of these components will undoubtedly deteriorate the precision of the solar irradiance forecasting models. Additionally, the application of WD can mitigate the above problems.

To summarize the information provided in Table 5, WD cannot effectively improve the forecasting performance of sunny days. Under the other three weather types, DWT-CNN-LSTM model performs best at WD level 2 when using the dataset of Elizabeth City State University. The results of performance comparison shown in Table 6 are different. Specifically speaking, DWT-CNN-LSTM model of cloudy days performs best at WD level 1 rather than WD level 2 when using the dataset of Desert Rock Station. Therefore, we can draw the conclusion that the influence of WD on forecasting performance, as well as the best WD level, generally varies under different weather types and validation datasets.

3.5. Performance Comparison Analysis of Different Solar Irradiance Forecasting Models

The proposed DWT-CNN-LSTM forecasting model is different from the previous traditional solar irradiance forecasting models. The key characteristics of the DWT-CNN-LSTM forecasting model are the perfect combination of the following parts: (1) DWT based solar irradiance sequence decomposition; (2) CNN based local feature extractor; and (3) LSTM based sequence forecasting model. In addition, the solar irradiance forecasting models are individually established under sunny, cloudy, rainy and heavy rainy days. Given this fact, the relevant performance comparison analysis is also shown and discussed under the above four weather types. The involved three error indexes (i.e., RMSE, MAE, and R) are considered as the basis of the following performance comparison analysis of different forecasting models.

3.5.1. Comparison Analysis of Sunny Days

As previously shown in Table 5, the DWT-CNN-LSTM forecasting model of sunny days performs best at WD level 1 among different WD levels. So in this part, the DWT-CNN-LSTM model at WD level 1 is compared with six solar irradiance forecasting models, namely CNN-LSTM (i.e., our proposed model without WD), artificial neural network (ANN), and manually extracted features (ANN, persistence forecasting, CNN and LSTM). As for the manually extracted features-ANN model, the relevant statistical features and their corresponding expressions are shown in Table 7.

Table 7. The list of manually extracted features.

Statistical Features	Expression
Variance [1]	$Z_{var} = (1/n)\sum_{i=1}^{n}(z_i - \mu)^2$
Maximum	$Z_{\max} = \max(z)$
Skewness	$Z_{skew} = E\left[((z - \mu)/\sigma)^3\right]$
Kurtosis	$Z_{skew} = E\left[((z - \mu)/\sigma)^4\right]$
Average	$Z_{aver} = (1/n)\sum_{i=1}^{n} z_i$

[1] z_i is the solar irradiance data point at time i during the whole day. z is the data point set of $\{z_1, z_2, \ldots, z_n\}$.

The performance comparisons of different sunny days' forecasting models using the dataset of Elizabeth City State University and Desert Rock Station are respectively shown in Tables 8 and 9. In Table 8, the prediction accuracy of DWT-CNN-LSTM (WD level 1) is worse than the single CNN-LSTM without WD. The corresponding conclusion can be drawn that the application of DWT based solar irradiance sequence decomposition does not improve the forecasting performance. The reason behind this phenomenon has already been explained in Section 3.5.

Table 8. The performance comparison of different sunny day's forecasting models using the dataset of Elizabeth City State University.

Forecasting Models	MAE	RMSE	R
DWT-CNN-LSTM (WD level 1)	23.174	36.548	0.991
CNN-LSTM	22.560	36.226	0.992
CNN	22.773	36.763	0.992
LSTM	24.497	37.049	0.990
Manually extracted features-ANN	43.045	54.796	0.985
ANN	23.533	36.888	0.989
Persistence forecasting	30.271	41.742	0.987
ARIMA	32.148	40.174	0.988

Table 9. The performance comparison of different sunny day's forecasting models using the dataset of Desert Rock Station.

Forecasting Models	MAE	RMSE	R
DWT-CNN-LSTM (WD level 1)	17.131	34.299	0.992
CNN-LSTM	16.573	32.411	0.993
CNN	16.222	33.178	0.993
LSTM	17.032	33.294	0.992
Manually extracted features-ANN	30.187	44.101	0.981
ANN	17.869	34.783	0.990
Persistence forecasting	21.034	38.341	0.984
ARIMA	20.433	37.781	0.987

As for our proposed model without WD (i.e., CNN-LSTM), it is superior to manually extracted features-ANN. This further verifies the ability of CNN to automatically and effectively extract representative and significant information from the raw input data. Additionally, ANN, persistence forecasting, and ARIMA models perform worse than CNN-LSTM, which also validates the advisability of applying the combined DL models in solar irradiance forecasting. By comparing among CNN-LSTM, CNN and LSTM, the comparing results also verify the reasonableness of the tandem connection of CNN and LSTM, because the performance evaluation (based on MAE, RMSE and R) results of CNN-LSTM are all better than those of CNN and LSTM. The above similar results can also be found in Table 9.

Figure 7 shows the actual and forecasted solar irradiance curve on sunny day pattern using dataset of Elizabeth City State University.

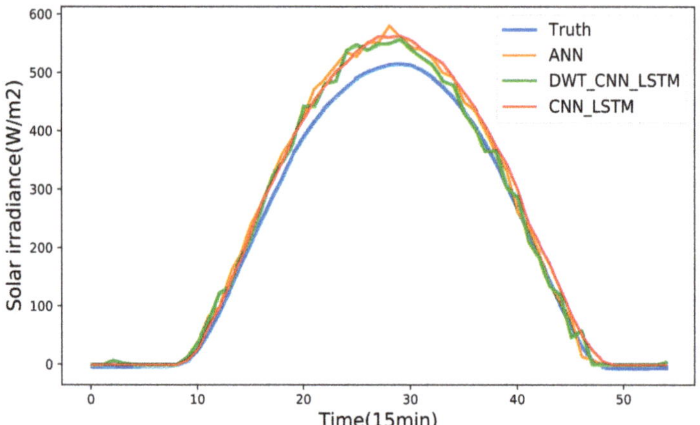

Figure 7. Actual and forecasted solar irradiance on sunny day pattern using dataset of Elizabeth City State University.

3.5.2. Comparison Analysis under Cloudy Day

Based on the dataset of Elizabeth City State University and Desert Rock Station, the performance comparisons among different cloudy day's forecasting models are presented in Tables 10 and 11, respectively. As previously discussed in Table 5, the DWT-CNN-LSTM model of cloudy days has the highest forecasting precision at WD level 2 when using the dataset of Elizabeth City State University. Therefore, as shown in Table 10, the proposed DWT-CNN-LSTM model with WD level 2 is selected to make comparisons with the other kinds of forecasting models.

First of all, it should be noted that all the error index values of DWT-CNN-LSTM (WD level 2) model is better than that of single CNN-LSTM. This result indicates that the DWT based solar irradiance sequence decomposition has the capability to further improve the forecasting performance of combined CNN-LSTM models. As discussed in Section 3.5, the obvious performance improvement can be attributed to the fact that the solar irradiance curve of cloudy days presents high volatility, variability and randomness. Therefore, the cloudy day's solar irradiance sequence includes nonlinear and dynamic components in the form of spikes and fluctuations. The existence of these components will undoubtedly deteriorate the precision of the solar irradiance forecasting models. Additionally, the application of WD could well mitigate the above problems.

Table 10. The performance comparison of different cloudy days' forecasting models using the dataset of Elizabeth City State University.

Forecasting Models	MAE	RMSE	R
DWT-CNN-LSTM (WD level 2)	81.466	118.645	0.928
CNN-LSTM	86.754	121.922	0.925
CNN	87.043	122.042	0.923
LSTM	87.997	122.479	0.921
Manually extracted features-ANN	90.310	125.871	0.905
ANN	89.743	123.532	0.917
Persistence forecasting	95.370	168.443	0.849
ARIMA	110.334	207.694	0.772

Table 11. The performance comparison of different cloudy days' forecasting models using the dataset of Desert Rock Station.

Forecasting Models	MAE	RMSE	R
DWT-CNN-LSTM (WD level 1)	62.761	91.098	0.965
CNN-LSTM	63.661	96.641	0.959
CNN	64.339	95.373	0.961
LSTM	66.752	97.523	0.954
Manually extracted features-ANN	128.06	165.98	0.817
ANN	69.522	100.811	0.950
Persistence forecasting	74.413	114.369	0.939
ARIMA	89.543	150.192	0.865

When compared to the manually extracted features-ANN, as well as the traditional forecasting models (i.e., ANN, persistence forecasting and ARIMA), the comparison results verify our proposed model's advantages in the following two respects. One is the ability to automatically extract representative and significant information from the raw input data, and the other is the ability to capture the long dependencies among the time series input data. In addition, the performance improvement of CNN-LSTM over CNN and LSTM also reveals the benefits of the combination of them. A similar discussion can also be made according to Table 11. Figure 8 shows the actual and forecasted solar irradiance curve on cloudy day pattern using dataset of Elizabeth City State University.

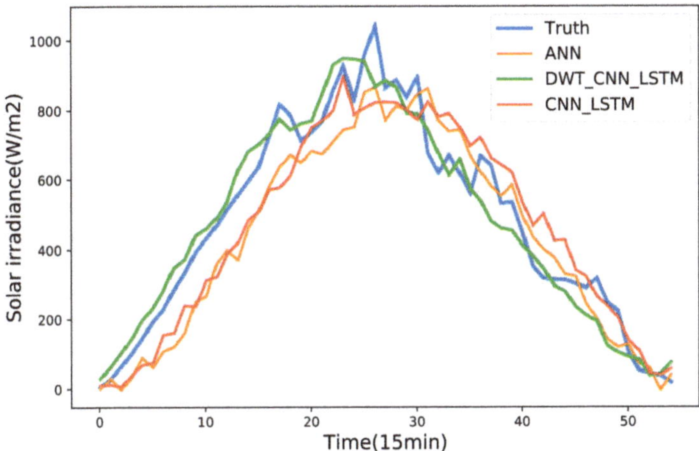

Figure 8. Actual and forecasted solar irradiance on cloudy day pattern using dataset of Elizabeth City State University.

3.5.3. Comparison Analysis under Rainy Days

In terms of the rain day, it is discussed in Section 3.5 that the corresponding DWT-CNN-LSTM model performs best at level 2 whether using the dataset of Elizabeth City State University or Desert Rock Station. Therefore, as shown in Tables 12 and 13, the DWT-CNN-LSTM (WD level 2) is compared with other forecasting models.

Table 12. The performance comparison of different rainy days' forecasting models using the dataset of Elizabeth City State University.

Forecasting Models	MAE	RMSE	R
DWT-CNN-LSTM (WD level 2)	89.503	139.133	0.757
CNN-LSTM	93.694	142.194	0.743
CNN	94.773	143.072	0.737
LSTM	95.089	142.877	0.741
Manually extracted features-ANN	132.321	189.842	0.639
ANN	97.894	147.818	0.736
Persistence forecasting	114.338	173.497	0.680
ARIMA	132.066	181.681	0.656

Table 13. The performance comparison of different rainy days' forecasting models using the dataset of Desert Rock Station.

Forecasting Models	MAE	RMSE	R
DWT-CNN-LSTM (WD level 2)	130.194	180.079	0.866
CNN-LSTM	132.831	181.973	0.857
CNN	132.755	183.076	0.857
LSTM	133.007	184.332	0.855
Manually extracted features-ANN	184.352	225.887	0.769
ANN	138.045	186.553	0.829
Persistence forecasting	155.661	205.340	0.788
ARIMA	177.053	210.119	0.772

When CNN-LSTM and DWT-CNN-LSTM (WD level 2) are compared, the results and the reasons for them are similar to those discussed in Section 3.5.3. Specifically, the MAE is lowered from 93.694 in CNN-LSTM to 89.503 in DWT-CNN-LSTM. The RMSE is lowered from 142.194 in CNN-LSTM to 139.133 in DWT-CNN-LSTM. At the same time, the R has also been improved from 0.743 in CNN-LSTM to 0.757 in DWT-CNN-LSTM. The lower MAE and RMAE denote smaller differences between forecasted and true solar irradiance data, and the higher R also represents that the forecasted solar irradiance curve is closer to the true one. Therefore, the application of the DWT based sequence decomposition also helps the improvement of forecasting performance. Additionally, the combined CNN-LSTM shows better forecasting performance than the rest models (i.e., single DL models and traditional forecasting models). This indicates that the reasonable combination of DL models can better take advantage of the CNN and LSTM.

In sum, the improved DL models (i.e., DWT-CNN-LSTM) not only leverages the advantages of DWT to obtain subsequences with good behavior (e.g., more stable variances and fewer outliers) in terms of regularity, but also absorbs the superiority of CNN-LSTM to automatically extract abstract features and find long dependencies. Similar results can also be found in Table 13. Figure 9 shows the actual and forecasted solar irradiance curve on rainy day pattern using dataset of Elizabeth City State University.

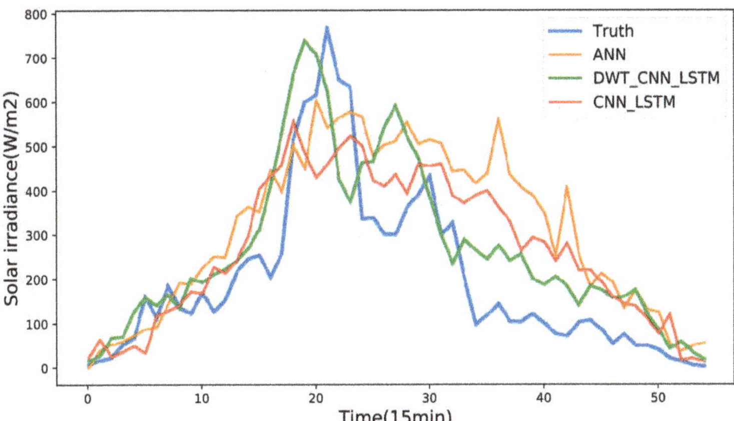

Figure 9. Actual and forecasted solar irradiance on rainy day pattern using dataset of Elizabeth City State University.

3.5.4. Comparison Analysis under Heavy rainy Days

Regarding the weather type of rainy days, the corresponding simulation result in Section 3.5 reveals that the DWT-CNN-LSTM model can reach the best precision at WD level 2. Therefore, the DWT-CNN-LSTM (WD level 2) is adopted once again to be compared with other forecasting models. Similar to the cloudy and rainy days, the solar irradiance data under heavy rainy days is also volatile and fluctuates. The introduction of DWT based sequence decomposition is able to mitigate the adverse influence of fluctuation on forecasting models. This idea is in accordance with comparison results shown in Tables 14 and 15.

Table 14. The performance comparison of different heavy rainy days' forecasting models using the dataset of Elizabeth City State University.

Forecasting Models	MAE	RMSE	R
DWT-CNN-LSTM (WD level 2)	38.642	67.574	0.641
CNN-LSTM	43.435	70.410	0.616
CNN	45.775	73.377	0.611
LSTM	44.373	74.086	0.611
Manually extracted features-ANN	54.580	120.495	0.354
ANN	48.956	77.034	0.589
Persistence forecasting	64.416	107.290	0.401
ARIMA	63.848	110.735	0.388

Table 15. The performance comparison of different heavy rainy days' forecasting models using the dataset of Desert Rock Station.

Forecasting Models	MAE	RMSE	R
DWT-CNN-LSTM (WD level 2)	62.160	94.977	0.680
CNN-LSTM	63.448	95.374	0.647
CNN	64.743	96.774	0.640
LSTM	65.014	97.096	0.641
Manually extracted features-ANN	81.249	138.689	0.454
ANN	66.312	99.863	0.615
Persistence forecasting	75.029	115.696	0.497
ARIMA	79.473	120.744	0.477

Additionally, the great performance improvement is also achieved via automatic feature extraction and long dependency identification, especially under unstable weather conditions. This can also be verified by the following results shown in Table 14. For example, the MAE is reduced a lot from 64.416 in persistence forecasting to 38.642 in DWT-CNN-LSTM (WD level 2). The RMSE is reduced a lot from 107.290 in persistence forecasting to 67.574 in DWT-CNN-LSTM (WD level 2). Additionally, the R is enhanced from 0.401 in persistence forecasting to 0.641 in DWT-CNN-LSTM (WD level 2). The performance improvement achieved by DWT-CNN-LSTM (WD level 2) can also be found when compared with other forecasting models shown in Table 14.

Moreover, it should be noted the applicability degree of DWT-CNN-LSTM model in different weather conditions is different. For instance, as mentioned in Section 3.5.1, the MAE of sunny days' forecasting is decreased little with 30.271 in the persistence forecasting model and 23.174 in the DWT-CNN-LSTM model. Nevertheless, in Table 12, the MAE of heavy rainy' forecasting is reduced a lot from 64.416 in the persistence forecasting model to 38.642 in the DWT-CNN-LSTM model. This further indicates that our proposed model is more applicable for the solar irradiance forecasting of extreme weather conditions. Similar results can also be found in Table 15. Figure 10 shows the actual and forecasted solar irradiance curve for rainy day pattern using dataset of Elizabeth City State University.

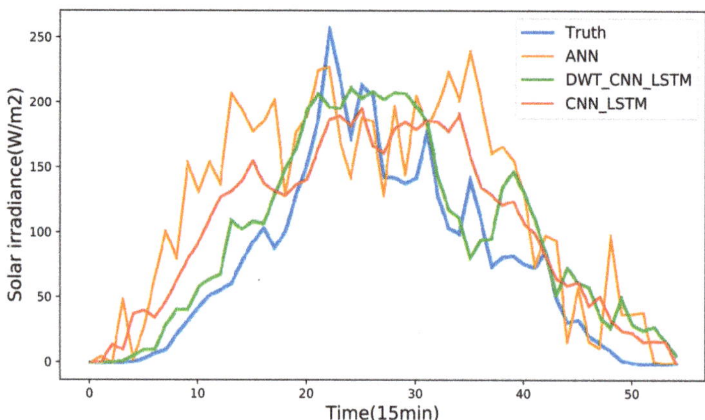

Figure 10. Actual and forecasted solar irradiance on heavy rainy day pattern using dataset of Elizabeth City State University.

3.6. Simulation Discussion

In this paper, an improved DL model (i.e., DWT-CNN-LSTM) based on WD, CNN, and LSTM is proposed for day-ahead solar irradiance forecasting. In the actual simulation based on two datasets, the model performance of DWT-CNN-LSTM model with Different WD Level is assessed for four general weather types (i.e., sunny, cloudy, rainy, and heavy rainy). At the same time, the DWT-CNN-LSTM model with certain WD Level is also compared with other DL models (e.g., CNN and LSTM) and traditional forecasting models (e.g., ANN, persistence forecast and ARIMA) for each weather type. The information previously shown in Tables 5–15 is vividly described in the following Figures 11–14, which is conducive to further summary. The changing trends of bars in these four figures are similar, which can be summarized as follows.

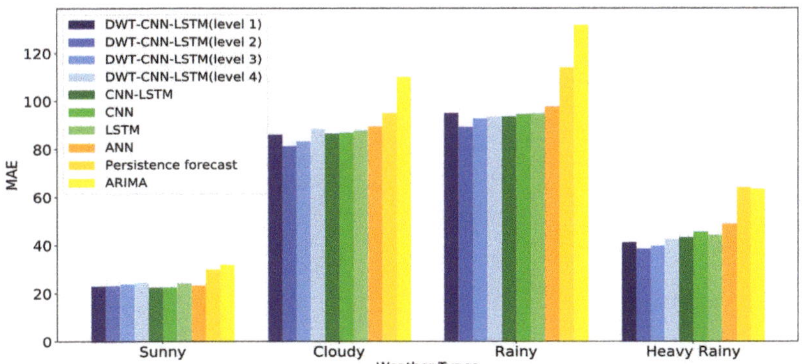

Figure 11. The MAE of different forecasting models for sunny, cloudy, rainy and heavy rainy days using the dataset of Elizabeth City State University.

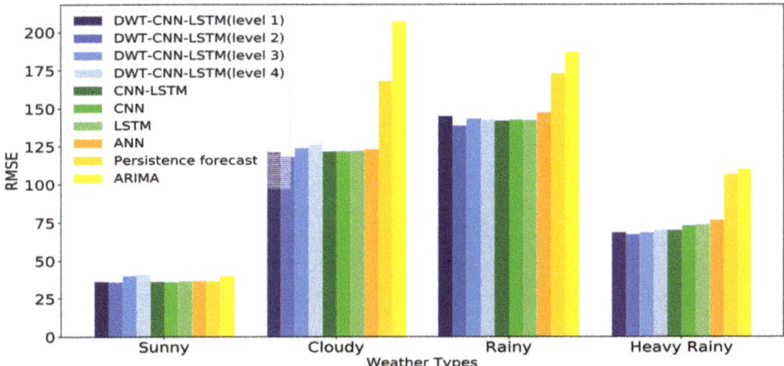

Figure 12. The RMSE of different forecasting models for sunny, cloudy, rainy and heavy rainy days using the dataset of Elizabeth City State University.

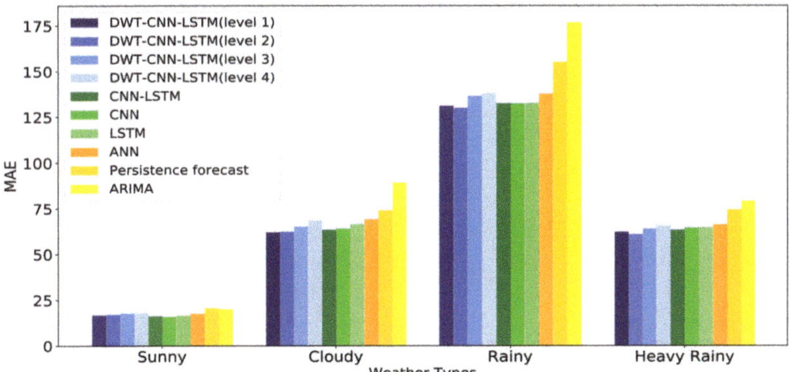

Figure 13. The MAE of different forecasting models for sunny, cloudy, rainy and heavy rainy days using the dataset of Desert Rock Station.

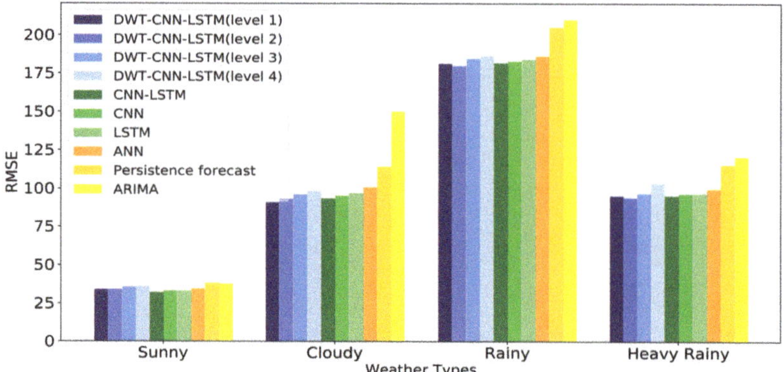

Figure 14. The RMSE of different forecasting models for sunny, cloudy, rainy and heavy rainy days using the dataset of Desert Rock Station.

First of all, it can be concluded that the influence of WD on forecasting performance, as well as the best WD level, generally varies under different weather types and validation datasets. Additionally, the introduction of certain WD level can improving the forecasting performance of DWT-CNN-LSTM model for cloudy, rainy and heavy rainy days, excluding sunny day. The conclusions are revealed by the fact in Figures 11–14 that the heights of all the blue bars (represent DWT-CNN-LSTM models with different WD Level) of sunny day are higher than the dark green bars (represents CNN-LSTM model). This can be explained by the fact that the solar irradiance curve of cloudy, rainy and heavy rainy days presents higher volatility, variability and randomness than that of sunny days. Therefore, the raw solar irradiance sequence of cloudy, rainy and heavy rainy days probably includes nonlinear and dynamic components in the form of spikes and fluctuations. The existence of these components will undoubtedly deteriorate the precision of the solar irradiance forecasting models. Additionally, the application of WD could mitigate the above problems.

Secondly, the proposed DWT-CNN-LSTM models with suitable WD Level are always superior to other DL models (e.g., CNN and LSTM) and traditional forecasting models (e.g., ANN, persistence forecast and ARIMA) for cloudy, rainy and heavy rainy days. For sunny days, the CNN-LSTM model without WD also performs better than other DL models and traditional forecasting models. The performance enhancement can be attributed to the application of WD and the reasonable tandem connection of CNN and LSTM. WD is used to decompose the raw solar irradiance sequence data of certain weather types into several subsequences with better behaviors (e.g., more stable variances and fewer outliers). CNN is good at automatically and effectively extracting representative and significant information from the raw subsequence data. As shown in Figure 15, the sequential characteristics with low and high frequency are well captured by CNN. LSTM is able to find the long dependencies of the time series input.

In the end, it should be noted that the applicability degree of DWT-CNN-LSTM model under the different weather is not the same. Specifically speaking, the height differences of bars under different weather types reveal that our proposed DWT-CNN-LSTM model obviously performs better than traditional forecasting models (e.g., ARMIA) under cloudy, rainy and heavy rainy days. In other words, our proposed model is more applicable for the solar irradiance forecasting of extreme weather conditions. However, as shown in Figures 7–10, there still exists a certain deviation between the actual solar irradiance value and the predicted value. This may be explained by the fact that the DWT-based decomposition of raw solar irradiance data may miss part of the information. It is an important problem needed be overcome in the next research stage.

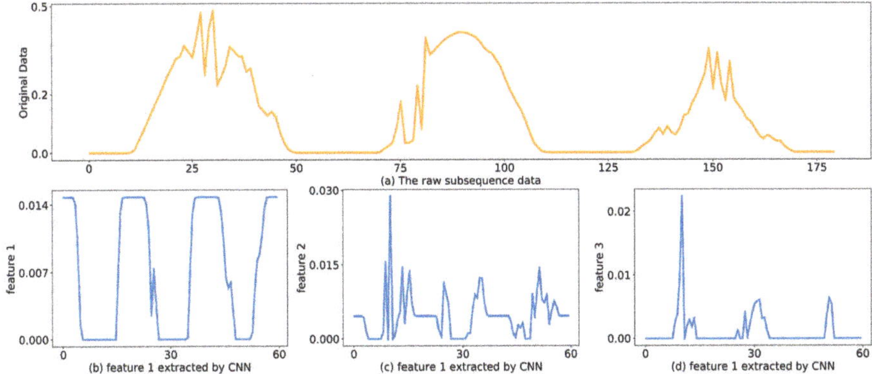

Figure 15. The visualization of feature maps extracted by CNN from the raw subsequence data. (**a**) the original data before convolution operation; (**b**) The first feature map yielded by convolution operation; (**c**) the second feature map yielded by convolution operation; and (**d**) the third feature map yielded by convolution operation.

4. Conclusions

The nature of the volatility and randomness characteristics of the output power of solar PV generation causes serious difficulty for the real-time power balance of the interconnected grid. This makes PV power forecasting become an important issue to the power grid in terms of the effective integration of large-scale PV plants. As the main influence factor of PV power generation, the solar irradiance and its accurate forecasting are prerequisites for solar PV power forecasting. Therefore, this paper proposes an improved DL model to enhance the accuracy of day-ahead solar irradiance forecasting. It should be noted that the DWT-CNN-LSTM model is individually established under four general weather types (i.e., sunny, cloudy, rainy and heavy rainy) due to the high dependency of solar irradiance on weather status.

The basic pipeline framework behind the data-driven DWT-CNN-LSTM model consists of three major parts: (1) DWT based solar irradiance sequence decomposition; (2) the CNN-based local feature extractor; and (3) the LSTM-based sequence forecasting model. In the solar irradiance forecasting under certain weather types, the raw solar irradiance sequence is decomposed into several subsequences via discrete wavelet transformation. Then each subsequence is fed to the CNN-based local feature extractor, which leverages the advantage of CNN to automatically learn the abstract feature representation from the raw subsequence data. Since the extracted features are also time series data, they are individually transported to LSTM to construct the subsequence forecasting model. In the end, the final solar irradiance forecasting results under certain weather types are obtained via the wavelet reconstruction of these forecasted subsequences.

In the case study using two datasets of Elizabeth City State University and Desert Rock Station, the performance of the proposed DWT-CNN-LSTM model is compared with another six solar irradiance forecasting models, namely, CNN-LSTM (i.e., our proposed model without WD), ANN, manually extracted features-ANN, persistence forecasting, CNN, and LSTM. Based on three error indexes (i.e., RMSE, MAE, and R), the simulation results indicate that DWT-CNN-LSTM model has high superiority in the solar irradiance forecasting, especially under extreme weather conditions. This mans the proposed DL technique-based day-ahead solar irradiance forecasting model has high potential for future practical applications.

Author Contributions: All authors have worked on this manuscript together and all authors have read and approved the final manuscript. F.W., Y.Y. and Z.Z. (Zhanyao Zhang) conceived and designed the experiments;

Y.Y. and Z.Z. (Zhanyao Zhang) performed the experiments; J.L., K.L., Z.Z. (Zhao Zhen) analyzed the data; F.W. and Y.Y. wrote the paper.

Funding: This work was supported by the National Key R&D Program of China (2018YFB0904200), the National Natural Science Foundation of China (51577067), the Beijing Natural Science Foundation of China (3162033), the Hebei Natural Science Foundation of China (E2015502060), the State Key Laboratory of Alternate Electrical Power System with Renewable Energy Sources (LAPS18008), the Science and Technology Project of State Grid Corporation of China (SGCC) (NY7117020), the Open Fund of State Key Laboratory of Operation and Control of Renewable Energy & Storage Systems (China Electric Power Research Institute) (5242001600FB), and the Fundamental Research Funds for the Central Universities (2018QN077).

Conflicts of Interest: The authors declare no conflict of interest.

Nomenclature

PV	photovoltaic
DL	deep learning
WD	wavelet decomposition
CNN	convolutional neural network
LSTM	long short-term memory
IEA	international energy agency
NWP	numerical weather prediction
TSI	total sky imagery
MOS	model output statistics
MA	moving average
AR	autoregressive
ARMA	autoregressive moving average
ARIMA	autoregressive integrated moving average
ANN	artificial neural network
SVM	support vector machine
SOM	self-organizing map
LVQ	learning vector quantization
SVR	support vector regression
A-P	Angstrom-Prescott
RNN	recurrent neural network
MFE	manual feature extraction
DWT	discrete wavelet transformation
LPF	low-pass filters
HPF	high-pass filters
ReLu	rectified linear units
RMSE	root mean squared error
MAE	mean absolute error
R	correlation coefficient
$\psi(t)$	mother wavelet function
$\varphi(t)$	scaling function
$\psi_{j,k}(t)$	a sequence of wavelet at time index t
$\varphi_{j,k}(t)$	binary scale-functions at time index t
$os(t)$	the original sequence at time index t
$c_{j,k}$	the approximation coefficient at scale j and location k
$d_{j,k}$	the detailed coefficient at scale j and location k
S	the raw sequential input
FS	the filter sequence
L	the length of raw sequential input
K	the number of total filters in the convolutional Layer
m	window size
$\oplus$	the concatenation operator

$s_{i:i+m-1}$	a concatenation vector representation of $s_i \oplus s_{i+1} \oplus s_{i+2} \oplus \cdots \oplus s_{i+m-1}$
f, σ	a nonlinear activation function
w_j	the j-th filter matrix
F_j	the feature map of j-th filter
p	pooling size
$F_{j-compress}$	the compressed feature vector from the pooling layer
X	the sequential vectors
$x(t)$	the input variable at time step t
$y(t)$	the expected output at time step t
W, U, V	weight matrixes
b, c	biases vectors
$h(t)$	a short-term state
$c(t)$	long-term state

References

1. Wang, F.; Zhen, Z.; Mi, Z.; Sun, H.; Su, S.; Yang, G. Solar irradiance feature extraction and support vector machines based weather status pattern recognition model for short-term photovoltaic power forecasting. *Energy Build.* **2015**, *86*, 427–438. [CrossRef]
2. World Energy Outlook 2016. Available online: https://www.iea.org/newsroom/news/2016/november/world-energy-outlook-2016.html (accessed on 20 June 2018).
3. Renewables 2017: Global Status Report. Available online: http://www.ren21.net/gsr-2017/ (accessed on 20 June 2018).
4. Inman, R.H.; Pedro, H.T.C.; Coimbra, C.F.M. Solar forecasting methods for renewable energy integration. *Prog. Energy Combust. Sci.* **2013**, *39*, 535–576. [CrossRef]
5. Yona, A.; Senjyu, T.; Funabashi, T.; Mandal, P.; Kim, C.-H. Optimizing Re-planning Operation for Smart House Applying Solar Radiation Forecasting. *Appl. Sci.* **2014**, *4*, 366–379. [CrossRef]
6. Sun, Y.; Wang, F.; Wang, B.; Chen, Q.; Engerer, N.A.; Mi, Z. Correlation feature selection and mutual information theory based quantitative research on meteorological impact factors of module temperature for solar photovoltaic systems. *Energies* **2017**, *10*, 7. [CrossRef]
7. Wang, J.; Li, P.; Ran, R.; Che, Y.; Zhou, Y. A short-term photovoltaic power prediction model based on the Gradient Boost Decision Tree. *Appl. Sci.* **2018**, *8*, 689. [CrossRef]
8. Baños, R.; Manzano-Agugliaro, F.; Montoya, F.G.; Gil, C.; Alcayde, A.; Gómez, J. Optimization methods applied to renewable and sustainable energy: A review. *Renew. Sustain. Energy Rev.* **2011**, *15*, 1753–1766. [CrossRef]
9. Sharma, A.; Kakkar, A. Forecasting daily global solar irradiance generation using machine learning. *Renew. Sustain. Energy Rev.* **2018**, *82*, 2254–2269. [CrossRef]
10. Christensen-Dalsgaard, J. Physics of Solar-Like Oscillations. *Highlights Astron.* **2005**, *13*, 397–402. [CrossRef]
11. Marquez, R.; Coimbra, C.F.M. Forecasting of global and direct solar irradiance using stochastic learning methods, ground experiments and the NWS database. *Sol. Energy* **2011**, *85*, 746–756. [CrossRef]
12. Li, J.; Ward, J.K.; Tong, J.; Collins, L.; Platt, G. Machine learning for solar irradiance forecasting of photovoltaic system. *Renew. Energy* **2016**, *90*, 542–553. [CrossRef]
13. Diagne, M.; David, M.; Lauret, P.; Boland, J.; Schmutz, N. Solar irradiation forecasting: state-of-the-art and proposition for future developments for small-scale insular grids. *Renew. Sustain. Energy Rev.* **2013**, *27*, 65–76. [CrossRef]
14. Reikard, G. Predicting solar radiation at high resolutions: A comparison of time series forecasts. *Sol. Energy* **2009**, *83*, 342–349. [CrossRef]
15. Lorenz, E.; Hammer, A.; Heinemann, D. Short term forecasting of solar radiation based on satellite data. In Proceedings of the EuroSun 2004 ISES Europe Solar Congress, Freiburg, Germany, 20–23 June 2004; pp. 841–848.
16. Arbizu-Barrena, C.; Ruiz-Arias, J.A.; Rodríguez-Benítez, F.J.; Pozo-Vázquez, D.; Tovar-Pescador, J. Short-term solar radiation forecasting by advecting and diffusing MSG cloud index. *Sol. Energy* **2017**, *155*, 1092–1103. [CrossRef]

17. Wang, F.; Zhen, Z.; Liu, C.; Mi, Z.; Hodge, B.M.; Shafie-khah, M.; Catalão, J.P.S. Image phase shift invariance based cloud motion displacement vector calculation method for ultra-short-term solar PV power forecasting. *Energy Convers. Manag.* **2018**, *157*, 123–135. [CrossRef]

18. Wang, F.; Li, K.; Wang, X.; Jiang, L.; Ren, J.; Mi, Z.; Shafie-khah, M.; Catalão, J.P.S. A Distributed PV System Capacity Estimation Approach Based on Support Vector Machine with Customer Net Load Curve Features. *Energies* **2018**, *11*, 1750. [CrossRef]

19. Verzijlbergh, R.A.; Heijnen, P.W.; de Roode, S.R.; Los, A.; Jonker, H.J.J. Improved model output statistics of numerical weather prediction based irradiance forecasts for solar power applications. *Sol. Energy* **2015**, *118*, 634–645. [CrossRef]

20. Bacher, P.; Madsen, H.; Nielsen, H.A. Online short-term solar power forecasting. *Sol. Energy* **2009**, *83*, 1772–1783. [CrossRef]

21. Huang, R.; Huang, T.; Gadh, R.; Li, N. Solar generation prediction using the ARMA model in a laboratory-level micro-grid. In Proceedings of the 2012 IEEE Third International Conference Smart Grid Communications, Tainan, Taiwan, 5–8 November 2012.

22. Perdomo, R.; Banguero, E.; Gordillo, G. Statistical Modeling for Global Solar Radiation Forecasting in Bogotá. In Proceedings of the 2010 35th IEEE Photovoltic Specialists Conference, Honolulu, HI, USA, 20–25 June 2010; pp. 2374–2379.

23. Wang, F.; Li, K.; Liu, C.; Mi, Z.; Shafie-khah, M.; Catalao, J.P.S. Synchronous Pattern Matching Principle Based Residential Demand Response Baseline Estimation: Mechanism Analysis and Approach Description. *IEEE Trans. Smart Grid* **2018**, *3053*, 1–13. [CrossRef]

24. Chen, Q.; Wang, F.; Hodge, B.-M.; Zhang, J.; Li, Z.; Shafie-Khah, M.; Catalao, J.P.S. Dynamic Price Vector Formation Model-Based Automatic Demand Response Strategy for PV-Assisted EV Charging Stations. *IEEE Trans. Smart Grid* **2017**, *8*, 2903–2915. [CrossRef]

25. Wang, F.; Xu, H.; Xu, T.; Li, K.; Shafie-Khah, M.; Catalao, J.P.S. The values of market-based demand response on improving power system reliability under extreme circumstances. *Appl. Energy* **2017**, *193*, 220–231. [CrossRef]

26. Wang, F.; Zhou, L.; Ren, H.; Liu, X.; Shafie-khah, M. Multi-objective Optimization Model of Source-Load-Storage Synergetic Dispatch for Building Energy System Based on TOU Price Demand Response. *IEEE Trans. Ind. Appl.* **2018**, *54*, 1017–1028. [CrossRef]

27. Maier, H.R.; Dandy, G.C. Neural networks for the prediction and forecasting of water resources variables: A review of modelling issues and applications. *Environ. Model. Softw.* **2000**, *15*, 101–124. [CrossRef]

28. Wang, F.; Mi, Z.; Su, S.; Zhao, H. Short-Term Solar Irradiance Forecasting Model Based on Artificial Neural Network Using Statistical Feature Parameters. *Energies* **2012**, *5*, 1355–1370. [CrossRef]

29. Zeng, J.; Qiao, W. Short-term solar power prediction using a support vector machine. *Renew. Energy* **2013**, *52*, 118–127. [CrossRef]

30. Shakya, A.; Michael, S.; Saunders, C.; Armstrong, D.; Pandey, P.; Chalise, S.; Tonkoski, R. Using Markov Switching Model for solar irradiance forecasting in remote microgrids. In Proceedings of the 2016 IEEE Energy Conversion Congress and Exposition, Milwaukee, WI, USA, 18–22 September 2016; pp. 895–905.

31. Wang, F.; Zhen, Z.; Wang, B.; Mi, Z. Comparative Study on KNN and SVM Based Weather Classification Models for Day Ahead Short Term Solar PV Power Forecasting. *Appl. Sci.* **2017**, *8*, 28. [CrossRef]

32. Gala, Y.; Fernández, Á.; Díaz, J.; Dorronsoro, J.R. Hybrid machine learning forecasting of solar radiation values. *Neurocomputing* **2016**, *176*, 48–59. [CrossRef]

33. Wang, F.; Zhou, L.; Ren, H.; Liu, X. Search Improvement Process-Chaotic Optimization-Particle Swarm Optimization-Elite Retention Strategy and Improved Combined Cooling-Heating-Power Strategy Based Two-Time Scale Multi-Objective Optimization Model for Stand-Alone Microgrid Operation. *Energies* **2017**, *10*, 1936. [CrossRef]

34. Yang, D.; Kleissl, J.; Gueymard, C.A.; Pedro, H.T.C.; Coimbra, C.F.M. History and trends in solar irradiance and PV power forecasting: A preliminary assessment and review using text mining. *Sol. Energy* **2018**, *168*, 60–101. [CrossRef]

35. Wan, C.; Zhao, J.; Song, Y.; Xu, Z.; Lin, J.; Hu, Z. Photovoltaic and solar power forecasting for smart grid energy management. *CSEE J. Power Energy Syst.* **2015**, *1*, 38–46. [CrossRef]

36. Ferlito, S.; Adinolfi, G.; Graditi, G. Comparative analysis of data-driven methods online and offline trained to the forecasting of grid-connected photovoltaic plant production. *Appl. Energy* **2017**, *205*, 116–129. [CrossRef]

37. Yang, H.-T.; Huang, C.-M.; Huang, Y.-C.; Pai, Y.S. A Weather-Based Hybrid method for one-day ahead hourly forecasting of PV power output. *IEEE Trans. Sustain. Energy* **2014**, *5*, 917–926. [CrossRef]

38. Gensler, A.; Henze, J.; Sick, B.; Raabe, N. Deep Learning for solar power forecasting-An approach using AutoEncoder and LSTM Neural Networks. In Proceedings of the 2016 IEEE International Conference on Systems, Man, and Cybernetics, Budapest, Hungary, 9–12 October 2016.

39. Hussain, S.; Alili, A. Day ahead hourly forecast of solar irradiance for Abu Dhabi, UAE. In Proceedings of the 2016 IEEE Smart Energy Grid Engineering (SEGE), Oshawa, ON, Canada, 21–24 August 2016.

40. Akarslan, E.; Hocaoglu, F.O.; Edizkan, R. Novel short term solar irradiance forecasting models. *Renew. Energy* **2018**, *123*, 58–66. [CrossRef]

41. Zhen, Z.; Wan, X.; Wang, Z.; Wang, F.; Ren, H.; Mi, Z. Multi-level wavelet decomposition based day-ahead solar irradiance forecasting. In Proceedings of the 2018 IEEE Power Energy Society Innovative Smart Grid Technologies Conference (ISGT), Washington, DC, USA, 19–22 February 2018; pp. 1–5.

42. Wang, F.; Zhen, Z.; Liu, C.; Mi, Z.; Shafie-Khah, M.; Catalão, J.P.S. Time-section fusion pattern classification based day-ahead solar irradiance ensemble forecasting model using mutual iterative optimization. *Energies* **2018**, *11*, 184. [CrossRef]

43. Qing, X.; Niu, Y. Hourly day-ahead solar irradiance prediction using weather forecasts by LSTM. *Energy* **2018**, *148*, 461–468. [CrossRef]

44. Llamas, J.; Lerones, P.M.; Medina, R.; Zalama, E.; Gómez-García-Bermejo, J. Classification of Architectural Heritage Images Using Deep Learning Techniques. *Appl. Sci.* **2017**, *7*, 992. [CrossRef]

45. Almeida, A.; Azkune, G. Predicting Human Behaviour with Recurrent Neural Networks. *Appl. Sci.* **2018**, *8*, 305. [CrossRef]

46. Yoo, Y.; Baek, J.-G. A Novel Image Feature for the Remaining Useful Lifetime Prediction of Bearings Based on Continuous Wavelet Transform and Convolutional Neural Network. *Appl. Sci.* **2018**, *8*, 1102. [CrossRef]

47. Panapakidis, I.P.; Dagoumas, A.S. Day-ahead natural gas demand forecasting based on the combination of wavelet transform and ANFIS/genetic algorithm/neural network model. *Energy* **2017**, *118*, 231–245. [CrossRef]

48. Mallat, S.G. A Theory for Multiresolution Signal Decomposition: The Wavelet Representation. *IEEE Computer Soc.* **1989**, *11*, 674–693. [CrossRef]

49. Zhao, R.; Yan, R.; Wang, J.; Mao, K. Learning to Monitor Machine Health with Convolutional Bi-Directional LSTM Networks. *Sensors* **2017**, *17*, 273. [CrossRef] [PubMed]

50. Gu, J.; Wang, Z.; Kuen, J.; Ma, L.; Shahroudy, A.; Shuai, B.; Liu, T.; Wang, X.; Wang, G.; Cai, J.; et al. Recent advances in convolutional neural networks. *Pattern Recognit.* **2018**, *77*, 354–377. [CrossRef]

51. Nair, V.; Hinton, G.E. Rectified linear units improve restricted boltzmann machines. In Proceedings of the 27th International Conference on Machine Learning, Haifa, Israel, 21–24 June 2010; pp. 807–814.

52. Längkvist, M.; Karlsson, L.; Loutfi, A. A review of unsupervised feature learning and deep learning for time-series modeling. *Pattern Recognit. Lett.* **2014**, *42*, 11–24. [CrossRef]

53. Hochreiter, S.; Schmidhuber, J. Long Short-Term Memory. *Neural Comput.* **1997**, *9*, 1735–1780. [CrossRef] [PubMed]

54. Bengio, Y.; Simard, P.; Frasconi, P. Learning Long-Term Dependencies with Gradient Descent is Dicfficult. *IEEE Trans. Neural Netw.* **1994**, *5*, 157–166. [CrossRef] [PubMed]

55. US Department of Energy, NREL, National Renewable Energy Laboratory. Available online: https://rredc.nrel.gov/solar/new_data/confrrm/bs/ (accessed on 20 June 2018).

56. US Department of Commerce, NOAA, Earth System Research Laboratory. Available online: https://www.esrl.noaa.gov/gmd/grad/surfrad/ (accessed on 20 June 2018).

57. Keras Documentation. Available online: https://keras.io/ (accessed on 20 June 2018).

58. Scikit-learn: Machine Learning in Python. Available online: http://scikit-learn.github.io/stable (accessed on 20 June 2018).

Article

Towards the Development of a Low-Cost Irradiance Nowcasting Sky Imager

Luis Valentín [1,*], Manuel I. Peña-Cruz [1], Daniela Moctezuma [2], Cesar M. Peña-Martínez [1], Carlos A. Pineda-Arellano [1] and Arturo Díaz-Ponce [1]

[1] CONACYT—Centro de Investigaciones en Optica, Unidad de Aguascalientes, Prol. Constitución 607, Reserva Loma Bonita, Aguascalientes 20200, Mexico; mipec@cio.mx (M.I.P.-C.); cesarmp@cio.mx (C.M.P.-M.); capia@cio.mx (C.A.P.-A.); adiaz@cio.mx (A.D.-P.)

[2] CONACYT—Centro de Investigación en Ciencias de Información Geoespacial A.C., 117 Circuito Tecnopolo Norte Col. Tecnopolo Pocitos II, Aguascalientes 20313, Mexico; dmoctezuma@centrogeo.edu.mx

* Correspondence: luismvc@cio.mx; Tel.: +52-449-442-8124 (ext. 126)

Received: 28 December 2018; Accepted: 1 February 2019; Published: 18 March 2019

Abstract: Solar resource assessment is fundamental to reduce the risk in selecting the solar power-plants' location; also for designing the appropriate solar-energy conversion technology and operating new sources of solar-power generation. Having a reliable methodology for solar irradiance forecasting allows accurately identifying variations in the plant energy production and, as a consequence, determining improvements in energy supply strategies. A new trend for solar resource assessment is based on the analysis of the sky dynamics by processing a set of images of the sky dome. In this paper, a methodology for processing the sky dome images to obtain the position of the Sun is presented; this parameter is relevant to compute the solar irradiance implemented in solar resource assessment. This methodology is based on the implementation of several techniques in order to achieve a combined, fast, and robust detection system for the Sun position regardless of the conditions of the sky, which is a complex task due to the variability of the sky dynamics. Identifying the correct position of the Sun is a critical parameter to project whether, in the presence of clouds, the occlusion of the Sun is occurring, which is essential in short-term solar resource assessment, the so-called irradiance nowcasting. The experimental results confirm that the proposed methodology performs well in the detection of the position of the Sun not only in a clear-sky day, but also in a cloudy one. The proposed methodology is also a reliable tool to cover the dynamics of the sky.

Keywords: irradiance nowcasting; Sun detection; sky dynamics; image processing; particle filtering

---

1. Introduction

The assessment of solar irradiance has become one of the most important topics in the field of solar energy, such as the photovoltaic industry (PV), concentrating solar power (CSP), and the solar chemistry process [1–4]. In the solar power industry, constant assessment of the solar resource is essential to establish the viability of exploiting a specific location and predicting the factors that may affect its energy production. Having an accurate assessment of the solar resource is an important advantage that can help in establishing solar energy as a more competitive alternative. Currently, a number of maps and databases describing the solar resource distribution can be found on the Internet; however, confidence intervals of data are not provided in any of them [5]. The proposal in [1] argued that the assessment and forecasting of the solar resource are essential tasks, since they allow better management of the solar field in real time, reducing the maintenance activities, thus improving the expected benefits. Currently, the installation of PV and CSP plants has generated an increase in the production of electricity with renewable energy, and it is to be expected that this trend will continue increasing. However, the fluctuating nature of solar energy poses particular problems in delivering the generated energy to the electricity grids, not to mention the high initial investment cost, which require a short-term forecast of the solar irradiance availability, also known as "nowcasting" [1,2,6,7]. This makes the construction and improvement of "nowcasting" systems very important.

There are two main approaches that allow the forecast of solar irradiance [8]. The first one is the physical-statistical method, which deals with the forecast of the availability of solar irradiance by means of the so-called numerical weather prediction models (NWP). These prediction models have a spatial resolution around of 100 km^2 and deliver resolutions up to 10 km^2 [9], with long-term irradiance prediction [10,11]. On the other hand, there are models based on real-time prediction by means of satellite images. The spatial resolution of the satellite prediction models is 1 km^2 with a medium temporal resolution greater than 6 h [2]. The need for low temporal resolution (minutes) has been addressed by the development of sky imagers. These systems provide a hemispherical view of the whole sky by using a fish-eye lens or a convex dome mirror and a camera; allowing the Sun motion and cloud vector to be estimated [12].

Generally, the forecast of the NWP has a better performance in the temporal perspective than satellite and sky imagers. However, the fluctuation in the initial conditions and the limited resolution mean that the NWP numerical models cannot predict the position of the clouds accurately, nor their effects on the solar irradiance in a specific location [2,9,13]. Therefore, it is inescapable to make better forecasts by short-term measurements with high temporal and spatial resolutions.

Most of solar power plants look for reliable long-, medium-, and short-term data to estimate more accurately the amount of energy they can produce daily and thus establish energy generation strategies. In general, for these processes, irradiance sensors are used to obtain global horizontal irradiance (GHI), direct normal irradiance (DNI), and diffuse horizontal irradiance (DHI) data. However, properly calibrated and well-maintained sensors have a high cost, not only for the communication networks required to centralize and store the data, but also for the installations that must be carried out, the acquisition, and maintenance of the sensors themselves. It could be said that the forecasting of solar irradiance is a technology that allows optimizing the costs, offering a better quality of energy supplied by the electrical network, as long as the solar irradiance variation can be predicted with great accuracy [14]. The combination of these two factors (costs and quality), has been the main motivation for the development of a complex field of research that aims to make better predictions of solar irradiance values at the ground level and, thus, be able to predict output values depending on the type of technology used. The observations with terrestrial instruments that use a sky camera and record complete images of the celestial dome are a powerful tool, since they represent a great opportunity to fill the prognostic gap at a low cost of computational processing [2,15].

In this sense, the development of a robust and automatic Sun detection algorithm becomes an important step in the process of the forecasting of the solar resource. By knowing the Sun position with precision (solar centroid) and the cloud trajectory through the sky dome and towards the Sun (cloud vector), irradiance variance can be forecasted due to possible Sun occlusion. In this work, a low-cost solar nowcasting system based on artificial vision is proposed. With the purpose of capturing the image of the full sky reflection, the system is equipped with a dome-type convex mirror and a Raspberry Pi camera module pointing downwards. The camera acquires images every minute with a resolution of 1024 × 780 pixels; while the mirror occupies 673 × 641 pixels. To process the acquired images, a methodology addressing several challenges related to the computation of the position of the Sun has been proposed. This methodology is based on the implementation of several algorithms, which help to achieve a robust detection; since knowing the correct position of the Sun allows other processes to be carried out. For instance, knowing the position of a cloud with respect to the Sun enables us to identify if the Sun will be occluded, as well as the time it will take. In order to test this methodology, the system was installed in Aguascalientes city, Mexico (21°51′ N 102°18′ W).

2. Methodology

System development was achieved through an artificial vision system that analyzes Earth-sky dynamics. The prototype was designed with the intention of capturing sky dome images for detecting the position of the Sun and analyzing sky dynamics, avoiding the use of high-cost fish eye lenses, as well as expanding CMOS sensor life by avoiding overexposure to the sunlight. A robust and automatic methodology based on a digital image processing algorithm capable of identifying the position of the Sun in the images throughout the day and under different conditions has been developed. Having this information, the system will be able to estimate DHI and DNI values.

2.1. System Design

The prototype consists of a low-cost image acquisition system, based on a Raspberry-Pi camera, mounted over a highly-convex mirror that reflects the Sun and sky dome over the CMOS IMX219PQ sensor (see Figure 1).

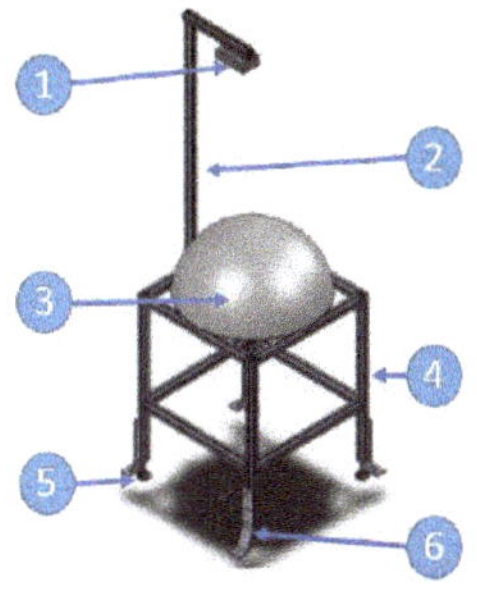

1. Camera.
2. Support.
3. Mirror.
4. , 5., and 6. Prototype structure.

Figure 1. System prototype.

2.2. Sun Detection

To build a system capable of estimating irradiance values based on sky images, as a first step, the correct detection of the Sun's position in the image throughout the day is necessary. To fulfill this purpose, an image processing algorithm capable of accurately calculating the position of the Sun, through the elimination of any element different from it in the image, has been implemented.

With the proposed methodology, the system is able to perform the detection, regardless of the conditions of the day. For the above, a completely clear sky (sunny day), a partially sunny day (partial occlusions of the Sun) and a cloudy day (where in a period of time, a complete occlusion of the Sun is presented) have been analyzed. Furthermore, the system is smart enough to perform the detection without any extra information other than the image by combining several image processing algorithms. The followed methodology, as well as the implemented algorithms that compose the core of the system are detailed below.

2.2.1. Framework

The schematic overview of the proposed methodology is shown in Figure 2.

The system begins by acquiring a color image, where the algorithm executes the first process that consists of the dome's detection, which is the region of interest (ROI), that is where the Sun should be in the image. Once this is done, the algorithm attempts to detect the Sun within the ROI, separating it from the rest of the elements in the image (also known as the segmentation process). Then, if the executed process finds the Sun, the system will store the position of the Sun in the image, the position that will be used as input to the tracking process through the particle filter. However, if the algorithm is not able to segment the Sun in the image, it is estimated through predefined and well-known equations, and this estimation is used as input to the particle filter. With this methodology, the proposed system is always able to return a position of the Sun in the image, no matter what conditions are present in the sky. This process is repeated automatically with each new image acquired by the sensor, except for the detection of the dome, which has already been performed on the first image, since it will not change its position while the system is running.

In the following sections, each of the processes followed by the proposed system are described in detail.

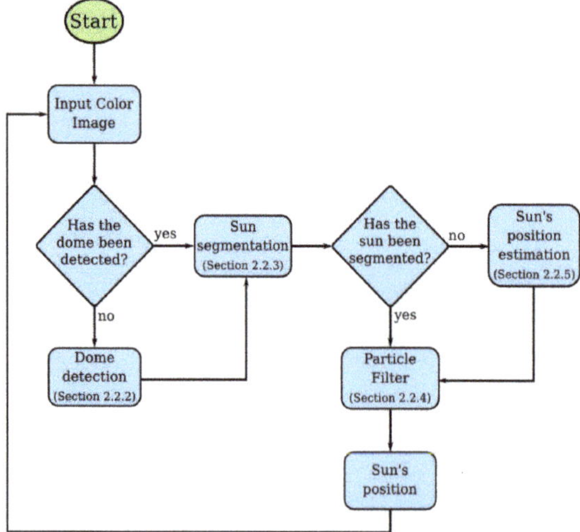

Figure 2. Sun position detection process.

2.2.2. Dome Detection

In this section, the dome detection method, for the region where the Sun should be, is described.

The dome detection is carried out as follows. Let $I(m, n, p)$ be the matrix, which represents an image with dimensions $m \times n \times p$, where $i = 1, 2, ..., m$ and $j = 1, 2, ..., n$ represent the spatial position of the pixel and $k = 1, ..., p$ is the channel. Hence, $I(i, j, k)$ is a function that returns the intensity value of image I in position (i, j) in the k channel; in this case, the image has three channels, red, green, and blue. The image I is converted to a gray scale image G. This conversion is given by $G(i, j) = 0.299 \cdot I(i, j, 1) + 0.587 \cdot I(i, j, 2) + 0.114 \cdot I(i, j, 3)$ [16].

Once the gray scale image has been computed, the system performs a noise reduction, resulting in a new image, denoted as $G_s(m, n)$. The noise reduction was done through the well-known Gaussian filter. This filter works with a convolution mask with dimensions of 11×11, which has the values of the center pixels weighted more (for more details about the Gaussian filter, see [17]). Then, $G_s(m, n)$ is binarized by an adaptive thresholding algorithm, which calculates the threshold for a small region of the image, getting different thresholds, according to its local neighborhood, for different regions of the same image. Image thresholding segments a digital image based on a certain characteristic of the pixels (for example, intensity value) based on a threshold (th); that means the new pixel value will depend on if the pixel value is below or above this threshold, as follows:

$$B(i, j) = \begin{cases} 0 & \text{if } G_s(i, j) < th \\ 1 & \text{if } G_s(i, j) \geq th \end{cases}$$

The goal is to create a binary representation of the image, let $B(m, n)$ be such a representation, classifying each pixel into one of two categories, such as "dark" ($B(i, j) = 0$) or "light" ($B(i, j) = 1$) [18].

In order to find the dome, the system uses $B(m, n)$ as the input of Algorithm 1. Here, the parameters m, n, and Δy are also needed, m and n are the image dimensions, and Δy is a slight vertical displacement that can take values in a range of 50–100 pixels with respect to the center of the image. What Algorithm 1 does is find the circle that best fits the dome; this circle is defined by three points, which are obtained through a search that starts at the center of the image and ends once a pixel, with a value of one, is found. The way to find the pixel is by moving along the image from the center, both to the left and to the right. As a result of this search, two of the three needed points are found. The third point is found by moving along the image to the left starting at the center, but adding a slight translation on the vertical axis (80 pixels). Having these three points, the parameters of the circle (center and radius) are computed through the procedure DomeParameters, which is composed of classical circle formulas. An example of dome detection is shown in Figure 3.

Algorithm 1 Dome detection algorithm.

1: **procedure** FINDDOME($B(m, n), \Delta y, m, n$)
2:     $w = n/2$
3:     $h = m/2$
4:     first $= True$
5:     second $= True$
6:     third $= True$
7:     **for** $j = 1$ to $(m/2)$ **do**
8:         **if** $B(h, w - j)$ **is equal to** 1 **and** first **then**
9:             $P_1 = \{h, w - j\}$
10:             first $= False$
11:         **end if**
12:         **if** $B(h, w + j)$ **is equal to** 1 **and** second **then**
13:             $P_2 = \{h, w + j\}$
14:             second $= False$
15:         **end if**
16:         **if** $B(h - \Delta y, w - j)$ **is equal to** 1 **and** third **then**
17:             $P_3 = \{h - \Delta y, w - j\}$
18:             third $= False$
19:         **end if**
20:         **if** first **is equal to** $False$ **and** second **is equal to** $False$ **and** third **is equal to** $False$ **then**
21:             **break**
22:         **end if**
23:     **end for**
24:     $(P_{dome}, r_{dome}) = $ DOMEPARAMETERS(P_1, P_2, P_3)
25:     **return** (P_{dome}, r_{dome})
26: **end procedure**

27: **procedure** DOMEPARAMETERS(P_1, P_2, P_3)
28:     $\{x_1, y_1\} = P_1$
29:     $\{x_2, y_2\} = P_2$
30:     $\{x_3, y_3\} = P_3$
31:     $A = x_1(y_2 - y_3) - y_1(x_2 - x_3) + x_2 y_3 - x_3 y_2$
32:     $B = (x_1^2 + y_1^2)(y_3 - y_2) + (x_2^2 + y_2^2)(y_1 - y_3) + (x_3^2 + y_3^2)(y_2 - y_1)$
33:     $C = (x_1^2 + y_1^2)(x_2 - x_3) + (x_2^2 + y_2^2)(x_3 - x_1) + (x_3^2 + y_3^2)(x_1 - x_2)$
34:     $D = (x_1^2 + y_1^2)(x_3 y_2 - x_2 y_3) + (x_2^2 + y_2^2)(x_1 y_3 - x_3 y_1) + (x_3^2 + y_3^2)(x_2 y_1 - x_1 y_2)$
35:     $x_d = -B/2A$
36:     $y_d = -C/2A$
37:     $r_d = \sqrt{(B^2 + C^2 - 4AD)/4A^2}$
38:     **return** $(x_d, y_d), r_d$
39: **end procedure**

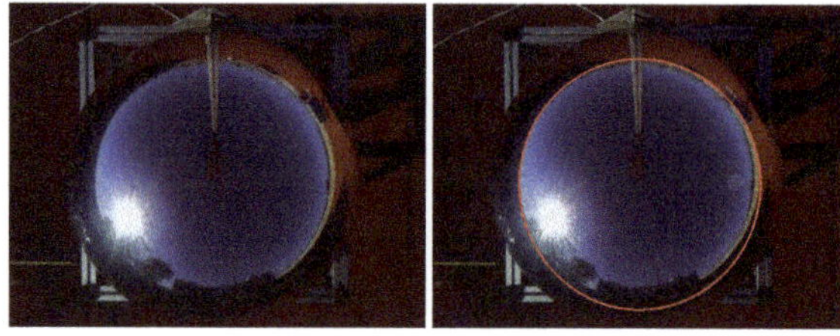

(**a**) Input image.   (**b**) Dome detection.

Figure 3. Dome detection example. The input color image, as well as the dome detection are shown in (**a**,**b**), respectively.

It is important to mention that this step is done only when the first image is acquired. Once the dome is detected, the system is ready to find the Sun within the ROI.

2.2.3. Clear Sky Sun Segmentation

The ideal case to make the detection of the position of the Sun is when a sunny day is presented, since there are no elements in the sky that interfere with the detection. This process is identified on the block called Sun segmentation from the diagram shown in Figure 2. The needed sequence of steps performed by the system to get the Sun's position in the image are shown in the diagram of Figure 4.

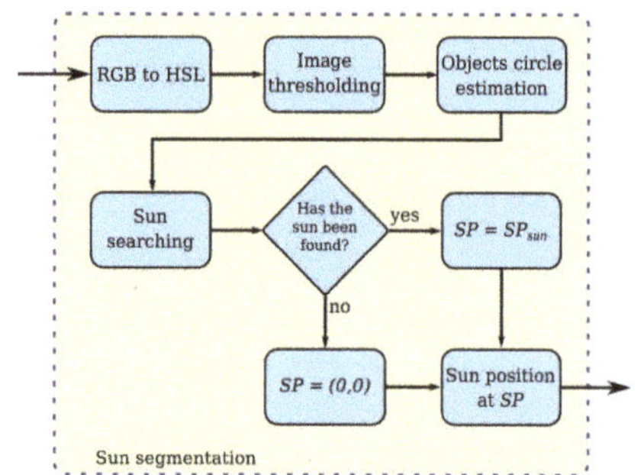

Figure 4. Sun segmentation based on image processing.

In order to compute the Sun's position, the system begins with a color space conversion from RGB to HSL; let $H(m, n, p)$ be the new image representation. HSL is another color space, which represents the color on three channels: hue, saturation, and brightness (how bright the color is). Thus, objects with very large brightness values can be easily separated from the background in the image. From the image $H(m, n, p)$, where in this case, $p = 3$, which represents the brightness channel in the image, the system performs thresholding over it and then a binarization. This threshold value is chosen based on the histogram of the image pixel intensities. From this operation, the matrix B_H is obtained, this matrix contains dark values in the object zone and white values in the background.

The resulting image is shown in Figure 5b, where the objects segmented are the those with the highest brightness.

From each object found, the system computes the minimum enclosing circle to get the center, as well as the radius; let $C = \{c_1(x_{c_1}, y_{c_1}, r_{c_1}), \ldots, c_n(x_{c_n}, y_{c_n}, r_{c_n})\}$ be the set of detected circles, where (x_{c_i}, y_{c_i}) represents the center coordinates and r_{c_i} the radius of the ith circle (c_i). To discriminate the Sun from all those objects found in the segmentation process, Algorithm 2 has been implemented.

(a) Input images.

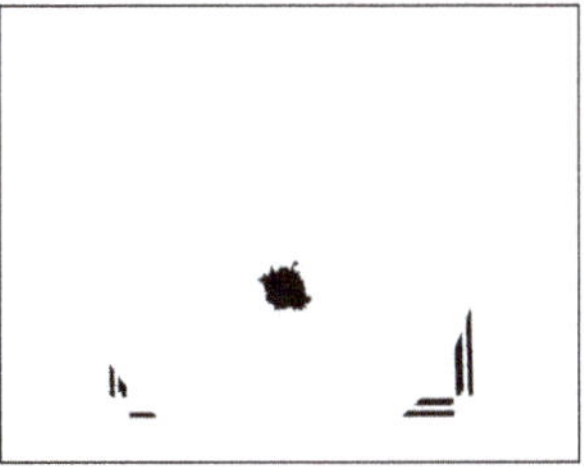

(b) Binary image.

Figure 5. Binarization of the image. (a) shows the input color image. (b) shows the result of the binarization.

Algorithm 2 Sun-searching algorithm.

1: **procedure** SUNSEARCHING(C, P_{dome}, r_{dome}, $MaxSunRadius$, $MinSunRadius$)
2:     $n = |C|$
3:     **for** $i = 1$ to n **do**
4:         $P_i, r_i = c_i(x_{c_i}, y_{c_i}, r_{c_i})$
5:         $d = \|P_i, P_{dome}\|$
6:         **if** $d < r_{dome}$ **and** ($MinSunRadius < r_i < MaxSunRadius$) **then**
7:             $c_{Sun} = c_i$
8:         **else**
9:             $c_{Sun} = (0,0,0)$
10:        **end if**
11:     **end for**
12:     **return** c_{Sun}
13: **end procedure**

The input parameters of Algorithm 2 are the detected circles C, the center and radius of the detected dome, and a max and min radius of the Sun, which have been determined experimentally.

From the input parameters, Algorithm 2 begins by computing the cardinality of the set C (detected circles), then the algorithm iterates over each element in C, computing the distance (d) of each element (centroid of the circle) to the dome center; if the distance of an element is lower than the dome radius and the radius is between values *MinSunRadius* and *MaxSunRadius*, then the system considers this element as one circle that could represent the Sun; otherwise, the system returns as the Sun position the point (0,0) with a radius of zero. That means, in the end, that the algorithm returns the last element that fulfills these conditions. As the output of this process, in Figure 6, a Sun detection example is shown.

Figure 6. Sun detection example.

2.2.4. Partially Sunny Day

The process described in Section 2.2.3 is repeated with each new image acquired by the camera.

However, it may be the case that a cloud with a brightness similar to the Sun is generated; if that happens, the system could segment two objects (the cloud and the Sun), and both objects could be within the circle that represents the dome; thus, they could be valid objects, and any of them could be considered as the Sun, or even worse, if a cloud is generated so close to the Sun such that the segmentation process generates an object that fuses these two elements, then the Sun's position in the image would have a translation relative to the actual position. An example of this possible scenario is shown in Figure 7.

(a) Input image.

(b) Segmentation result.

(c) Sun detection.

Figure 7. Sun and cloud segmented as a single object. (**b**) shows the output of the segmentation process, while in (**c**), the resulting Sun position (red dot), as well as the enclosing circle are shown.

In order to deal with this problem, an object tracking algorithm based on particle filters was implemented.

Particle Filter

Particle filters are based on probabilistic representations of states by a set of samples (particles), with the advantage of making possible the representation of a non-linear system and measurement models and multi-modal non-Gaussian density states [19,20]. The main idea of particle filters is to represent the required posterior density function by a set of random samples with associated weights and to compute estimates based on these samples and weights.

Let $\{\mathbf{x}_{0:k}^{i}, \omega_{0:k}^{i}\}_{i=0}^{N}$ denote the set of samples that characterizes the a posteriori probability density function (PDF) $p(\mathbf{x}_{0:k}|\mathbf{z}_{0:k})$, where $\{\mathbf{x}_{0:k}^{i}, i = 1,\ldots,N\}$ is a set of points with associated weights $\{\omega_{0:k}^{i}, i = 1,\ldots,N\}$ and $\mathbf{x}_{0:k} = \{\mathbf{x}_{j}, j = 0,\ldots,k\}$ and $\mathbf{z}_{1:k} = \{\mathbf{z}_{j}, j = 0,\ldots,k\}$ are the set of all states and measurements up to time k, respectively.

The weights are normalized such that $\sum_{i} \omega_{k}^{i} = 1$, and they can be chosen using the principle of importance sampling [21].

In the particle filter implementation, a re-sampling step has been applied. This step allows the reduction of the effects of degeneracy, observed in the basic particle filter algorithm. The basic idea is to eliminate particles that have small weights and to concentrate on particles with large weights [21].

Particle filtering provides a robust tracking framework, as it models uncertainty; besides, particle filters can deal with occlusions.

Sun's Position Based on the Particle Filter

After the segmentation process, the position of the Sun in the image is computed, allowing the system to use this information as an input parameter for the particle filter. By doing this, the system is able to track the Sun in each frame; then, if this position suffers an unexpected variation because it has been merged with a very close cloud, the system is capable of mitigating this change.

In Figure 8, the detection (red dot), as well as the tracking (green dot) positions of the Sun are shown.

Figure 8. Sun detection (red dot) and tracking (green dot).

On the other hand, from the case shown in Figure 7, in which the Sun and a cloud (that has suddenly formed very close to it) have been merged as a single object, causing the enclosing circle of this object to be such that its center moves away from the circumsolar area, the system is capable of mitigating the change, keeping the tracking position close to the circumsolar area, due to the use of the particle filter, as shown in Figure 9.

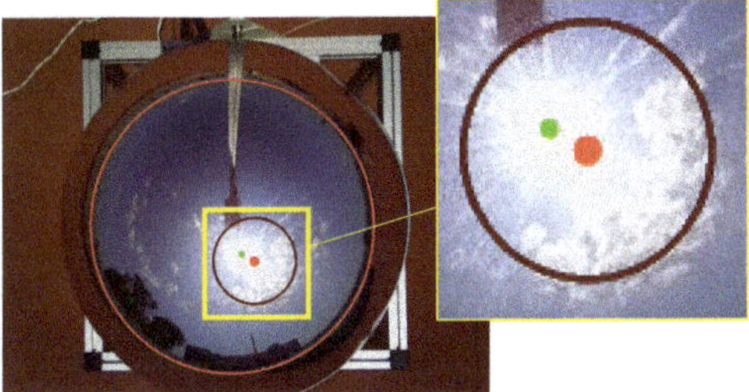

Figure 9. Correction of the Sun's position through tracking.

At this point, with the proposed approach, it is possible to determine the position of the Sun even in cases where there are elements in the scene that may interfere, as is the case of clouds whose intensity in the image could have values similar to the intensity of the Sun. However, it may be the case that the Sun cannot be detected because it is totally occluded by a cloud during a long period of time, for instance on a very cloudy day. To address this problem, a solution based on computing the position of the Sun through the geographical position of the prototype has been implemented.

2.2.5. Detection of the Sun on Cloudy Days

When the position of the Sun, in the image, cannot be computed due to an occlusion commonly caused by a cloud, the system attempts to approximate it through the geographical position (latitude and longitude) of the prototype. In order to do that, the Pysolar library (available at pysolar.org) has been used.

Pysolar is an online library of Python codes that implement an algorithm developed by the National Renewable Energy Laboratory (NREL) for solar position calculation [22]. The implemented algorithm is a set of fairly straightforward equations that calculate the Sun's location in the sky at a specific time of a specific day.

In Figure 10, a set of positions of the Sun (red dots) computed by the Pysolar library on a unit sphere throughout the day are shown.

When the Sun is totally occluded by some clouds and it cannot be segmented in the image, which means that it is not possible to obtain its position, the position computed by the Pysolar library (for that specific day and time) is used. The computed point is projected onto the image and becomes the input parameter of the particle filter instead of the output of the segmentation process.

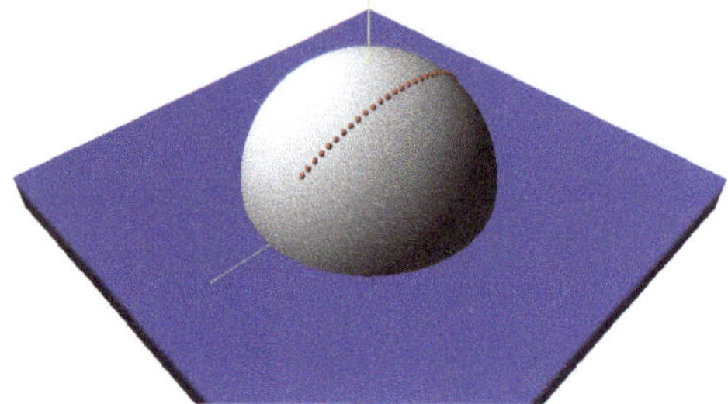

Figure 10. Solar position calculation made by Pysolar.

An example of the case where the Sun is occluded by some clouds is shown in Figure 11. Due to fact that the Sun's position cannot be segmented in the image, the system estimates its position by the Pysolar library; however, once the segmentation process is capable of detecting the Sun, the system improves that approximation, since the position of the Sun given by the segmentation is much more precise.

(**a**) Cloudy day.

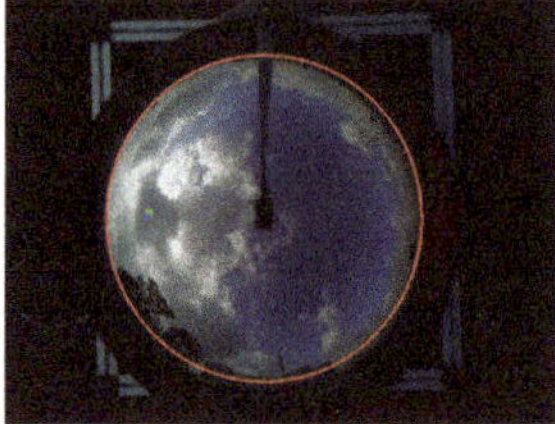

(**b**) Estimated position of the Sun.

Figure 11. Estimation of the Sun's position when it is occluded by a cloud. (**a**) shows the input image where the Sun is occluded by some clouds, while in (**b**), the position estimation made by the Pysolar library (blue dot), as well as the position computed by the particle filter (green dot) are shown.

By using this methodology, the proposed system is capable of detecting the Sun in the image, regardless of the day's conditions. Knowing the position of the Sun at all times is a fundamental step, since from this information, a system like the one that has been proposed would be able to compute or infer other information, for instance an irradiance value.

3. Results and Discussion

To test the proposed approach, three different scenarios have been selected. These three scenarios are considered as the set of scenarios that cover all the possible conditions that may occur. In the first scenario, a full clear-sky day has been selected, in the second one, a day where a partial occlusion of the Sun in some instant of the day is present has been analyzed, and finally, for the third one, a day where in a period of time, a complete occlusion of the Sun is present, has been selected.

3.1. Scenario 1 (Sunny Day)

In this scenario, a full clear-sky day has been selected. For this scenario, a set of images taken every 15 min, from 9:00–17:00 on 8 November (i.e., from sunrise to sunset), has been analyzed. In Figure 12, three representative images of this first scenario are shown. In this case, since it is a clear-sky day, the segmentation process was able to segment the Sun in each frame.

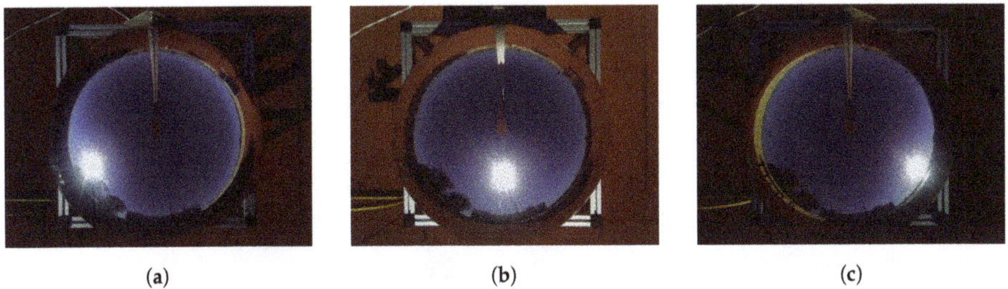

(a)   (b)   (c)

Figure 12. Representative examples with full clear-sky day conditions. (**a**) shows an image taken at 9:00, (**b**) shows an image taken at 13:00, and finally, (**c**) shows an image taken at 17:00.

In Figure 13, the Sun's path throughout the day is shown, where red dots represent the position given by the segmentation process, while green dots represent the position given by the particle filter.

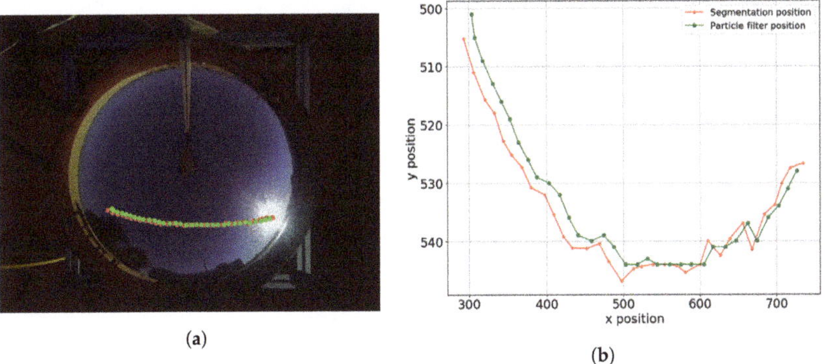

(a)   (b)

Figure 13. Path followed by the Sun. (**a**) Projection of the trajectory of the Sun. (**b**) Trajectory of the Sun.

In order to compare these two trajectories, the distance between the resulting positions of both the segmentation process and the particle filter has been computed. Figure 14 shows a graph of the computed distance, in which is shown the average distance between these two trajectories (dashed black line), the value of which is about 8.03 pixels, as well as the standard deviation with a value that is around 2.05 pixels. Considering that the average radius of the Sun in the image was about 45.5 pixels, then it can be said that this error is not significant.

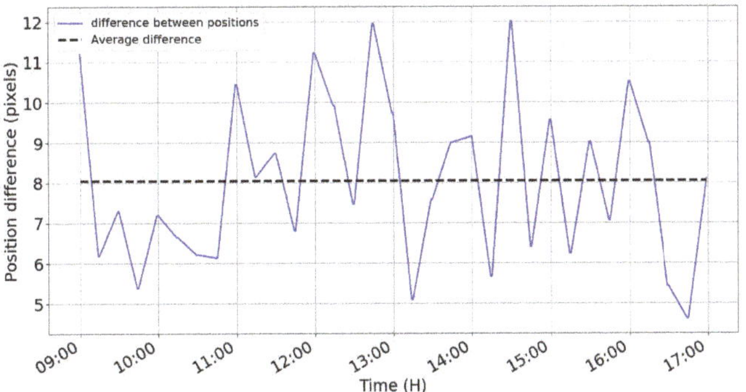

Figure 14. Difference of positions between segmentation and particle filter.

3.2. Scenario 2 (Partially Sunny Day)

In this scenario, a day in which a partial occlusion of the Sun is present has been selected. The set of analyzed data have been acquired by the sensor every 15 min, from 9:00 to 18:00 on 13 September. Figure 15 shows, as in the first case, three representative images of this scenario.

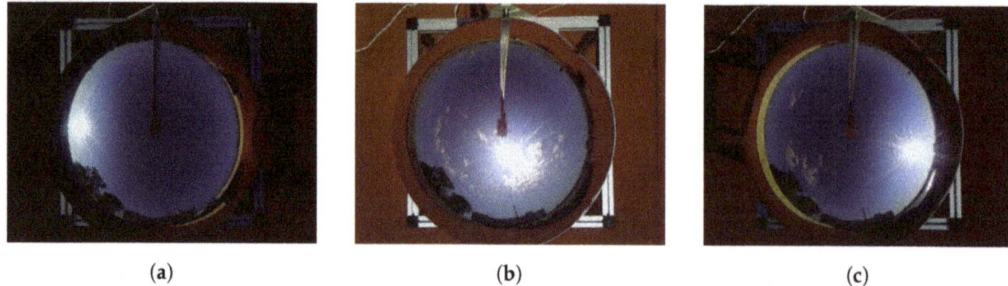

(a)   (b)   (c)

Figure 15. Representative examples where a partially sunny day is presented. (**a**) shows an image taken at 9:00, (**b**) shows an image taken at 14:30, and (**c**) shows an image taken at 18:00.

The partial occlusions of the Sun presented in this scenario are due to a cloud that formed very close to it, which caused the segmentation process to fuse the Sun and a part of the cloud as a single object; then, the calculated position is such that it is between the Sun and the cloud. In Figure 16, the trajectory during the day, as well as the partial occlusions of the Sun are shown.

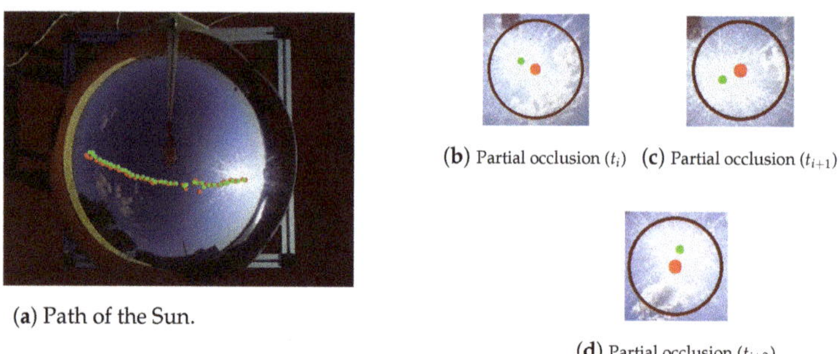

(**b**) Partial occlusion (t_i)   (**c**) Partial occlusion (t_{i+1})

(**a**) Path of the Sun.

(**d**) Partial occlusion (t_{i+2})

Figure 16. Path followed by the Sun (Set 2). In (**b**–**d**), partial occlusion of the Sun are shown.

As in Section 3.1, the distance between the positions given by the segmentation process and the particle filter have been computed.

In Figure 17a, three instants during the day where there is a clear difference in the position of the Sun computed by the segmentation process and the particle filter are shown. These differences become more evident when the distance between these positions are plotted, as shown in Figure 17b, in which an upper limit given by two-times the standard deviation of the distance above the mean value has been marked, and as can be seen, these points are above this limit; this means that these three points moved too far away from the average distance, an effect caused by inaccurate detection in the segmentation process (see Figure 16b–d).

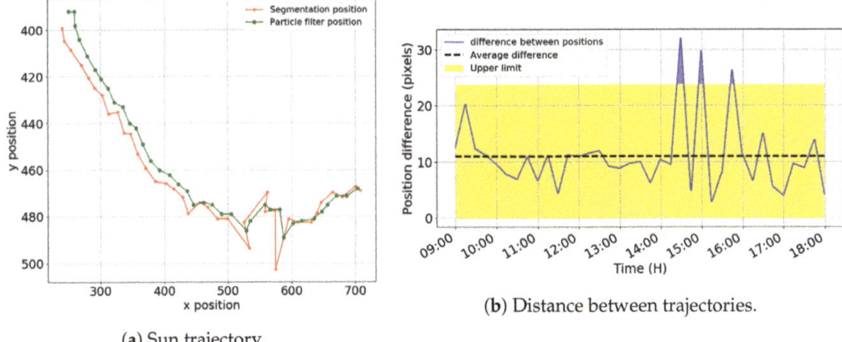

(a) Sun trajectory.

(b) Distance between trajectories.

Figure 17. Sun trajectory and distance between trajectories.

3.3. Scenario 3 (Cloudy Day)

Figure 18 show the representative images of the third test scenario.

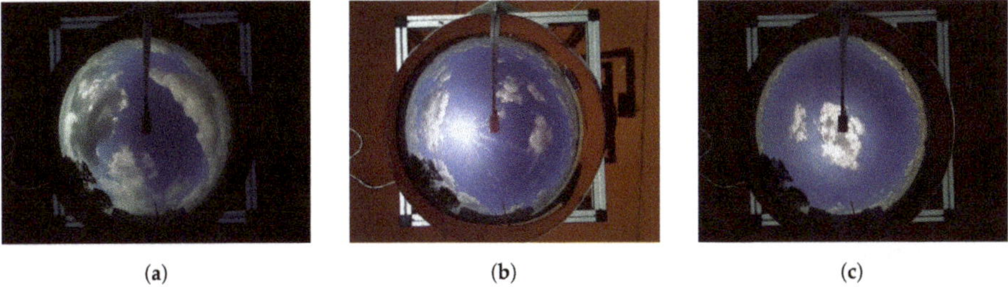

(a)

(b)

(c)

Figure 18. Representative examples of cloudy day conditions. (**a**) shows an image taken at 10:00, while (**b**,**c**) show an image taken at 12:15 and 14:00, respectively.

For this scenario, the images were taken from 10:00–14:00, every 15 min, on 17 August. Along this period of time, the Sun was completely occluded at several moments; one of these moments was at the beginning, as shown in Figure 18a.

In order to compute an approximate position of the Sun, the system uses the the geographical position (latitude and longitude) of the prototype to then, project this position in the image (see Figure 19).

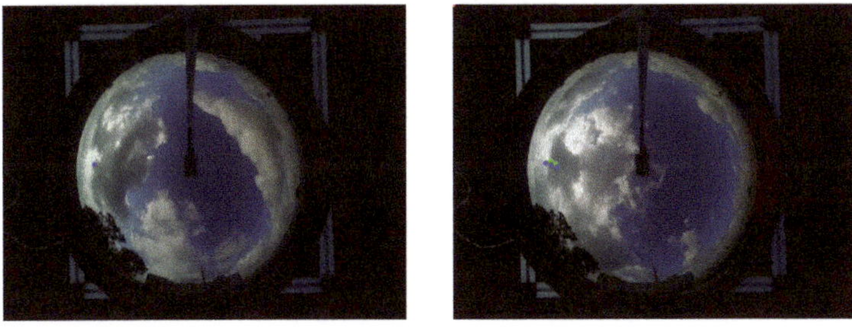

(a) Estimated position of the Sun.

(b) Estimated Sun trajectory.

Figure 19. Estimated position of the Sun. (**a**) shows the Sun position estimation at the beginning, and (**b**) shows an example of the trajectory of the Sun given by the Pysolar model (blue dots), as well as the trajectory given by the particle filter (green dots).

The system attempts to follow the Sun using the computed estimated position until the system is able to segment the Sun. This is done because it is very important to know the position at any moment. However, given that the projected position is just an approximation, we improve it once the system is able to detect the Sun in the segmentation process, as is shown in Figure 20.

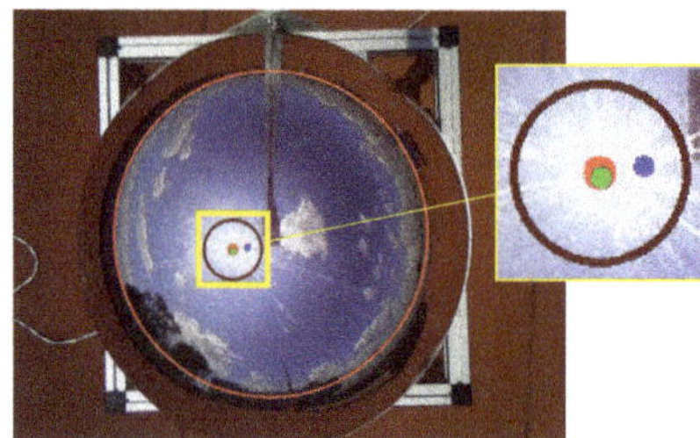

Figure 20. Sun detection (red dot), estimation (blue dot), and tracking (green dot).

By adding this strategy to the system, it is possible to have an estimation of the position of the Sun in the image regardless of the occlusions generated by the clouds.

In Figure 21, the complete trajectory during the evaluated period of time is shown. It can be observed that there are some positions away from the trajectory; this is due to the cloud that was initially generating an occlusion of the Sun begins to move, so there is an instant where the Sun is partially occluded, and as a consequence, the segmentation process considers these two elements as a single one; however, by using the particle filter, the system avoids this abrupt change.

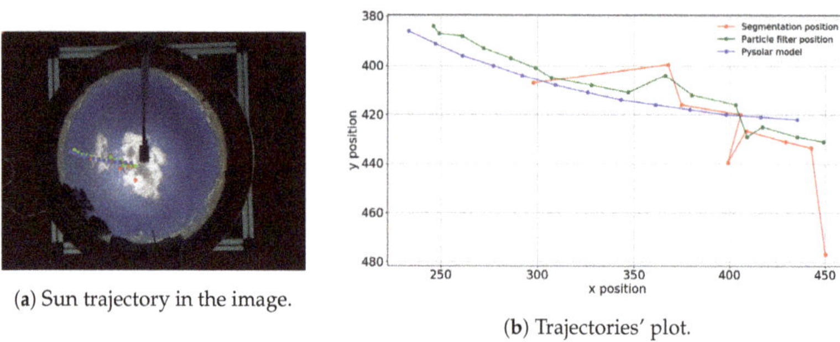

(**a**) Sun trajectory in the image.

(**b**) Trajectories' plot.

Figure 21. Trajectory of the Sun during the evaluated period of time detection.

To form a better view of the performance of the system, a video in the three scenarios described above, can be seen online at Supplementary Materials https://youtu.be/HY2plwtBR-4.

3.4. Detection Accuracy

In order to test how accurate the system is, a day in which the system can always detect the Sun has been chosen, then, as reference the position given by the segmentation process, the positions generated by both the particle filter and the Pysolar model position have been compared. In Figure 22, the trajectories of the Sun given by the three proposed strategies are shown.

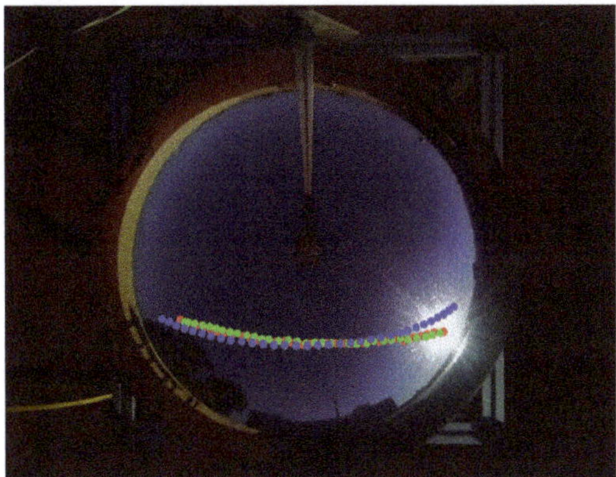

Figure 22. Trajectory of the Sun given by the segmentation process (red dots), the particle filter (green dots), and the estimated position (blue dots), during the day.

In Figure 23, each of the three trajectories have been plotted to clearly show their behavior. It can be observed that the estimated position (position given by the Pysolar model) diverges considerably from the trajectory of the Sun in the image, while the position given by the particle filter remains close. Nevertheless, as was mentioned before, when the system cannot segment the Sun, the only information it has is the result of the position model.

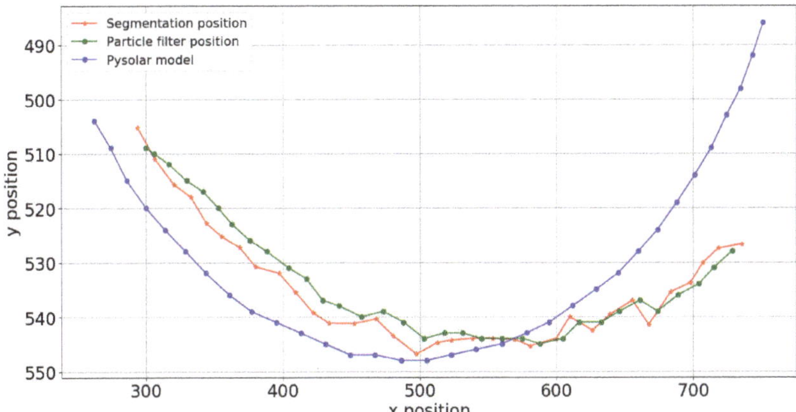

Figure 23. Trajectory of the Sun during a day where we can always detect it.

To evaluate the system quantitatively, the solar zenith angle (SZA) has been calculated as follows. From each point (P_{dome} with coordinates (x, y)) of the positions of the Sun in the dome given by the segmentation process, particle filter and Pysolar model, the distance to the center of the dome has been calculated (this distance is denoted as d_p in Figure 24).

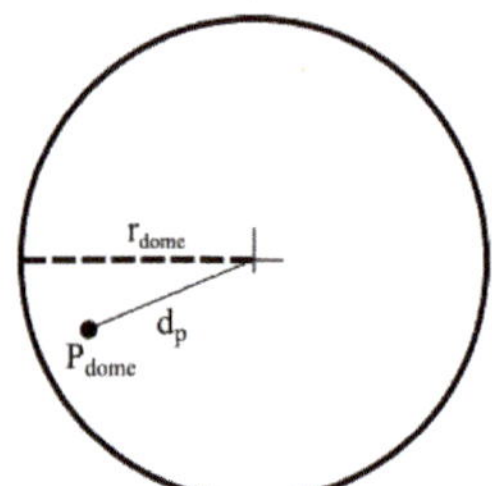

Figure 24. Sun position in the dome.

Then, each of these distances is normalized by dividing them by the radius (defined as r_{dome} in Figure 24) of the dome. Thus, it is possible to compute the solar zenith angle on a unit sphere according to the following expression,

$$SZA_i = \arctan\left(\frac{d_{p_i}/r_{dome}}{\sqrt{1 - ((x_i/r_{dome})^2 + (y_i/r_{dome})^2)}}\right), \qquad i \in \{1, \ldots, N\}$$

By doing this for the N Sun's positions given by the segmentation process, the particle filter and the Pysolar model, the set of plots shown in the Figure 25 results.

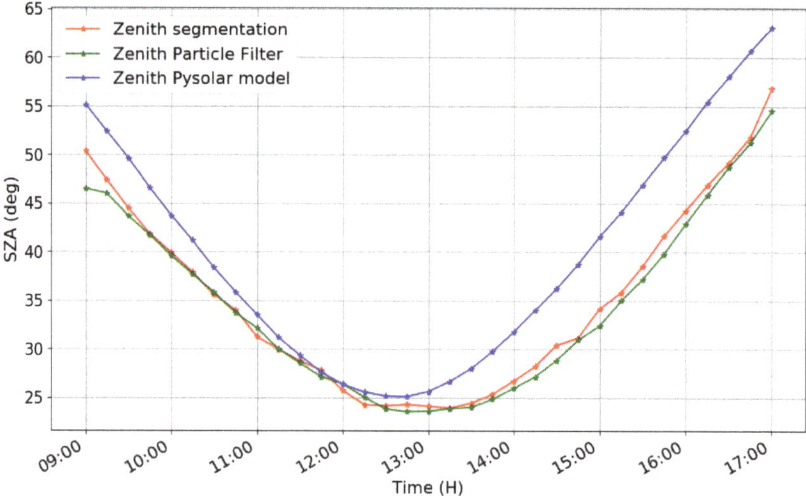

Figure 25. Solar zenith angle (SZA).

Using the SZA of the segmentation process as the reference, the error from the SZA for both the particle filter and Pysolar model have been calculated.

As can be seen in Figure 26, the SZA given by the particle filter deviated about $0.86°$ away on average from the SZA given by the segmentation process; this indicates that the position of the particle filter remained practically in the center of the circumsolar area of the Sun. On the other hand, the SZA given by the Pysolar model was around $4.51°$ away on average from the SZA given by the segmentation process; thus, using only this information can result in significant errors in later processes, for instance the irradiance estimation.

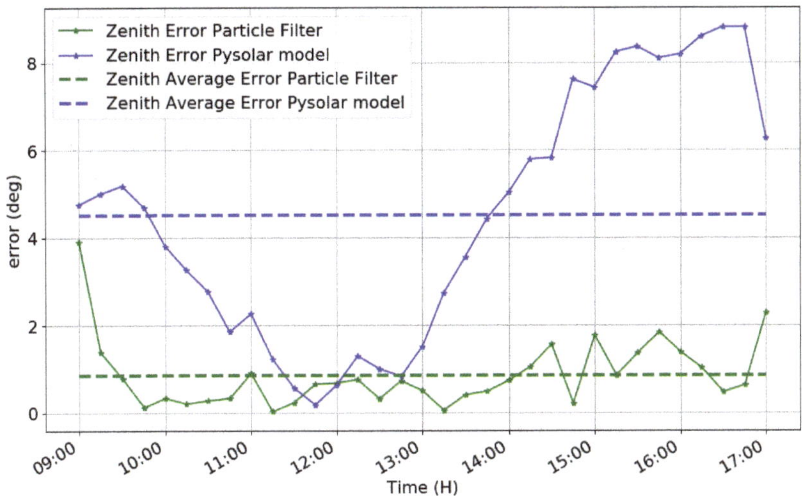

Figure 26. Solar zenith error.

Besides of the SZA calculation, the solar azimuth angle (SAA) has also been calculated using the following expression:

$$SAA_i = \arctan\left(\frac{y_i/r_{dome}}{x_i/r_{dome}}\right), \qquad i \in \{1,\dots,N\}$$

Figure 27 shows the set of plots of the azimuth angles of the segmentation process, the particle filter, and the Pysolar model.

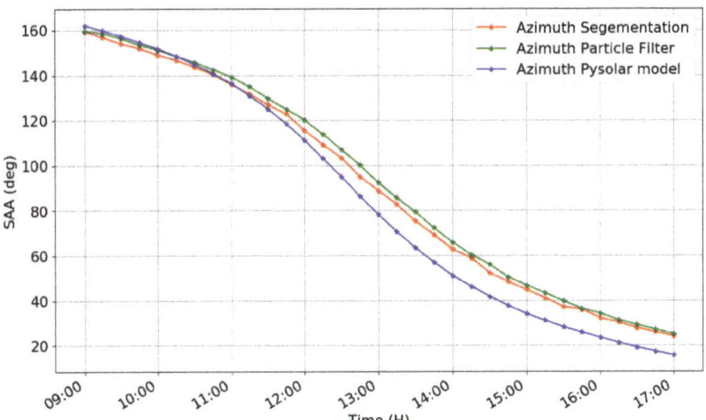

Figure 27. Solar azimuth angle (SAA).

Similar to the previous case, taking as a reference the SAA of the segmentation process, the error of both the SAA of the particle filter and the SAA of the Pysolar model have been computed; in Figure 28, these errors are shown.

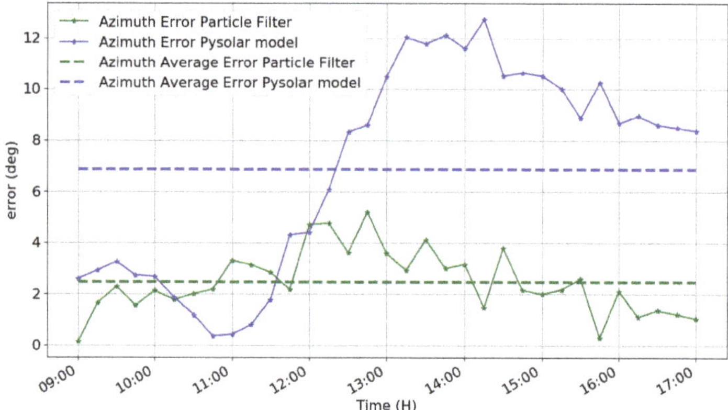

Figure 28. Solar azimuth error.

The behavior of the SAA errors was similar to the behavior of the SZA errors, the SAA of the particle filter having an average error of 2.47° while the SAA of the Pysolar model was around of 6.88° on average.

From the results obtained, it can be observed that clouds occluding the Sun generated the perception of changing the size of the Sun depending on the water vapor concentration of the cloud. "Thin" clouds (low water vapor concentration) tended to generate a perception of growth in the Sun size. On the other hand, when "heavy" clouds were present (high water vapor concentration), the sunlight was obstructed completely, and the Sun position could not be determined. Nevertheless, having the position of the Sun, at any moment, was crucial, and this could only be accomplished by a combination of algorithms that was ready for all the sky scenarios. In addition, mapping sky coordinates to image coordinates allowed working in a new reference system that avoided hardware errors.

4. Conclusions

A novel methodology based on the implementation of several image processing techniques to achieve a robust and automatic detection of the position of the Sun from a set of images, acquired with a low-cost artificial vision system, was proposed. The methodology allowed not only detecting the position of the Sun in a clear-sky day, but also in a cloudy one, even if the Sun was completely occluded. The proposed methodology has been tested using a set of images of three different scenarios, clear-sky day, a partially sunny day, and a cloudy day. Experimental results demonstrated that regardless of the conditions of the day, the computed position of the Sun was within the circumsolar area. It is important to mention that knowing an accurate position of the Sun at any moment allows performing other processes, for instance knowing not only the position of a cloud with respect to the position of the Sun to identify if an occlusion will occur, but also the time it will take, which is fundamental in the implementation of better solar resource assessment systems.

Supplementary Materials: A supporting video is available at: https://youtu.be/HY2plwtBR-4.

Author Contributions: Conceptualization and methodology, L.V. and M.I.P.-C.; software, L.V. and C.M.P.-M.; formal analysis, L.V., M.I.P.-C., and Daniela Moctezuma; writing, original draft preparation, L.V., D.M., and M.I.P.-C.; writing, review and editing, C.A.P.-A. and A.D.-P.

Funding: This research was partially funded by project CONACyT-Problemas Nacionales, grant number 2015-01-1651.

Acknowledgments: The authors acknowledge CONACyT for the partial financial support to the project Problemas Nacionales 2015-01-1651.

Conflicts of Interest: The authors declare no conflict of interest.

References

1. Chauvin, R.; Nou, J.; Thil, S.; Grieu, S. Modelling the clear-sky intensity distribution using a sky imager. *Sol. Energy* **2015**, *119*, 1–17. [CrossRef]
2. Yang, H.; Kurtz, B.; Nguyen, D.; Urquhart, B.; Chow, C.W.; Ghonima, M.; Kleissl, J. Solar irradiance forecasting using a ground-based sky imager developed at UC San Diego. *Sol. Energy* **2014**, *103*, 502–524. [CrossRef]
3. Long, C.N.; Sabburg, J.M.; Calbó, J.; Pagès, D. Retrieving Cloud Characteristics from Ground-Based Daytime Color All-Sky Images. *J. Atmos. Ocean. Technol.* **2006**, *23*, 633–652. [CrossRef]
4. Meza, F.; Varas, E. Estimation of mean monthly solar global radiation as a function of temperature. *Agric. For. Meteorol.* **2000**, *100*, 231–241. [CrossRef]
5. Valdes-Barrón, M.; Riveros-Rosas, D.; Arancibia-Bulnes, C.; Bonifaz, R. The Solar Resource Assessment in Mexico: State of the Art. *Energy Procedia* **2014**, *57*, 1299–1308. [CrossRef]
6. Igawa, N.; Koga, Y.; Matsuzawa, T.; Nakamura, H. Models of sky radiance distribution and sky luminance distribution. *Sol. Energy* **2004**, *77*, 137–157. [CrossRef]
7. Florita, A.; Hodge, B.; Orwig, K. Identifying Wind and Solar Ramping Events. In Proceedings of the 2013 IEEE Green Technologies Conference (GreenTech), Denver, CO, USA, 4–5 April 2013; pp. 147–152.
8. Lorenz, E.; Kühnert, J.; Heinemann, D. Short Term Forecasting of Solar Irradiance by Combining Satellite Data and Numerical Weather Predictions. In Proceedings of the 27th European Photovoltaic Solar Energy Conference and Exhibition, Frankfurt, Germany, 24–28 September 2012; pp. 4401–4405.
9. Chow, C.W.; Urquhart, B.; Lave, M.; Dominguez, A.; Kleissl, J.; Shields, J.; Washom, B. Intra-hour forecasting with a total sky imager at the UC San Diego solar energy testbed. *Sol. Energy* **2011**, *85*, 2881–2893. [CrossRef]
10. Reikard, G. Predicting solar radiation at high resolutions: A comparison of time series forecasts. *Sol. Energy* **2009**, *83*, 342–349. [CrossRef]
11. Marquez, R.; Coimbra, C.F. Forecasting of global and direct solar irradiance using stochastic learning methods, ground experiments and the NWS database. *Sol. Energy* **2011**, *85*, 746–756. [CrossRef]
12. Chu, Y.; Pedro, H.T.; Li, M.; Coimbra, C.F. Real-time forecasting of solar irradiance ramps with smart image processing. *Sol. Energy* **2015**, *114*, 91–104. [CrossRef]
13. Marquez, R.; Coimbra, C.F. Intra-hour DNI forecasting based on cloud tracking image analysis. *Sol. Energy* **2013**, *91*, 327–336. [CrossRef]
14. Diagne, M.; David, M.; Lauret, P.; Boland, J.; Schmutz, N. Review of solar irradiance forecasting methods and a proposition for small-scale insular grids. *Renew. Sustain. Energy Rev.* **2013**, *27*, 65–76. [CrossRef]
15. Kurtz, B.; Kleissl, J. Measuring diffuse, direct, and global irradiance using a sky imager. *Sol. Energy* **2017**, *141*, 311–322. [CrossRef]
16. Nguyen, R.M.H.; Brown, M.S. Why You Should Forget Luminance Conversion and Do Something Better. In Proceedings of the 2017 IEEE Conference on Computer Vision and Pattern Recognition (CVPR), Honolulu, HI, USA, 21–26 July 2017; pp. 5920–5928.
17. Deng, G.; Cahill, L.W. An adaptive Gaussian filter for noise reduction and edge detection. In Proceedings of the 1993 IEEE Conference Record Nuclear Science Symposium and Medical Imaging Conference, San Francisco, CA, USA, 31 October–6 November 1993; Volume 3, pp. 1615–1619.
18. Bradley, D.; Roth, G. Adaptive Thresholding using the Integral Image. *J. Graph. Tools* **2007**, *12*, 13–21. [CrossRef]
19. Almeida, A.; Almeida, J.; Araujo, R. Real-Time Tracking of Moving Objects Using Particle Filters. In Proceedings of the IEEE International Symposium on Industrial Electronics, ISIE 2005, Dubrovnik, Croatia, 20–23 June 2005; Volume 4, pp. 1327–1332.
20. Nummiaro, K.; Koller-Meier, E.; Gool, L.V. An adaptive color-based particle filter. *Image Vis. Comput.* **2003**, *21*, 99–110. [CrossRef]

21. Arulampalam, M.S.; Maskell, S.; Gordon, N.; Clapp, T. A tutorial on particle filters for online nonlinear/non-Gaussian Bayesian tracking. *IEEE Trans. Signal Process.* **2002**, *50*, 174–188. [CrossRef]
22. Reda, I.; Andreas, A. Solar position algorithm for solar radiation applications. *Sol. Energy* **2004**, *76*, 577–589. [CrossRef]

 *applied sciences*

Article

Validation of All-Sky Imager Technology and Solar Irradiance Forecasting at Three Locations: NREL, San Antonio, Texas, and the Canary Islands, Spain

Walter Richardson Jr. [1,*], **David Cañadillas** [2], **Ariana Moncada** [1], **Ricardo Guerrero-Lemus** [2], **Les Shephard** [3], **Rolando Vega-Avila** [4] **and Hariharan Krishnaswami** [5]

[1] Department of Mathematics, University of Texas at San Antonio, San Antonio, TX 78219, USA; arianam.moncada@gmail.com
[2] Departamento de Física, Universidad de La Laguna, S/N 38206 S/C de Tenerife, Spain; dcanadil@ull.edu.es (D.C.); rglemus@ull.edu.es (R.G.-L.)
[3] Department of Civil Engineering, University of Texas at San Antonio, San Antonio, TX 78219, USA; Les.Shephard@utsa.edu
[4] CPS Energy, San Antonio, TX 78219, USA; RVega-Avila@cpsenergy.com
[5] Department of Electrical and Computer Engineering, University of Texas at San Antonio, San Antonio, TX 78219, USA; harikrismn@gmail.com
* Correspondence: Walter.Richardson@utsa.edu; Tel.: +01-210-458-4760

Received: 30 December 2018; Accepted: 8 February 2019; Published: 17 February 2019

Featured Application: Validation of all-sky imager technology at 3 locations: NREL, Joint Base San Antonio, and the Canary Islands, Spain.

Abstract: Increasing photovoltaic (PV) generation in the world's power grid necessitates accurate solar irradiance forecasts to ensure grid stability and reliability. The University of Texas at San Antonio (UTSA) SkyImager was designed as a low cost, edge computing, all-sky imager that provides intra-hour irradiance forecasts. The SkyImager utilizes a single board computer and high-resolution camera with a fisheye lens housed in an all-weather enclosure. General Purpose IO pins allow external sensors to be connected, a unique aspect is the use of only open source software. Code for the SkyImager is written in Python and calls libraries such as OpenCV, Scikit-Learn, SQLite, and Mosquito. The SkyImager was first deployed in 2015 at the National Renewable Energy Laboratory (NREL) as part of the DOE INTEGRATE project. This effort aggregated renewable resources and loads into microgrids which were then controlled by an Energy Management System using the OpenFMB Reference Architecture. In 2016 a second SkyImager was installed at the CPS Energy microgrid at Joint Base San Antonio. As part of a collaborative effort between CPS Energy, UT San Antonio, ENDESA, and Universidad de La Laguna, two SkyImagers have also been deployed in the Canary Islands that utilize stereoscopic images to determine cloud heights. Deployments at three geographically diverse locations not only provided large amounts of image data, but also operational experience under very different climatic conditions. This resulted in improvements/additions to the original design: weatherproofing techniques, environmental sensors, maintenance schedules, optimal deployment locations, OpenFMB protocols, and offloading data to the cloud. Originally, optical flow followed by ray-tracing was used to predict cumulus cloud shadows. The latter problem is ill-posed and was replaced by a machine learning strategy with impressive results. R^2 values for the multi-layer perceptron of 0.95 for 5 moderately cloudy days and 1.00 for 5 clear sky days validate using images to determine irradiance. The SkyImager in a distributed environment with cloud-computing will be an integral part of the command and control for today's SmartGrid and Internet of Things.

Keywords: distributed PV generation; microgrid; irradiance forecasting; all-sky imager; Raspberry Pi; optical flow; machine learning; cloud-computing; SmartGrid; Internet of Things (IoT)

---

1. Introduction

The Department of Energy (DOE) estimates that PV power will grow to 14% of the electricity supply by 2030 as the price of solar electricity reaches a point at which it is cost-competitive with cogen ($0.06/kwh by 2020). It is imperative that power grid reliability and stability be maintained under this high penetration of low carbon energy [1]. Organizations such as North American Electric Reliability Corporation (NERC) and California Energy Commission (CEC) [2] have formulated several requirements needed in a "grid-friendly" PV power plant. For instance, CEC has developed a set of several smart inverter functionalities such as dynamic volt/var operation and ramp rate control. Existing PV plants do not have these functionalities even though the inverters are capable, due to the lack of communications standards and dynamic control. For PV power plants to participate in energy markets and ancillary services markets, they need to be considered "dispatchable" power plants.

High penetration of PV systems can be achieved if PV inverters [3] participate in the grid frequency regulation by active power control. Currently, frequency disturbances in the grid are handled by load curtailment. The disadvantage of this methodology is that it can cause voltage stress on the distributed generation. The alternative is to operate the PV system below its MPP (Maximum Power Point) to provide active power control. This can be done by modifying the MPP algorithm in such a way that it can track the next MPP point while working in the reduced power mode (RPM). A critical component of coordinated inverter control is forecasted solar power output or forecasted MPP at the array level. Having accurate intra-hour solar forecasts can enable implementation of a coordinated inverter control strategy capable of regulating a set-point, which may be a signal from a utility requiring either power curtailment or frequency regulation. The electric utility industry has yet to see an integrated solution to the dispatchability problem of PV plants, a system solution that effectively integrates intra-hour solar forecasting and smart control of inverters to achieve not only a grid-friendly plant, but also one that provides monetary and efficiency benefits to the PV plant operator.

Microgrids lack the stabilizing effects present in a large urban macrogrid that itself is joined to an interconnect; these issues are critical when a microgrid is operated in islanded mode. The Energy Management Systems (EMS) that balance PV output, load, and battery storage require accurate intra-hour irradiance forecasts to solve the control problem by shedding non-critical load when power generated is predicted to drop significantly below the load. An increasingly important problem for utilities is optimal scheduling and dispatch of a microgrid [4–7], both when connected to the macrogrid and when operated as an island. This task is divided into Day-Ahead Scheduling, which finds optimal schedules for the next operating day and focuses on energy markets, and Intra-day Dispatching and Scheduling in which schedules are continuously updated during the current day. Both cases follow these steps: (1) forecast day-ahead load, (2) forecast day-ahead renewable power (solar, wind), (3) Micro-Grid Management System (MGMS) optimizes the day-ahead plan, produces schedules for flows within the microgrid and to the PCC, (4) MGMS transmits schedules to utility control center.

This article describes a four-year research effort to develop hardware and software with an aim to solve the intra-hour solar forecasting problem for electric utilities. It was a collaborative effort between many groups, including national labs and research institutes (NREL and the Texas Sustainable Energy Research Institute TSERI), two universities (The University of Texas at San Antonio and Universidad de La Laguna in the Canary Islands), public and private utilities (CPS Energy, ENDESA, and Duke Energy), and a private company, Siemens-Omnetric. It serves as a case study in managing research in a joint theatre of operations and integrating the efforts of researchers and engineers who came from very different university and industrial cultures. Details of our research and technology development have been presented in journal articles [8,9], conference proceedings [10–13], and technical reports [14]. While this paper gives a detailed overview of that research, our primary goal is to describe how the

UTSA SkyImager was validated at three geographically diverse locations, the pitfalls encountered, the lessons learned, and the outlook for future research efforts.

The SkyImager evolved from a realization that existing all-sky imaging systems were too expensive to be deployed in large numbers, suffered from data-loss issues caused by the shadow band and camera arm, and used proprietary software. The Raspberry Pi single board computer ($35) and programable high-resolution Pi-Cam ($20) with a fisheye lens ($20) formed the heart of the new system. The most expensive component was the all-weather security camera enclosure ($350). In addition, General Purpose IO (GPIO) pins would allow a variety of external sensors to be connected to the Pi. Low cost and ease of use were essential if the SkyImager were to be deployed in a rural sustainable development microgrid. In contrast to some commercial all-sky imaging systems, only open source software would be used. The Pi accommodates several operating systems (OS) including Raspbian, a Linux-based derivative of Debian. It can be operated with a monitor or in "headless" mode, and once deployed can be accessed remotely with SSH/SFTP. Code for the SkyImager would be written in Python and allow calls to libraries such as Open Computer Vision, Scikit-Learn, SQLite, and Mosquito. In the summer of 2014 it was an open question whether the proposed imager could deliver the functionality of much more expensive systems and be thoroughly tested before its deployment.

As part of the DOE microgrid INTEGRATE program, the first deployment of the SkyImager occurred in Fall 2015 at the ESIF building at NREL. INTEGRATE aggregated sustainable generation and loads into microgrids controlled by an Energy Management System with the OpenFMB protocol. In 2016 a second SkyImager was installed at the CPS Energy microgrid at Joint Base San Antonio. A multi-year collaboration between CPS-UTSA and the Universidad de La Laguna resulted in the deployment of two SkyImagers in the Canary Islands. These utilize stereoscopic images to determine heights of the bases of cumulus clouds. Deployment of SkyImagers in three diverse locations provided not only big data, but operational experience in harsh extremes of climate. This resulted in improvements and additions to our original design: weatherproofing, new environmental sensors, the need for scheduled maintenance, optimal positioning of the camera, communications with OpenFMB publish-subscribe protocols, and using WiFi and cloud computing. The SkyImager will be an integral part of the command and control for microgrids, both as part of the larger SmartGrid in an urban environment or in an islanded mode in a military or rural setting.

Solar forecasting is widely considered a key means of integrating solar power efficiently and reliably into the electric grid. For a utility to meet projected customer demand with electricity from sustainable resources, high-accuracy global horizontal irradiance (GHI) forecasts must be available over widely different time and space scales. A convenient separation of this forecasting problem into two parts is as follows: (1) *intra-hour forecasts* of the ramp events that are caused when cumulus clouds move between the sun and the solar panels, and (2) *day-ahead forecasts* for 12, 24, and perhaps 36 h into the future. There is overlap between the two parts, but this taxonomy is convenient not only because the physics and forecasting techniques are generally different, but also the way in which the utility makes use of the forecasts. A single ramp event on a microgrid powered primarily by PV arrays can result in over/under voltages, as well as frequency deviations and may require secondary spinning reserves to be brought on line. Day-ahead irradiance forecasts are useful in predicting surplus/deficit generation capacity that can then be augmented or sold in the day-ahead electricity market.

1.1. Day-Ahead GHI Forecasting

For day-ahead GHI forecasting, both numerical weather prediction (NWP) [15] and satellite imagery provide effective tools for forecasting irradiance. The National Center for Environmental Prediction (NCEP), a part of NOAA, runs two versions of the Rapid Refresh (RAP) numerical weather model to predict environmental data. The first version generates weather data on a 13-km (8-mile) resolution horizontal grid and the second, the High-Resolution Rapid Refresh (HRRR), generates data on a 3-km (2-mile) grid. RAP forecasts use multiple data sources: commercial aircraft weather data, balloon data, radar data, surface observations, and satellite data to generate forecasts with hourly resolution in time and forecast lengths of 18 hours. For further details, consult the RAP website [16]. RAP data are available for download through the National Model Archive and Distribution System (NOMADS). Several benefits accrue from using NWP for irradiance forecasting: NOAA incurs much of the computational burden and these models incorporate first-principles physics such as the Navier-Stokes equations, thereby allowing for the dynamic formation of clouds. Satellite technology is advancing rapidly with GOES-16 (Geostationary Operational Environmental Satellite) pictures being updated every 5 min with maximum resolution of 5000 by 3000 pixels. Figure 1 shows such an image cropped to the central Texas region. The temporal sampling of the data is still insufficient to support optical flow predictions 15-min ahead. Continued improvements in GOES-R (geostationary satellite with high spatial and temporal resolution) may well make the satellite approach to minutes-ahead irradiance prediction more attractive in the future [17–20]. Statistical methods based on historical time-series data and climatology are also useful for day-ahead PV forecasting.

Figure 1. GOES high-resolution satellite image cropped to show the central Texas region.

1.2. Intra-Hour Solar Forecasting

The energy alliance between UTSA and CPS Energy has as one of its primary goals the development of new solar forecasting technologies that combine inexpensive all-sky imaging cameras with sophisticated image processing techniques and artificial intelligence software to produce high-accuracy 15-min ahead solar irradiance forecasts. GHI consists of two components, the Direct Normal Irradiance (DNI) caused by sunlight traveling in a direct path from sun to PV array and the Diffuse Horizontal Irradiance (DHI), background illumination that is due to secondary reflections and absorption/re-radiation. The formula is $GHI = \cos(\theta_z)\,DNI + DHI$ where θ_z is the zenith angle. Shadows cast by low-level cumulus clouds significantly impact the DNI but have little effect upon the background DHI. While it is possible to predict DNI separately [21], for verifying PV power output forecasts GHI is used. Figure 2 displays the three quantities: GHI, DNI, and DHI, on the date 27 October 2015 at the NREL ESIF facility in Golden, CO. It shows that moderately cloudy conditions occasion multiple ramp events.

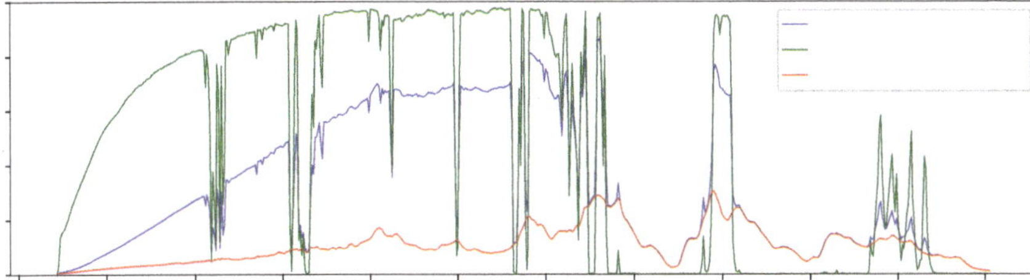

Figure 2. Daily solar irradiance (W/m^2) for 27 October 2015, Golden, CO.

Figure 3 displays a sequence of eight pictures taken by the SkyImager at the NREL site, one every two minutes starting (upper left) at 12:31pm MST. At that time the sun is not obscured, but cumulus clouds are moving in from the left. At 12:35 the cloud begins to enter the solar disk and by 12:37 the sun is completely occluded. This continues until 12:44 when the cloud has moved past the sun and the DNI recovers. This ramp event is seen in the DNI oscillations that occur around the noon hour in Figure 2. While this example considers a single ramp event, it strongly suggests that the correlation between measured GHI and the presence of clouds obscuring the sun in the SkyImager pictures could be learned by AI models.

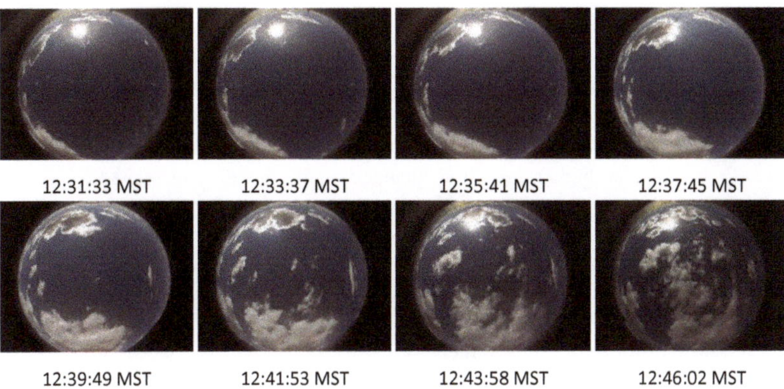

Figure 3. SkyImager sequence (27 October 2015, Golden, CO) showing a ramp event.

The challenge in short term prediction of PV power is simply "Where will cumulus cloud shadows be fifteen minutes from now?" Our approach incorporates as much of the physics as possible, but is an idealization necessitated by the requirement to produce GHI forecasts in real time for the MGMS. Solving Navier-Stokes for the true dynamics of the atmosphere is not feasible on a Raspberry Pi. If GHI can be accurately forecast, then predicting PV power output is straightforward. The evolution of clouds and irradiance shown in Figures 2 and 3 is even more striking when video of the images is viewed, confirming that SkyImager pictures are highly correlated with the observed GHI time series. Moreover, it suggests the camera sensor could be used to measure irradiance as well as predict it. Once GHI has been accurately forecast, it is usually straightforward to assign a corresponding PV power output, which is what the MGMS requires.

Figure 4 shows the relationship between GHI (Watts/m^2) and PV power (Watts) from the RSF2 PV arrays located at NREL. The relationship is almost linear with a slight hysteresis effect that reflects the differences in morning versus afternoon irradiance. The task of predicting PV power output over multiple temporal and spatial scales, and from a variety of different equipment is a challenging one [22–24].

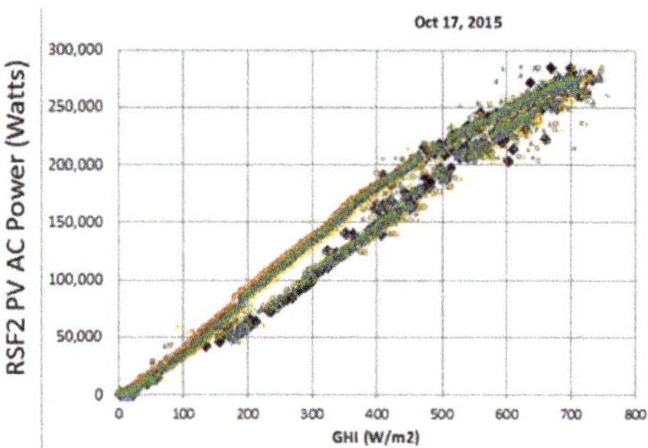

Figure 4. Relationship between GHI (W/m^2) and PV Power (Watts) determined at NREL.

State of the Art in Solar Forecasting

As photovoltaics achieve greater penetration, the SmartGrid will demand accurate solar forecasting hence a network of low cost, distributed sensors to acquire large amounts of image and weather data for input to the forecasting algorithms. Commercial sky imaging systems often prove too costly and have proprietary software, leading several research groups to develop their own systems. The solar forecasting research group at UC San Diego [25–29] has done pioneering work in this field for many years. For one example, Coimbra et al. [30] proposed DNI forecasting models using images from a Yankee TSI 880 with hemispherical mirror as inputs to Artificial Neural Networks (ANN). The TSI has high capital and maintenance costs, uses a shadow band mechanism, and requires proprietary software. The UCSD Sky Imager described in Yang et al. [31] captures images with an upward-facing charge-coupled device (CCD) Panasonic sensor and a 4.5 mm focal length fisheye lens. Compared with the TSI it has higher resolution, greater dynamic range, and lossless PNG compression. The Universitat Erlangen-Nurnberg [32] group used a five-megapixel C-mount camera equipped with a fisheye lens. They implemented the Thirions '"daemons"' algorithm for image registration and cloud-motion estimation similar to optical flow. In Australia, West et al. [33] used off-the-shelf IP security cameras (Mobotix Q24, Vivotek FE8172V) for all-sky imaging. Inexpensive compared to the TSI, the cost of such systems still is ~800 euros. Rather than a feature-tracking strategy, they used dense optical flow to estimate cloud movement. See also Wood-Bradley [34]. Several research groups in China are working on the irradiance forecasting problem [35,36] generally using a TSI imager, but in one case Geostationary Statellite data [37]. As mentioned before dramatic improvements in GOES-R technology and resolution (spatial and temporal) will make this approach more attractive for intra-hour forecasting. See also the historical review [38] of irradiance and PV power forecasting that was produced using text mining and machine learning.

A recurring theme in the INTEGRATE project was that while all-sky imaging was a critical component of microgrid stability and control, it could not be developed in a stand-alone fashion but must be fully integrated into the microgrid management system (MGMS). Uriate et al. [39] discuss the importance of Ramp Rates (RR) on the inertial stability margin of a microgrid deployed at the Marine Corp Base at Twentynine Palms. They show that a large ramp in PV power can destabilize frequency when the PV load is suddenly transferred to the cogen. The inertia constant H of a generator is the ratio of stored kinetic energy to system capacity. Microgrids usually have $H < 1$ s compared to 2–10 s for large generation plants. Frequency stability is defined by the condition $\left| \Delta f_{pu} \right| < \Delta f_{max}^{pu}$ where allowable frequency deviation Δf_{max}^{pu} in p.u. is typically 0.01–0.05 per unit. In [39] the authors derive the ODE $\omega_{mech} J d\omega_{mech}/dt + \omega_{mech}^2 D = P_{accel}$ where ω_{mech} is the mechanical speed of the generator in rad/s, J the moment of inertia (kg·m^2), D a damping coefficient (Nm/s), and P_{accel} the power imbalance exerted on a generator rotor, to model the microgrid stability control problem. The NREL microgrid has a 300 kW Caterpillar diesel generator, whereas JBSA has no cogen. However, the same issues of stability and frequency control apply when there are no spinning resources. Some electric codes are specifying ancillary control must be added to the EMS in order to handle ramp events of a certain magnitude and duration. See [40–42] for details.

1.3. Climatology and Microgrid Architectures at the Three Locations

As shown in Figure 5 the UTSA SkyImager has been deployed at 3 geographically diverse locations: Golden, Colorado on the rooftop of the ESIF building at NREL, in San Antonio, Texas at the CPS Energy microgrid facility at Joint Base San Antonio (JBSA) and the Engineering Building at UTSA, as well as in the Canary Islands, Spain at Tenerife and Caleta de Sebo. Each location presented unique challenges in terms of local climate, physical and cyber access, and microgrid design, equipment, operation, and customer needs.

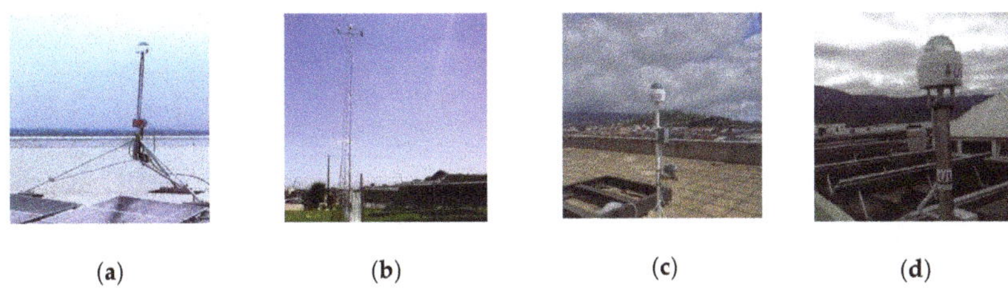

(a)   **(b)**   **(c)**   **(d)**

Figure 5. SkyImager at (**a**) NREL, (**b**) JBSA, (**c**) Tenerife and (**d**) Caleta de Sebo.

The UTSA SkyImager was first conceived as a technology for providing accurate intra-hour irradiance forecasts as inputs to a microgrid management system that would then provide the utility with command and control of the microgrid in either connected or islanded mode. The Department of Energy INTEGRATE project [43] lasted for 18 months beginning on 6 March 2015 and partnered NREL, Omnetric-Siemens, CPS Energy, Duke Energy, and UTSA. The project goal was to increase the capacity of the electric grid to incorporate renewables by upgrading and optimizing architectures for control and communication in microgrids. There were three major components: (1) OpenFMB, a reference architecture that allows real time interaction among distributed intelligent nodes, (2) optimization with the Spectrum Power Microgrid Management System based on the Siemens SP7 Platform, and (3) PV and Load Forecasting using UTSA's applications for both intra-hour and day-ahead irradiance and building load forecasts. The OpenFMB framework leverages existing standards such as IEC's Common Information Model (CIM) semantic data model and the Internet of Things (IoT) publish/subscribe protocols (DDS, MQTT, and AMQP) to allow flexible integration of renewable energy and storage into the existing electric grid. The OpenFMB standard was ratified by the North American Energy Standards Board (NAESB) in March of 2016 and allows communication between diverse grid devices–meters, relays, inverters, capacitor bank controllers, etc. It allows federated message exchanges with readings such as kW, kVAR, V, I, frequency, phase, and State of Charge (SOC) published every 2 seconds as well as data-driven events, alarms, and control in near-real-time.

1.3.1. SkyImager at National Renewable Energy Laboratory in Golden, CO

The site of the first SkyImager deployment was NREL in the Rocky Mountains. Golden's high elevation and mid-latitude interior continent geography results in a cool, dry climate. There are large seasonal and diurnal swings in temperature. At night, temperatures drop quickly and freezing temperatures are possible in some mountain locations year-round. The thin atmosphere allows for greater penetration of solar radiation. As a result of Colorado's distance from major sources of moisture (Pacific Ocean, Gulf of Mexico), precipitation is generally light in lower elevations.

Eastward-moving storms from the Pacific lose much of their moisture falling as rain or snow on the mountaintops. Eastern slopes receive relatively little rainfall, particularly in mid-winter. The SkyImager enclosure came equipped with a heater/fan that performed well at NREL. Given the climate, it proved useful in keeping frost off the plastic dome. It adds to the expense and complexity of the technology and may not be required at other locations. Most installations of the security camera enclosure would be facing downward and perhaps under a building overhang. Used facing upward and exposed to the sky, there were issues with water getting inside the enclosure. A simple solution was silicon caulk applied at the base of the dome. In a typical security installation, a green tinted plastic dome is used with the enclosure to protect components from UV radiation. For all-sky imaging a clear plastic dome is a necessity. With any plastic material on a bright sunny day there can be issues with glare caused by the dome, but this was minor. The alternative is a glass dome but that has it own set of problems.

As shown in Figure 6, the microgrid at NREL was already well established and the process of deploying the SkyImager went relatively smoothly. Denver International Airport is located some 36 miles from Golden; this distance introduces some error in the Cloud Base Height for the ray-tracing algorithm originally used in the SkyImager. The ESIF building at NREL had the infrastructure necessary for easy installation of both the SkyImager and the Hardkernel Odroid C1 single board computer (SBC) used for load and day-ahead PV forecasts. Information was transferred using a Wi-Fi network on a LAN system. NREL also provided un-interruptible 120 VAC power, ample Ethernet connectivity, and excellent on-site weather and irradiance data. In addition to solar PV arrays, generation included a 500 kW wind power simulator and a 300 kW caterpillar diesel. A 300 kWh battery system provided energy storage and the load was separated into a controllable component (250 kW) and a critical load (250 kW).

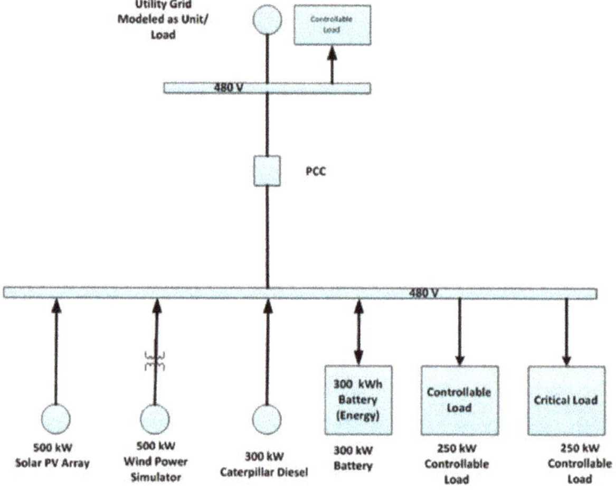

Figure 6. Microgrid at NREL ESIF Building where the first SkyImager was deployed in 2015.

1.3.2. SkyImager at San Antonio, TX, USA

Texas produces more electricity than any of the other 49 states, and as a result has its own interconnect ERCOT. In 2017, power statewide was generated by a variety of sources: natural gas (45%), coal (30%), wind (15%), and nuclear (9%). In 2014, wind replaced nuclear as the third-largest source of power and Texas now produces more wind power than any other state. Solar generation is increasing, but still relatively small for a state with abundant annual sunshine. Located in central Texas some 200 miles from the Gulf of Mexico, San Antonio is home to almost 1.5 million people and several military bases. CPS Energy serves San Antonio and is the nation's largest public power, natural gas and electric company. They are committed to renewables, funding a 400 MWac project with multiple PV plants (Alamo 1–7) close to San Antonio, and wind farms in West and South Texas. CPS Energy is among the top public power wind energy buyers in the nation and number one in Texas for solar generation. In keeping with this commitment, TSERI was formed in 2001 as an alliance between CPS Energy and UTSA.

For San Antonio, the most significant local weather issue is low-level Gulf stratus [44]. Elevations of the terrain increase from sea level at the Gulf coast to almost 800 ft at San Antonio, and a moist air mass over the Gulf of Mexico will cool adiabatically to saturation as it moves upslope. Nocturnal radiational cooling causes cloud formation before midnight, resulting in a ceiling of 500–1000 feet. A solid cloud deck will cover much of central Texas and remain in place until late morning when the sun burns off the stratus and cumulus clouds begin to form. Forecasting Gulf stratus is an important problem for aviation; it is a matter of accurately predicting low-level wind flow (<5000 ft) with the most favored wind direction for stratus formation from 90° to 180°. It is important to address these local weather conditions that occur below the spatial and temporal resolution of NWP, but are crucial for both inter-hour and day-ahead irradiance forecasts. Use of machine learning using local datasets and climatology will allow the information and intelligence of a study such as [44] to be incorporated in site-specific irradiance forecasts.

The Fort Sam Houston Library location at JBSA presented several unique challenges for the deployment of the UTSA hardware and software, challenges that provide valuable insights for other researchers. Many of the issues that arose were heavily dependent on the specific location. At JBSA, the Sky Imager was deployed using an edge-computing configuration with a wired Ethernet connection for cyber security. The JBSA microgrid is shown in Figure 7 and includes the Base Library building, solar arrays, inverters, and the pod housing the battery energy storage system (ESS). The need for accurate on-site meteorological observations necessitated installation of a complete MET Station atop a 10m antenna tower. A Campbell Scientific weatherproof instrument box at the tower base contained a National Instruments MyRio computer, a transformer, backup battery, and an Odroid C2 single board computer (SBC) for calculating the day-ahead load/PV forecasts. Atop the tower sat the SkyImager, a WXT520 Vaisala weather transmitter, and a pyranometer.

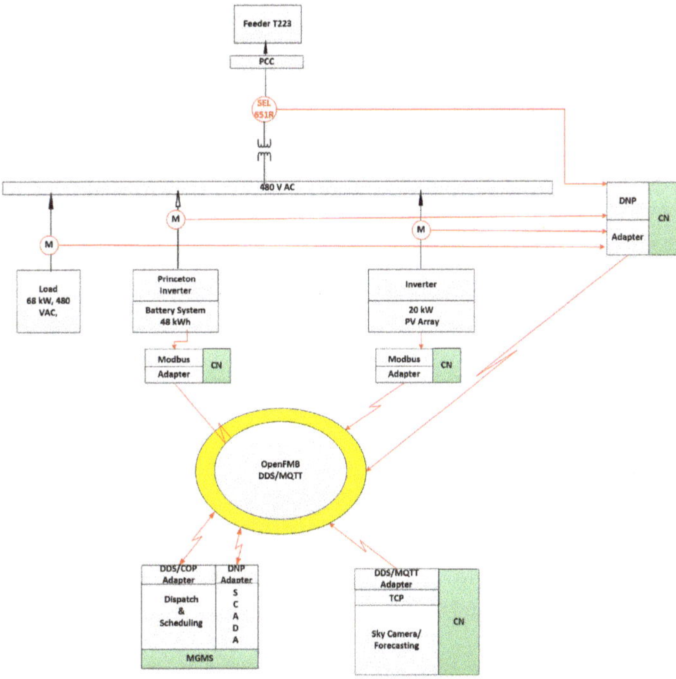

Figure 7. The Microgrid at Joint Base San Antonio, TX, USA.

In July of 2018 another SkyImager was deployed in San Antonio at the location of a university PV generation project. Funded by a DOE-SECO grant [45] in 2014, solar panels were installed on the Engineering Building, HEB University Center III, and Durango buildings at UTSA. In addition, equipment was installed to record measurements from 4 Combiners, 4 Inverters, 2 Kipp & Zonen CMP11 pyranometers, and a WXT520 Vaisala Weather Transmitter at the UCIII. Figure 8a displays the SkyImager and PV panels and Figure 8b shows combiners/inverters atop the Engineering Building. The only ingredient lacking to make this a research microgrid was energy storage.

(a)

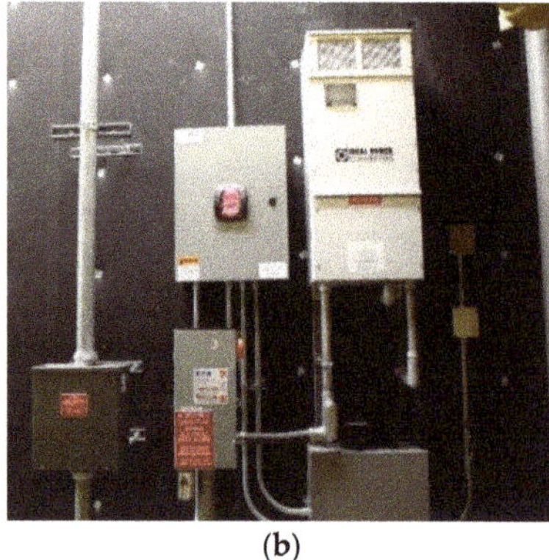

(b)

Figure 8. (**a**) PV panels and SkyImager at UTSA; (**b**) Inverters & combiners, Engineering Bldg.

1.3.3. SkyImager in the Canary Islands

Six insular power grids comprise the electrical network in the Canary Islands. Conventional generation costs more here than PV technologies, and savings can be shared between the PV system owners and the Spanish utility ENDESA. Penetration of renewables varies among these grids, from a high of 60% penetration of wind energy in El Hierro (after Gorona del Viento hydro-wind power plant is operational) to Lanzarote-Fuerteventura which achieves single-digit integration of renewables because of strong environmental regulations, a weak power grid, and an unstable regulatory environment in Spain for renewable energy infrastructure during the period 2011-15. However, new regulatory policies will provide a more attractive framework for investment. The Canary Islands Government plans to avoid ground-based renewable facilities with a large environmental footprint in favor of smaller rooftop plants close to electricity users. As in Hawaii, the existing distribution grid is not prepared for a large penetration of residential PV systems with resulting reverse flows, voltage and frequency instabilities, and drops at the end of long lines. ENDESA has built a testbed smart grid in a village at the north end of the Fuerteventura-Lanzarote insular power system (La Graciosa).

La Graciosa is the smallest island in the Canary Archipelago with a surface area of 29 km^2. It is in a marine nature reserve north of Lanzarote and home to about 700 people in the island capital of Caleta de Sebo. Average global irradiation is 5.157 kWh/ kW·day (1883 kWh/kW·yr) while the average monthly high temperature is 20.8 °C. Located a few kilometers away from the African coast, its proximity to the Sahara Desert gives to La Graciosa particularly stable atmospheric characteristics due to a quasi-permanent subsidence thermal inversion. Constant north trade winds, along with the high content of aerosols and dust in the atmosphere, have a large influence over the cloud dynamics and therefore, the irradiance in the region. As shown in Figure 9, the La Graciosa grid is supplied by three 20/0.4 kV transformers (600, 400, and 400 kVA) and tied by a 20 kV seabed cable to Lanzarote. The island has two PV generation plants (5 kW and 30 kW), but recently La Graciosa PV capacity was increased, enhancing the attractiveness of a smart grid energy management system.

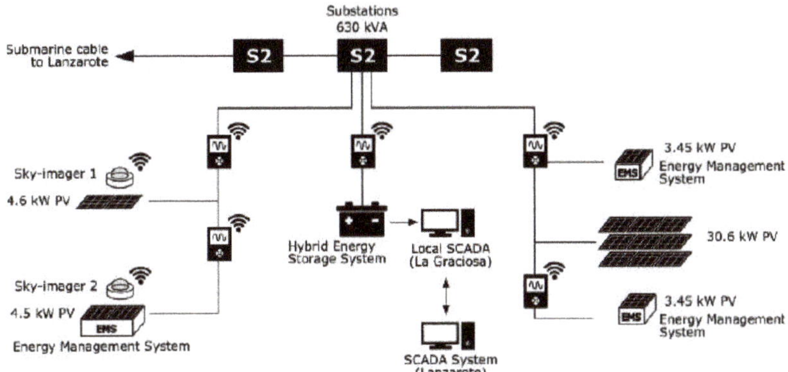

Figure 9. Microgrid in the Canary Islands.

One of the main differences between La Graciosa project and the prior two experiences in the USA, is that in the island, a system composed of two sky-imagers was installed, as can be seen in Figure 9. The reason behind this was to give the forecasting system the ability to estimate cloud base height (CBH) making use of stereoscopic techniques as in [46,47]. This provides the system with added value in terms of functionality and gives extra data to incorporate in the next steps of the image processing and forecasting pipeline. A recent paper comparing the use of different instruments to measure the CBH concluded that using a pair of inexpensive cameras was the most cost-effective alternative in comparison with other methods such as a ceilometer or LIDAR [48]. In fact, cloud base height is quite important to estimate the position of the shadows if a ray tracing approach is taken and it can also be included as a feature if Machine Learning methods are preferred, as it correlates with the position of the clouds in the image and the recorded irradiance or PV production.

In this case, the device falls away from the Internet of Things (IoT) concept, since two cameras are involved in the system, and some computation must be done either in one of the devices or (as it was done in the project) on a dedicated server. Of course, there are some advantages and drawbacks for using either method, but we found particularly easy the connection between the sky-imagers and the server, and we could exploit the higher computational capabilities of the dedicated server. The main requirement to work this way is to have a robust internet access, which fortunately was granted by the owners of the buildings where the sky-imagers were installed. The network speed can also influence the way of operating, as it can act as a bottleneck in the data stream (due to the relatively large size of images compared to other types of files). Also, in the future the use of Machine Learning algorithms could be done on the server, which is expected to perform better than computing directly on the device.

2. Materials and Methods

In the original configuration of the SkyImager, a security camera enclosure housed a Raspberry Pi single board computer with programmable Pi camera. The enclosure contained a small circuit board with heater and fan that runs off a supplied 24V AC power supply, standard with many security cameras. The 12V AC output from this board is input to an AC-DC converter which supplies 12V DC to a TOBSON converter which supplies 5VDC at 3A for the Raspberry Pi.

2.1. SkyImager Hardware

Figure 10a displays the original SkyImager hardware. At NREL it was found necessary to add an extra SBC for increased computational power–the Odroid C1 by Hardkernel. Heat dissipation is an issue with SBC in Texas summers. One C1 was destroyed by heat and as result a cooling fan was added to the design. The new C2 Odroid has a heat sink to eliminate the overheating issue.

Acquiring and fusing the 3-exposure images could be done with just the Raspberry Pi 2. The Pi 3 model is 50% faster than its predecessor; careful optimization of the workflow will allow acquisition, processing, and forecasting with just a Pi 3. This would reduce cost and greatly simplify network connections. Figure 10b shows this configuration with a single Pi 3, plastic case, cooling fan, camera, and WeatherBoard. Images could be pushed to the cloud for processing, however, the necessary bandwidth would be substantial. The sky imagers in La Graciosa are built upon a Raspberri Pi 3 model B with no ancillary boards, and a super wide fish-eye lens (field of view over 180°). An inexpensive mini PV module was added to record irradiance at the camera locations there.

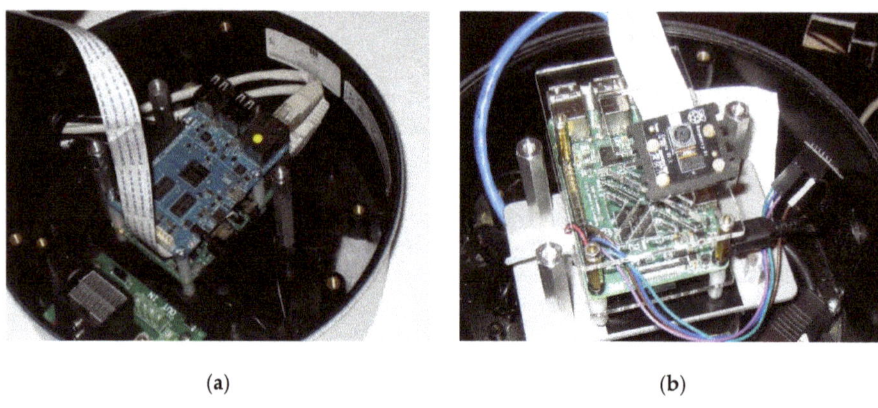

(a)   (b)

Figure 10. (**a**) Original configuration Raspberry Pi 2/Odroid C1, (**b**) Current hardware with a single Pi 3 Model B.

Figure 11 shows the equipment deployed on the MET tower at Ft. Sam: a Kipp&Zonen CMP11 pyranometer, the UTSA SkyImager, and a Vaisala WXT520 weather transmitter. Each of the commercial devices costs several thousand dollars; this cost prompted us to search for low-cost substitutes.

Figure 11. CMP11 pyranometer, UTSA SkyImager, Vaisala Weather Station.

Several inexpensive alternatives to a commercial pyranometer exist [49]. Devices can be added to the GPIO pins on the Raspberry Pi. The Hardkernel Weather-Board 2 shown in Figure 12a can take not only temperature, humidity, and pressure readings (bme280 Application-Specific Integrated Circuit ASIC), but also measures light in the Visible, Infra-Red, and Ultra-Violet bands (si1132 ASIC). There is a Python interface for data retrieval. After calibration and conversion of Lux to W/m^2, this provides irradiance measurements and limited weather data in real time. Another ancillary device that will be useful during initial deployment of the SkyImager is a GPS locator. At $20, it looks like a small mouse for a desktop computer and plugs into a USB port. The Linux programs gpsmon and cgps can be installed on Raspbian and used to take readings of the exact position using the latest GPS satellite data. These are just two of many environmental sensors that can be connected to the Raspberry Pi. Figure 12b shows the new PiNoIR camera, which captures infrared light as well as visible. This would allow for increased contrast between low-level cumulus and high-level cirrus clouds composed of ice crystals. The SkyImager can be used for additional tasks such as air quality monitoring. Figure 12c shows the $25 MQ-131 ozone detection sensor, for example.

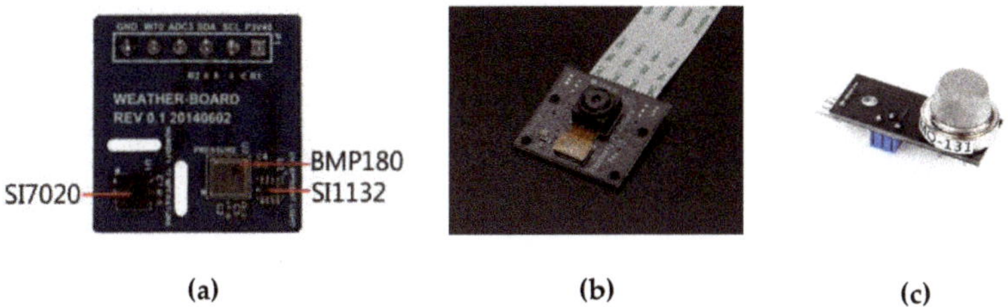

(a)   **(b)**   **(c)**

Figure 12. (**a**) Hardkernel's WeatherBoard, (**b**) PiNoIR camera, and (**c**) MQ 131 Ozone Sensor.

2.2. Image Processing Pipeline

Several additional external inputs were required for our forecasting algorithms: distortion parameters for the fish eye lens, zenith angle, True North, and most importantly the cloud base height (CBH). These inputs are used in the image processing pipeline (Figure 13) to output real-time GHI forecasts for the MGMS. A summary of the pipeline is included here, for details see [8]. (1) Distortion Removal (due to fish eye lens), (2) Cropping and Masking, (3) Calculation of "Red-to-Blue Ratio" (RBR), (4) Apply Median Filter to remove impulsive noise, (5) Thresholding to determine cloud presence, (6) Compute Cloud Cover percentage (clear/moderately-cloudy/overcast), (7) Project Clouds to height of CBH, (8) Use Optical Flow to move clouds forward in time, (9) Ray-Tracing to locate cloud shadows, and finally (10) Calculate GHI using shadow locations. Although physically correct, Step (9) Ray-Tracing is an inverse problem mathematically, hence "ill-posed". Small errors in locating shadows can produce significant errors in the forecast irradiance. To address this issue, we investigated using artificial intelligence and neural works to predict GHI values directly from forecast cloud locations.

At NREL another raw image was acquired every 15-seconds. The pipeline described above must be fine-tuned if the processing SBC is to achieve the necessary throughput. An SBC is much more limited than a desktop server as regards CPU speed and available memory/storage, which is provided by a 32 Gb micro-SD card. The usual tradeoffs between keeping a large array in memory versus writing it to disk, are still present even though there is no disk. Efficient programming constructs are required if the goal of low cost is to be achieved. The EMS may run on a military grade RuggedCom server but the SkyImager software is constrained run on an ARM architecture. Steps in the pipeline that have little effect on the overall forecast accuracy can be eliminated. Profiling/timing runs on the optical flow algorithms will show bottlenecks that can be addressed. This is important whether a ray-tracing approach or a machine learning strategy is employed.

The goal in intra-hour solar forecasting is real time PV power predictions. Those forecasts result from a two-step process: predicting cumulus cloud locations 15-min in the future and using projected cloud locations to forecast irradiance. Each step introduces errors. Work is ongoing for *Step 1 - Optical Flow*: compute error metrics of the 15-min ahead image versus the actual image. *Step 2 - Machine Learning* takes the predicted image and computes GHI. This approach separates optical flow [50] from machine learning (ML) and allows *GHI to be predicted directly from the image itself*. The SkyImager can be used as a pyranometer for measuring/observing irradiance.

The training datasets for the neural networks are region-specific, if not site-specific, and require many all-sky images taken on moderately cloudy days. The weights determined for Golden, CO, will have to be fined tuned for deployment in the San Antonio area for example. Training is computationally expensive, whereas the inference or prediction is very fast and can be handled by the Pi. Research on massive deep learning networks requires Graphical Processing Units (GPU) for training and software such as Theano, Keras, or Tensorflow.

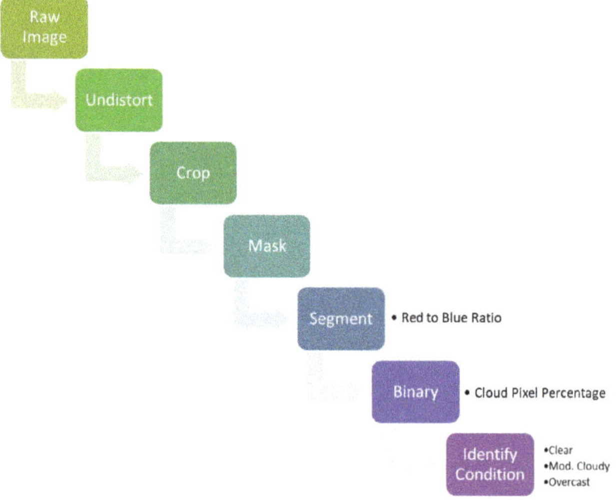

Figure 13. Image Processing Pipeline.

2.3. Machine Learning for Irradiance Forecasting

Machine Learning (ML) is now ubiquitous in all areas of engineering and data science. It has been used in many different ways to help solve the solar power forecasting problem, as described in [51,52]. Another area where ML is widely used is forecasting building load [53,54] which includes methods that are physics-based, statistics-based (Gaussian Process, Linear Regression), and use machine learning (Artificial Neural Network, Support Vector Machine, Deep Learning). Classic references for Deep Learning include [55–57] and for Convolutional Neural Networks, [58].

AI software for data mining has evolved dramatically over the last few years. In data analytics Python is the premier programming language [59] and this fully validated our decision to use it for the SkyImager project. Rapidminer (Version 8.1.001) [60] is a machine learning platform with a point and click interface. As shown in Figure 14, a data flow pipeline is established that permits the user to input a data set, select attributes to analyze, determine target and predictor variable roles, partition the data into training, validation, and sometimes testing subsets, create a logical fork to apply different subprocess models such as Random Forests or Deep Learning, run the model(s), and assess error metrics and overall performance. It is a proprietary package, but a version with somewhat reduced functionality is available for educational use. For some models Rapidminer utilizes the H2O machine learning modules (Version 3.8.2.6) [61,62]. Specifically, our Deep Learning (DL) model is found in H2O as the Python function H2ODeepLearningEstimator(). An open source Python package Scikit-Learn (Version 0.19.0) [63] allows a user to prototype and compare a variety of classification, clustering, and regression models. Neural networks and deep learning has seen the evolution of specialized software such as Keras, Theano, and Google's Tensorflow, which recently became open source. The computational demands of training networks on big data are extreme, and this has resulted in a hardware evolution from central processing units (CPU) to graphical processing units (GPU) to special purpose tensor processing units (TPU).

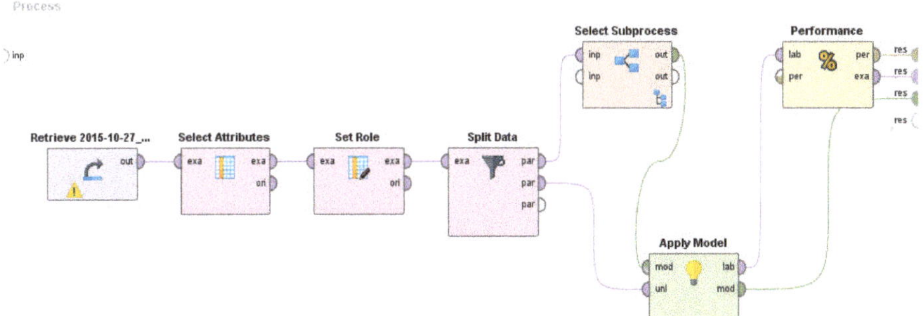

Figure 14. Point and click GUI for Rapidminer allows easy model development.

Before predicting irradiance for cloudy days, consider the much simpler problem of forecasting on a clear day. The Haurwitz analytic model performs well on days with no clouds. Using physics, one can derive a closed-form functional relationship [64]: $GHI_{clr} = 1098\ [cos\theta_z\ exp(-0.057/cos\theta_z)]$, where θ_z is the solar zenith angle. Can neural networks learn this relationship, given a large enough dataset on which to train? The actual forecasting problem on a cloudy day is of course much harder. Information in each all-sky image is used to locate and track low-level cumulus clouds as they move between the sun and PV-arrays. It is the difference between using ML to recognize machine-written (or even hand-written) digits versus recognizing and identifying faces in a crowd of people moving down the street. Work is ongoing to identify the best features to extract from the images, to efficiently solve the intra-hour solar forecasting problem and to predict very short-term ramp events.

A critical component of any machine learning strategy is deciding which features or input variables are most strongly correlated with the labels or output variables. A second aspect involves finding a representation of the data that is compressed or sparse in some basis. This dimensionality reduction [65] can be achieved through principal component analysis (PCA) or by the simple process of discarding unimportant features in the inputs. In our studies, the label or target variables were scalars: GHI values measured atop the ESIF Building at NREL. Other choices are possible such as the value $GHI_{clr} - GHI_{mea}$, the deviation from the clear sky value.

Each input or example is a 3-channel RGB image from the Pi camera. In the current configuration, 3 images taken 5 seconds apart at low, medium, and high exposure times are fused using the Mertens algorithm into one raw image. As mentioned, this approach reduces over-exposure and washout in the circumsolar region. The 1024×768 JPEG forms the basic input measurement for both the optical flow and machine learning algorithms. Low level cumulus clouds between the sun and the PV arrays have the greatest effect on the DNI, hence on GHI. For that reason, and to satisfy the need for dimensionality reduction, the first preprocessing step is to locate the area in the image that surrounds the sun and extract a 128×128 subimage.

Several approaches to locating the sun in the image are possible. Calculating the zenith angle from the SOLPOS program, finding true North, and then mapping physical to pixel coordinates would require extensive calibration. It was decided to use a simple robust image processing approach that finds the maximum intensity in the image. On a clear day, this always locates the sun, but occasionally when the sun is totally obscured by broken clouds, the brightest point in the picture is actually sunlight reflected off of a nearby cloud. This can be observed in a time lapse video clip in which for a few frames the sun is not at the center of the sub-image. While the cause of some transient errors, it never lasts long and does not happen when the sun is totally obscured by a uniform cloud deck without breaks. Figure 15 shows the low exposure image used to locate and center the sun and the resulting raw fused image that will ultimately become the input to the neural networks. Lastly the subimage will be resized to a point (8×8) where the neural networks will train in a reasonable amount of time. In supervised learning, the neural networks require labeled training examples: ordered pairs (x, y)

where x is the input vector, in this case the extracted subimage *img*, and y is the measured irradiance in units of Watts/m^2 at the time the picture was taken. Since new images are fused every 15 seconds, GHI values were treated as constant on a 60 second interval for the purposes of assigning labels for the neural network images.

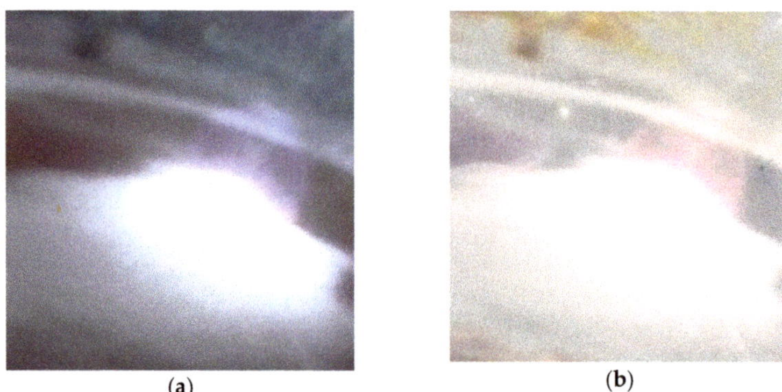

(a) **(b)**

Figure 15. Sub-images of circumsolar region: (**a**) low exposure and (**b**) raw fused.

Many different metrics are used in data analysis and ML for evaluating model performance. Table 1 lists common ones for solar forecasting. A_t is the actual value, F_t the forecast value, and μ the mean; the summation over t can be over all observations in the ML context or over the values of a time series. The metrics provide a posteriori error bounds upon which utilities can make economic decisions. MAPE is relative L_1 error, normalized for number of observations and converted to percent. When the denominator of a fraction is close to zero, μ is used instead of A_t, a common practice when predicting spot electricity prices. Some metrics such as L_1 are more robust–less sensitive to outliers–than the classical RMS error norms.

Table 1. Regression Metrics: A actual, F forecast, μ mean.

Metric	Definition
Mean Squared Error	$\text{MSE} = (1/n) \sum_t (A_t - F_t)^2$
Normalized RMS Error	$nRMSE = \sqrt{MSE}/\rho, \rho = (A_{max} - A_{min})$
Explained Variance	$R^2 = 1 - \sum_t (A_t - F_t)^2 / \sum_t (A_t - \mu)^2$
Mean Absolute Error	$MAE = (1/n) \sum_t |A_t - F_t|$
Mean Absolute Percent Error	$\text{MAPE} = \frac{100\%}{n} \sum_t |(A_t - F_t)/A_t|$

2.4. Stereographic Method for CBH Estimation

Obtaining CBH from paired sky images has been done by several authors in the past. For instance, the method proposed in [46] generated blocks of clouds which are computed to make a 3D reconstruction of the clouds. Then, the authors were able to obtain the height of the clouds by using geometric computation. On the other hand, the method in [66] used the cross correlation of non-projected saturation images to find all possible combinations that yield feasible heights, selecting the most correlated one (or the one with the minimum error).

In the GRACIOSA project, a pure geometric method based on the relative position of the sky-imagers and the clouds was implemented. First, the algorithm looks for the same cloud feature in the images coming from the two sky-imagers. This task can be extremely difficult due to the chaotic nature of clouds and slight changes in image properties such as luminosity. But fortunately, a Scale-Invariant Feature Transform (SIFT) algorithm [67], can handle this situation. This algorithm performs excellently, identifying features in a constantly changing shape (as clouds), since it considers possible changes in scale and orientation. SIFT is applied to pairs of simultaneous images from both cameras (Figure 16). Once the features from both images have been paired up, the best matches are selected to continue the calculations. Valid features are then transposed from the image (pixels) to real space (azimuth and zenith). With real space coordinates defined and the projection matrix of the lenses known, geometric computation is used to obtain the length of the vectors containing each feature and the geographical position of each camera in real space, from which the height of the evaluated feature can be derived.

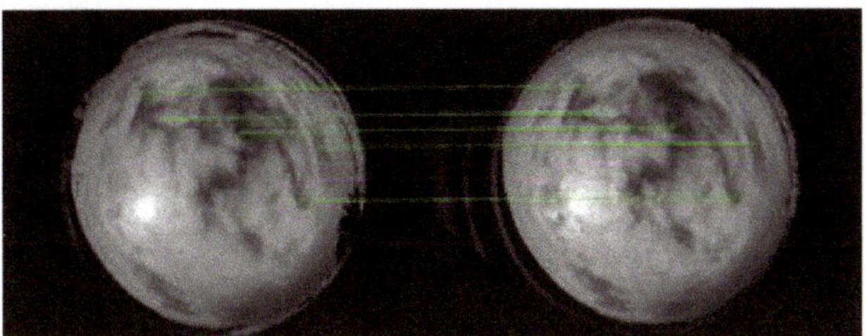

Figure 16. Feature matching by SIFT algorithm in a pair of images from both cameras.

3. Results

Almost a terabyte of image data was collected at NREL from 15 October 2015 through 16 April 2016. The measured DNI values were recorded using NREL's CHP1-L pyranometer with units of Watts/m^2. Days were grouped into three categories: (1) Clear Sky; consisting of predominantly clear days with little or no cloud cover; (2) Overcast; large masses of clouds that obscure the sun for most of the day; and finally (3) Moderately Cloudy; characterized by large variation in irradiance values and multiple ramp events. For Clear Sky and Overcast conditions there is no forecasting to do –persistence can't be beat. Other cloud cover classifications are possible. One could use unsupervised ML clustering algorithms on the raw irradiance data to find other breakdowns. Standard METAR cloud classification separates clouds into low, middle, and high level. All clouds affect measured irradiance [68]. We focus on cumulus clouds because they have the greatest effect on ramp events.

Prediction of intra-hour GHI can be partitioned into several distinct sub-tasks. (1) Acquiring a time series of all-sky images. Every 15 s a new raw image is fused [69] from three different exposure times to allow for High Dynamic Range (HDR) [70]. (2) Using the recent past images and optical flow to extrapolate cloud locations 15 min into the future. It is possible to enhance the algorithm, but it must not hamper production of real time forecasts. (3) Using the predicted image and the weights from training the neural network, a predicted GHI value is output to the microgrid management system.

In the original configuration, our software used optical flow to track movement of cumulus clouds and then ray tracing to predict cloud shadow locations. A better methodology would utilize artificial intelligence (AI) to classify the reduction in GHI that will result when cumulus clouds are predicted in the circumsolar region. It is expensive to train neural networks, but this is done offline for a given location. Once optimal weights are determined, the calculation of a single GHI value is very fast–amounting to an inner product. If the optical flow calculation requires too much time, a second single board computer can be added to the hardware as was employed at NREL.

Details of our research on machine learning to predict solar irradiance are described in the paper [9]. Our intent here is to provide the reader with a concise summary of that research. A critical outcome was verification that the SkyImager with the Pi camera can measure GHI in real time. Lacking the accuracy of an expensive pyranometer, this approach would use the image sequence acquired to solve the forecasting problem, in order to simultaneously estimate GHI. This data could be incorporated with readings from the WeatherBoard to provide additional inputs to the MGMS using MQTT or DDS protocols. Variables that are considered deterministic should be treated as stochastic random variables. Convolutional neural networks [58] which preserve spatial information offer the best performance for image datasets. Expensive offline training of the networks is normally done one time but it is possible to do continuous learning where the networks use feedback in the form of newly acquired data to refine the learned weights.

In the field of machine learning, there are standard ways of visualizing both the input dataset consisting of vectors in a high dimensional space, as well as targets and predicted outputs. The UTSA SkyImager collected all images in this study on the ESIF building rooftop at NREL as part of the INTEGRATE project. During 147 days from October, 2015 until May, 2016 there were 14 days of no data collected (technical reasons) and 27 days of partial data collection. The remaining 106 days of no missing data formed the inputs for training and testing the neural networks. This yielded 156,495 observations (examples or rows) for input to the neural networks. We used the standard split (70% − 30%) of the data into training and testing subsets: 109,547 examples for training and 46,948 for testing.

3.1. Comparing 4 Different ML Models

Each input example is uniquely associated with one of the SkyImager pictures taken every 15 s. The normalized pixel values for the Red-Green-Blue (RGB) channels of an 8×8 resized subimage centered about the sun are flattened into one row vector. Note that other color spaces such as HSV or HSL could also be used. Resizing provided dimensionality reduction and reduced runtimes, but with substantial computer resources the 128×128 sub-images could be used for training. Average values of each channel were included as additional features for a total of $3 \times 64 + 3 = 195$ features. The first entry in each row is the measured GHI in W/m^2. To show how well random variables X and Y are correlated, one uses a scatter diagram. Figure 17 shows scatter diagrams of measured GHI versus predicted GHI for four ML models: Multi-Layer Perceptron (MLP), Random Forest (RF), Deep Learning (DL), and Gradient Boosted Trees (GBT). While all models perform well (tight clustering around $y = x$, perfect correlation), DL and GBT have fewer outliers and visually outperform the other two models.

Note that the MLP and RF models were run using the Scikit-Learn ML software package while DL and GBT were run on the Rapidminer platform. This validated our results on different ML packages—results should depend on the algorithms, not the platform on which they are implemented. Currently, Scikit-Learn does not offer a deep learning model. Using *Rapidminer* is very convenient on a powerful desktop PC, our ultimate goal is to use the trained weights on a Raspberry Pi 3 computer for real time forecasting. Scikit-Learn with its open source Python interface should prove valuable for that task.

MAE, MAPE, nRMSE, and R^2 error metrics are given in Table 2 and run times in minutes. The explained variance (R^2) in the last column is very significant. Both DL and GBT achieve values of 0.87, while MLP and RF are 0.1 less. The other error metrics closely track the R^2 values. In DL the extra accuracy is at the expense of much longer run times, but GBT gets the highest accuracy and is very fast: 10 min faster than RF.

Table 2. Evaluation of four machine learning models.

ML Model	Platform	Runtime (min)	MAE	MAPE	nRMSE	R^2
Multilayer Perceptron	Scikit-Learn	3:05	81.21	33.05%	32%	0.71
Random Forest	Scikit-Learn	14:53	66.86	28%	29%	0.76
Deep Learning	Rapidminer	43:52	50.992	27.13%	21.6%	0.871
Gradient Boosted Trees	Rapidminer	4:50	47.072	22.73%	21.1%	0.875

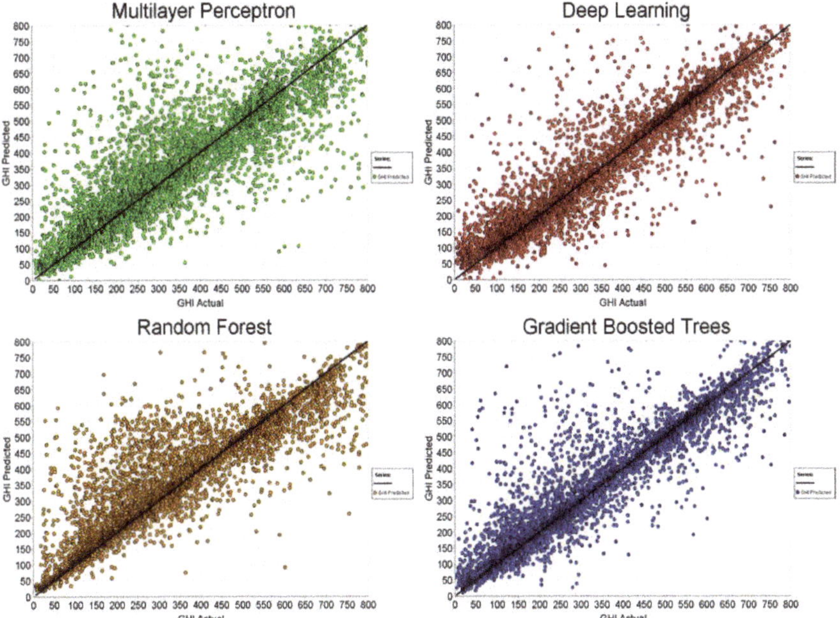

Figure 17. GHI actual vs GHI predicted on 24 October 2015 using different AI models.

A time series is another approach to visualizing the results: GHI values are plotted on the *y*-axis and time on the *x*-axis. Figure 18 compares measured and predicted GHI for one day, 3 October 2015. Although there are differences in the two curves, they track each other well. More illuminating is a time series display for the entire group of testing days shown in Figure 19, where actual GHI is blue and forecast values are red. It is difficult to distinguish the two curves because they track each other so closely. For both figures the deep learning model was used for prediction. Observe in the center of Figure 19 a group of five consecutive clear sky days that are easy to predict, as are completely overcast days.

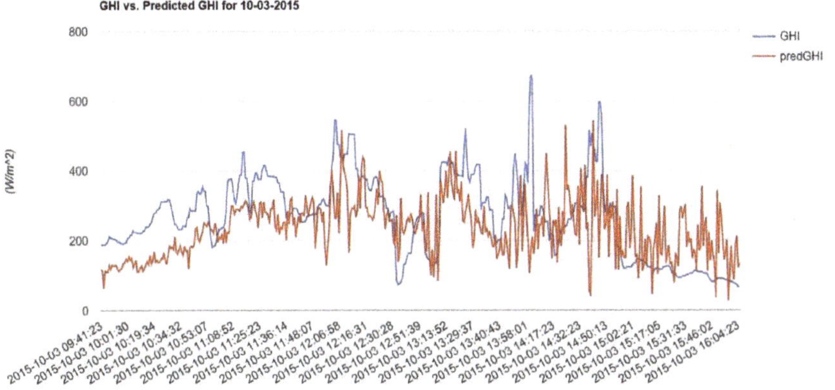

Figure 18. Time series of actual versus predicted GHI for single day of data.

The two curves differ most on moderately cloudy days with air mass cumulus clouds present.

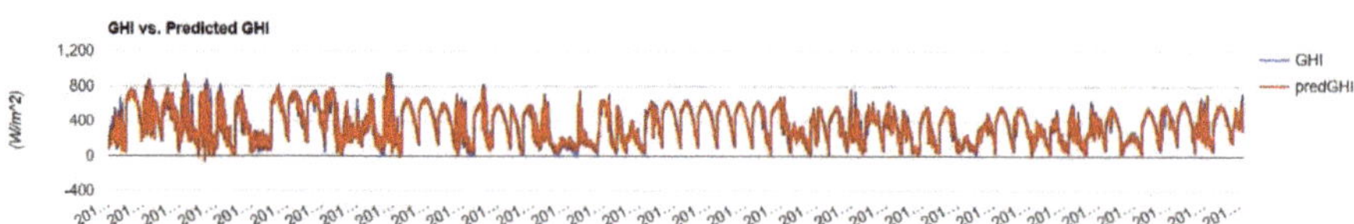

Figure 19. Time series of measured and predicted GHI for all testing days.

3.2. Different Deep Learning Model Results

Machine learning algorithms have many hyper-parameters that can be optimized to improve accuracies and reduce run times. Table 3 shows how changing the number of hidden layers for the DL model, nodes in the layers, and number of epochs (complete passes through the training dataset) affects results. Model 1 has 2 hidden layers each with 50 nodes; it requires ten epochs to train with a run time of ~2 min and R^2 = 0.815. Model 2 also has two hidden layers (195,195) and 10 epochs, but runs 3 times longer and only improves R^2 to 0.824. To achieve R^2 = 0.871 Model 3 (195,195,195) needs 500 epochs and ~45 min. A point of diminishing returns is reached: Model 4 (195,195,97, 195,195) takes more than an hour to run on a desktop PC. Further improvements in accuracy would require larger input images or tuning DL parameters.

Table 3. Comparison of 4 Deep Learning Models.

Hid. Layer	Nodes per H. Layer	# Epochs	Run Time	MAE	MAPE	nRMSE	R^2
2	50,50	10	1:55 min	65.807	32.73%	25.8%	0.815
2	195,195	10	7:10 min	66.872	40.76%	25.6%	0.824
3	195,195,195	500	43:52 min	50.992	27.13%	21.6%	0.871
5	195,195,97,195,195	100	67:42 min	48.519	26.09%	21.6%	0.871

3.3. Cloudy Versus Clear Sky Days

From each 1024 × 768 fused raw image, the algorithm extracts a 128 × 128 pixel subimage centered on the sun. Using the *transform.resize* function from Skimage, this is resized to 8 × 8 pixels to achieve dimensionality reduction. This idea is critical to successful machine learning: each example in the dataset is a vector in a high-dimensional space and there are many examples. Principal Component Analysis and Linear Discriminant Analysis are other techniques for reduction, but our approach is simple and has proved to be effective.

In the following case study ten days of SkyImager data acquired at NREL were used to synthesize two datasets. Five moderately cloudy days comprised the first dataset, October 16, 17, 18, 19, 20 in 2015. Five clear sky days of data from November 10, 11, 12, 13, 14 of 2015 made up the second dataset. The neural networks were fed 32 × 32 pixel resized images and 4 ML models from Scikit-Learn were compared: Generalized Linear Regression Model (GLM), Multi-Layer Perceptron (MLP), Random Forest Regressor (RFR), and Gradient Boosted Trees (GBT). Table 4 and Figure 20 show the results. GLM and GBT have much shorter runtimes in both cases, while MLP and RFR achieve higher R^2 values. Maximum accuracies are achieved with MLP but at a cost of increased runtimes. Extreme accuracy in R^2 values for clear sky days (0.97, 1.0, 1.0, 0.99) indicates the networks are learning the analytic form of the Haurwitz clear-sky GHI model well.

Table 4. Results for two datasets of 32 × 32 resized sub-images.

Model Name	5 Moderately Cloudy Days				5 Clear Sky Days			
	GLM	MLP	RFR	GBT	GLM	MLP	RFR	GBT
MAPE	39.75	13.53	15.79	20.93	12.0	2.44	2.66	3.79
Explained Variance R^2	0.70	0.95	0.93	0.90	0.97	1.00	1.00	0.99
Mean Absolute Error	86.43	30.66	33.30	44.68	20.43	5.30	5.84	8.57
Elapsed Time (Sec)	0.1	513.1	69.9	12.4	0.1	512.8	60.5	12.0

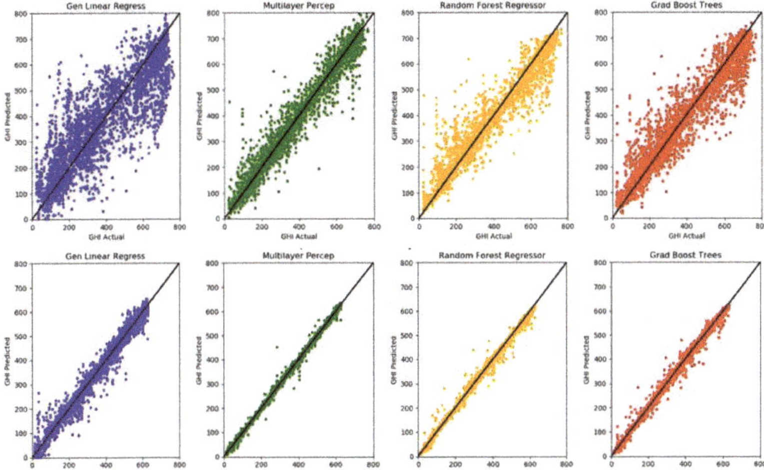

Figure 20. Scatterplots of Actual GHI versus Predicted GHI for 4 ML Models on moderately cloudy days (top row) and clear sky days (bottom row).

Using still finer sampling for the resized sub-images should yield better results at the cost of larger data files and runtimes. At some point however, statistics suggests diminishing returns. In addition to detailed descriptions of ML models and software our article [9] presents another case study. It uses only one moderately cloudy day (17 October 2015) of observations and runs the four ML models with 8 × 8, 32 × 32, and 64 × 64-pixel sub-images. The size of the CSV input data file increases quickly: 3 Megabytes, 111 Mb, and 442 Mb, as do runtimes 139 s, 444 s, and 1309 s. Accuracies improve, but not beyond a certain point.

3.4. JBSA Microgrid Data

The JBSA microgrid was built as a testbed for the CPS Energy Grid Modernization Laboratory. Management and control of a microgrid must address many factors including cybersecurity, data acquisition, data management, real-time computation, storage, bandwidth, interoperability, and usability requirements. In addition to the data acquired by the UTSA equipment–SkyImager, WXT520 Vaisala weather station, and pyranometer; this includes 36-hour ahead hourly weather forecasts scraped from the web, the day-ahead Load/PV forecasts, battery State-Of-Charge (SOC) readings, actual load for the base library building, and control data from the Siemens MGMS. The goal is to use all available data in order to refine site-specific solar irradiance forecasts for improved operation and control of the microgrid. Non-UTSA data from the JBSA microgrid was acquired from Itron MV-90 xi meters. This is a system used for the collection and management of interval data consisting of time stamped readings taken every x minutes where x can be 5, 15, 30, or 60. A large electric utility may acquire a billion interval readings in a single year and use them in a variety of ways including billing (demand response, real time pricing, curtailable rates), open market operations, and load/market research.

Analyses of the JBSA MV90 data, all of which were taken at 15 min intervals, demonstrated that much finer temporal resolution would be required to capture details of ramp events and provide accurate irradiance forecasts to the MGMS. While 1 min resolution was provided by the UTSA equipment such as the pyranometer and Vaisala weather station, the cost of this equipment precluded widespread deployment in a distributed environment. Similar equipment at the UTSA solar testbed provided a wealth of 1 min data, but we envisioned a network of hundreds of low cost SkyImagers spread across the city of San Antonio, and for this scenario low cost was an essential requirement. As previously discussed, there are a plethora of low-cost sensors that can be connected to the GPIO pins on a SBC, including the WeatherBoard2 (WB2), air quality and ozone sensors, and even devices to measure dust. Costing tens of dollars they provide a cost-effective way of environmental sensing that can easily be incorporated with the SkyImager. Figure 21 shows observations of irradiance, temperature, humidity, and pressure taken every minute on 26 November 2018 with the WB2 sensor mounted on a SkyImager.

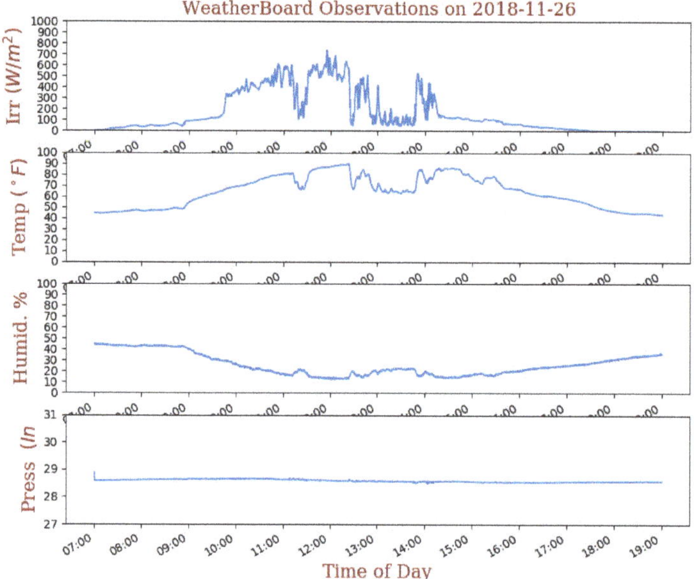

Figure 21. Data from the WeatherBoard sensor on the UTSA SkyImager.

The mini photovoltaic module used at ULL is an off-the-shelf PV module, made of c-Si, with open circuit voltage of 6V and a short circuit current of 200 mA, with a maximum DC output of 1.1 W. The module was connected to a resistive load to dissipate the heat, and the data was registered by a INA219 DC Current Sensor able to measure very small currents, attached to the Raspberry Pi 3 model B.

3.5. One Second Minimodule Data from La Graciosa

The Universidad de La Laguna in the Canary Islands provided 1 sec data from the minimodule at the La Graciosa microgrid operated by ENDESA. How does the spectral content of the voltage signal change when moving from 1 s, 5 s, 15 s, to 60 s sampling? Figure 22 shows the effect of sub-sampling on the time-series. Some of the noise present in the data might be due to voltage fluctuations or seagulls (the location was the Fisherman's Guild building). Certainly the area of the minimodule is very small, so it behaves as a point measurement where small occulding objects can drastically the affect irradiance measurements. Still, the analysis strongly suggests that to resolve the frequency and voltage swings that occur during a sudden ramp event, PMU measurements may be a necessity.

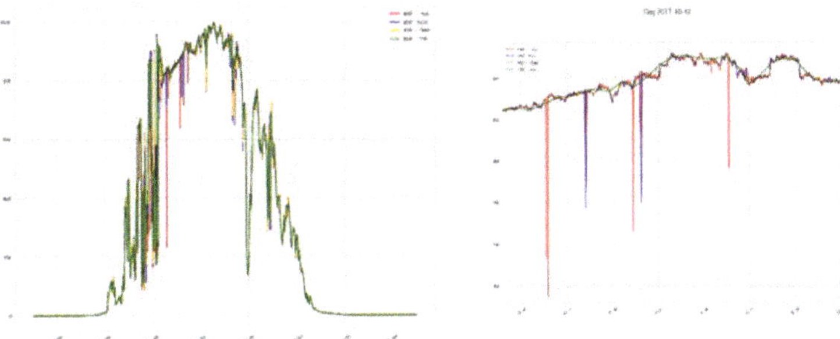

Figure 22. Effect of subsampling on the 1-sec minimodule data from La Graciosa.

3.6. CBH Estimations

The results of the CBH estimations are presented in Figure 23. The left boxplot shows the statistical distribution of the heights obtained with the stereographic method, while the right boxplot shows the statistical distribution of a weather station located in Arrecife, Lanzarote, which belongs to the network of the University of Wyoming.

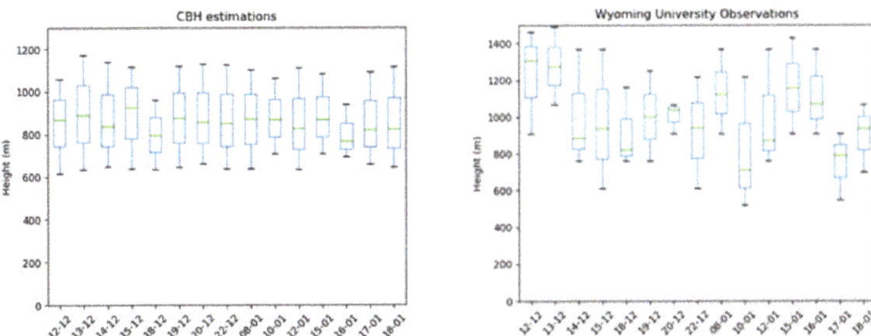

Figure 23. Statistical distribution of the heights obtained by two-camera stereographic method (**left**) and from a weather station located in Arrecife, Lanzarote (**right**).

There are several reasons for the apparent mismatch in the data. First, the weather station is located 30 km south of the position of the cameras, which undoubtedly has a significant effect taking into account how the atmospheric conditions develop in the region (with the thermal inversion steadily rising its level from the Sahara Desert). Second, the data of the weather station is obtained using a punctual measurement such as LIDAR, while the CBH estimations of the cameras cover a larger area of the sky (mainly the central part of the fish-eye image, since the distortion on the borders makes it almost impossible to compute the height). Finally, the temporal resolution of the weather station data is up to 1 hour, while the estimation of the CBH by the stereographic approach is done every minute. Likely the most important conclusion here is that the estimations made by the systems are coherent with the previous knowledge of the atmospheric conditions, with a stable thermal inversion ranging from 600 to 2000 m depending on the season of the year, which prevents clouds to rise over a certain height.

4. Discussion

Our future research efforts will be directed in several areas. Optical flow is a critical area for the success of intra-hour solar forecasting. IoT and cyber-security also form a critical component. Solving the AC Optimal Power Flow equations with GHI forecasts from the SkyImager will be important for solving the energy storage and microgrid control problems. Using the Raspberry Pi additionally as a multiple-sensor platform will be investigated. Environmental studies of the effects of dust and bird feces on the solar panels may well utilize SkyImager technology.

It was on the INTEGRATE project that a synergism developed between researchers and engineers at the national lab, universities, utilities, and private industry that continued after the project ended. Management styles are quite different in academia and private industry with national labs somewhere in between. Software version control was critical, as was careful documentation of all work. Both for the utility where engineers would use the hardware/software and for the university where graduate students and faculty would move on to other projects this was very important. Our decision to use Python was an excellent one. Increasingly, both documentation, tutorials, and example programs are being delivered in the form of IPython notebooks (.IPNB files) as, for example, Google's Tensorflow. Even a package such as Open Computer Vision [71,72] that is written in C++ for efficiency has Python bindings that allow easy access to routines for image fusion and optical flow.

4.1. Lessons Learned at the 3 Deployment Locations

4.1.1. SkyImager at NREL

The NREL microgrid was located at a specialized research facility, but every attempt was made to simulate conditions at a utility. The communications network was well established; there was abundant state-of-the-art ancillary equipment such as pyranometers and an on-site weather station, and the process of deployment went relatively smoothly. Still there were important lessons to be learned. The SkyImager was initially configured with a single Raspberry Pi 2, which proved insufficient for both acquiring images and processing them through the pipeline to produce irradiance forecasts. This problem was solved by adding an Odroid C1, but this made the design more complex and required bridging between the two SBC using a USB-internet connection. The plethora of operating systems for both the SBC and EMS servers provided still another challenge. There are differences in the way open source packages such as OpenCV and Mosquitto install and operate on Raspbian, Ubuntu 14, and Ubuntu 16. In some cases, there are compiled binaries available and in others software must be compiled from source files. From a solar forecasting perspective, NREL was where the best, most complete data was acquired: over six months of daily images and ground truth pyranometer observations. It took researchers over a year to analyze the data and the process is ongoing.

4.1.2. SkyImager at San Antonio, TX, USA

Two critical applications of islanded microgrids are remote installations in developing countries and power systems for the military that must be entirely stand alone. While this made a military base the perfect site for testing a microgrid EMS, it also meant that obtaining base access for UTSA researchers was an issue. For safety purposes, it requires at least two people to lower the 10 m MET tower. Beyond the initial installation, access to the tower is required every month to inspect the instruments and clean the surface of the plastic dome that covers the camera enclosure. The cost of the pyranometer and weather station exceeded that of the SkyImager by a factor of 20 and prompted adding low-cost sensors to the SBC for measuring temperature, humidity, and light.

Occasional loss of power to the SBC as the battery went through its initial testing phase was a minor issue which required tinkering so that the forecasting software was immediately brought back on line during a reboot. Initially the Mosquitto MQTT broker was chosen for the UTSA software, but because of compatibility issues with the MyRio software, HiveMQ proved to be a better choice. Direct internet access to our SBC from outside the corporate network was available only through a WebEx session that required close coordination between the utility and university personnel. Lesson learned: when initially transitioning hardware/software from a research environment to a production one, it is imperative to have physical/cyber access to the equipment and network. Several Odroid's were damaged by high temperatures in the enclosure boxes and had to be replaced. MV90 meter readings every 15 min are clearly insufficient for the intra-hour forecasting problem. We are currently working with a group in Austin to add inexpensive Phasor Measurement Units (PMU) to acquire detailed frequency and phase information in conjunction with weather, irradiance, and sky images. Cloud computing [73] and 5G networks offer unique opportunities to move most of the computations off the Raspberry Pi to the cloud.

4.1.3. SkyImager at La Graciosa, Canary Islands

Some valuable lessons learned from the Graciosa project have to do with the performance and durability of the sky-imagers in a harsh, dusty, and salty environment such as La Graciosa. The closeness of the island to the Sahara Desert makes the dust content in the atmosphere high. Quality of the images is severely affected by the deposition of dust on the enclosure, as seen in Figure 24, showing that scheduled cleanings of the enclosure are necessary to ensure the proper function of the devices. Our first estimation is that cleaning is necessary at least twice a year, but it is highly dependent on the climatologic and atmospheric conditions. Besides the deposition of dust on the enclosure, water infiltrations in the interior of the enclosure have been a frequent problem, even if the equipment was specially selected to have a high degree of protection to water and dust (IP67). Closeness to the sea, as well as the strong rains that occurred in December of 2017, appear to be the main source of this problem.

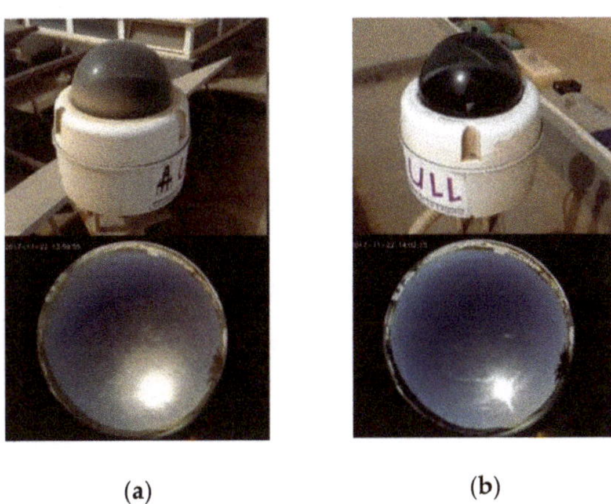

(a)   (b)

Figure 24. (a) SkyImager/image with dust on enclosure, (b) after cleaning it.

5. Conclusions

In March of 2018 the California Energy Commission mandated that beginning in 2020, all new home and apartment construction must include solar generation. When this level of distributed generation becomes part of the electric grid, ISO are faced with new challenges in terms of frequency and voltage control. Indeed, in Hawaii these issues resulted in a temporary hold on rebates for new residential solar installations.

In a macrogrid extending over hundreds of square kilometers, integrating a mix of generation conventional power plants, solar, and wind, and functioning as part of a larger interconnect such as ERCOT, there is an inherent inertia that works on the side of the utility. In microgrids, however, this inertia is lacking, and the control problem becomes much more difficult to solve when in islanded mode [4,7]. There are also different temporal scales involved. Optimal day-ahead scheduling and control of a microgrid is a distinct problem from hour-ahead control to ensure frequency and voltage do not vary outside prescribed limits. Using equipment such as the OP4500 RT-LAB/RCP/HIL real-time power grid digital simulator by Opal RT, we hope to use Hardware-in-the-Loop equipment to analyze the microgrid at JBSA. PMU measurements also need to be incorporated.

All-sky imaging technology will be a critical component in the overall solution strategy to predict solar irradiance 15 min ahead, and to take corrective measures during ramp events. It must, however, be fully integrated with NWP and satellite-based approaches for day-ahead load forecasting and optimal control of a microgrid. Optimal use of this technology will encompass a diverse group of specializations, including IoT and edge-computing, cyber-security, machine learning, and image processing.

For example, the characteristics and statistics of the all-sky imager must be included in the stochastic optimization programming for risk neutral and risk adverse operational control of a microgrid [6]. An holistic R&D approach is required. While a Raspberry Pi is the essence of plug-N-play and it is relatively straightforward to build a SkyImager, integration into the IoT and field deployment will remain an active area of research. Imagers will range the gamut in cost, accuracy, and interoperability. MGMS will integrate forecasts from imagers, NWP, and satellites, as well as hundreds of other meters and devices to solve the microgrid control problem. While physics-based methodology will continue to be important, machine learning and IoT technology will play an increasingly critical role. Development of standards such as OpenFMB for interoperability of thousands of devices will also be a necessary component.

6. Patents

A provisional US patent application "distributed solar energy prediction imaging" has resulted from the work reported in this manuscript.

Author Contributions: Formal analysis, A.M.; Funding acquisition, L.S.; Methodology, R.V.-A.; Writing—original draft, W.R.J. and David Canadillas; Writing—review & editing, R.G.-L. and H.K. All authors contributed to the final version of the manuscript.

Funding: This project and the preparation of this paper were funded in part by monies provided by CPS Energy through an Agreement with The University of Texas at San Antonio. The Universidad de La Laguna acknowledges support from ENDESA.

Acknowledgments: The authors acknowledge support from James Boston, Jorge Deleon, Michael Sparkman, Gaelen McFadden, and Michael Cervantes at CPS Energy, Greg Martin and Mike Simpson at NREL, Jim Waight and Shailendra Grover at Siemens-Omnetric, as well as collaboration with Bing Dong, Zhaoxuan Li, Brian Kelley, Brian Bendele, Jonathan Esquivel, and Marzieh Jafary at UTSA.

Conflicts of Interest: The authors declare no conflict of interest. The funders had no role in the design of the study; in the collection, analyses, or interpretation of data; in the writing of the manuscript, or in the decision to publish the results.

References

1.   Guerrero-Lemus, R.; Shephard, L. *Low Carbon Energy in Africa and Latin America: Renewable Technologies, Natural Gas and Nuclear Energy (Lecture Notes in Energy)*, 1st ed.; Springer: Berlin, Germany, 2017.
2.   Available online: https://uc-ciee.org/downloads/appendixA.pdf (accessed on 16 February 2019).
3.   Janssen, T.; Krishnaswami, H. Voltage and Current Control of a Multi-port NPC Inverter Configuration for a Grid-Connected Photovoltaic System. In Proceedings of the IEEE 17th Workshop on Control and Modeling for Power Electronics, Trondheim, Norway, 27–30 June 2016.
4.   Olivares, D.; Lara, J.; Canizares, C.; Kazerani, M. Stochastic-Predictive Energy Management System for Isolated Microgrids. *IEEE Trans. Smart Grid* **2015**, *6*, 2681–2693. [CrossRef]
5.   Canizares, C.; Palma-Behnke, R.; Olivares, D.; Mehrizi-Sani, A.; Etermadi, A.; Iravani, R.; Kazerani, M.; Hajimiragha, A.; Gomis-Bellmunt, O.; Saeedifard, M.; et al. Trends in Microgrid Control. *IEEE Trans. Smart Grid* **2014**, *5*, 1905–1919.
6.   Farzan, F.; Jafari, M.; Masiello, R.; Lu, Y. Toward Optimal Day-Ahead Scheduling and Operational Control of Microgrids Under Uncertainty. *IEEE Trans. Smart Grid* **2015**, *6*, 499–507. [CrossRef]
7.   Michaelson, D.; Mahmood, H.; Jiang, J. A Predictive Energy Management System Using Pre-Emptive Load Shedding for Islanded Photovoltaic Microgrids. *IEEE Trans. Ind. Electron.* **2017**, *64*, 5440–5448. [CrossRef]

8.   Jr, W.R.; Krishnaswami, H.; Vega, R.; Cervantes, M. A low cost, edge computing, all-sky imager for cloud tracking and intra-hour irradiance forecasting. *Sustainability* **2017**, *12*, 482.

9.   Moncada, A.; Richardson, W., Jr.; Vega-Avila, R. Deep Learning to Forecast Solar Irradiance Using a Six-Month UTSA SkyImager Dataset. *Energies* **2018**, *11*, 1988. [CrossRef]

10.   Nummikoski, J.; Manjili, Y.S.; Vega, R.; Krishnaswami, H. Adaptive Rule Generation for Solar Forecasting: Interfacing with A Knowledge-Base Library. In Proceedings of the IEEE 39th Photovoltaic Specialists Conference (PVSC), Tampa, FL, USA, 16–21 June 2013.

11.   Cervantes, M.; Krishnaswami, H.; Richardson, W.; Vega, R. Utilization of Low Cost, Sky-Imaging Technology for Irradiance Forecasting of Distributed Solar Generation. In Proceedings of the IEEE GreenTech Conference, Kansas City, MO, USA, 6–8 April 2016. [CrossRef]

12.   Cañadillas, D.; Richardson, W., Jr.; González-Díaz, B.; Shephard, L.; Guerrero-Lemus, R. First Results of a Low Cost All-Sky Imager for Cloud Tracking and Intra-Hour Irradiance Forecasting serving a PV-based Smart Grid in La Graciosa Island. In Proceedings of the IEEE PVSC-44, Washington, DC, USA, 25 June 2017.

13.   Richardson, W.; Krishnaswami, H.; Shephard, L.; Vega, R. Machine Learning versus Ray-Tracing to Forecast Irradiance for an Edge-Computing SkyImager. In Proceedings of the 19th International Conference on Intelligent System Application to Power Systems (ISAP), San Antonio, TX, USA, 17–20 September 2017.

14.   Waight, J.; Grover, S.; Laval, S.; Shephard, L.; Boston, J.; Lui, R.; Mathew, J.; Bradley, D.; Lawrence, D.; Sparkman, M.; et al. NREL Integrate: RCS -4-42326: Topic Area 3 OpenFMB Reference Architecture Demonstration Final Report, Minneapolis, 2017.

15.   Mathiesen, P.; Kleissl, J. Evaluation of numerical weather prediction for intra-day solar forecasting in the continental United States. *Solar Energy* **2011**, *85*, 967–977. [CrossRef]

16.   Available online: https://www.ncdc.noaa.gov/data-access/model-data/model-datasets/rapid-refresh-rap (accessed on 20 December 2018).

17.   Xia, S.; Mestas-Nuñez, A.M.; Xie, H.; Vega, R. An Evaluation of Satellite Estimates of Solar Surface Irradiance Using Ground Observations in San Antonio, Texas, USA. *Remote Sens.* **2017**, *9*, 1268. [CrossRef]

18.   Mueller, R.; Trentmann, J.; Träger-Chatterjee, C.; Posselt, R.; Stöckli, R. The Role of the Effective Cloud Albedo for Climate Monitoring and Analysis. *Remote Sens.* **2011**, *3*, 2305–2320. [CrossRef]

19.   Urbich, I.; Bendix, J.; Muller, R. A Novel Approach for the Short-Term Forecast of the Effective Cloud Albedo. *Remote Sens.* **2018**, *10*, 995. [CrossRef]

20.   Perez, R.; Ineichen, P.; Moore, K.; Kmiecik, M.; Chain, C.; George, R.; Vignola, F. A New Operational Model for Satellite-Derived Irradiances: Description and Validation. *Solar Energy* **2002**, *73*, 307–317. [CrossRef]

21.   Law, E.; Prasad, A.; Kay, M.; Taylor, R. Direct normal irradiance forecasting and its application to concentrated solar thermal output forecasting: A review. *Solar Energy* **2014**, *108*, 287–307. [CrossRef]

22.   Raza, M.Q.; Nadarajah, M.; Ekanayake, C. On recent advances in PV output power forecasting. *Sol. Energy* **2016**, *136*, 125–144. [CrossRef]

23.   Barbieri, F.; Rajakaruna, S.; Gosh, A. Very short-term photovoltaic power forecasting with cloud modeling: A review. *Renew. Sustain. Energy* **2017**, *75*, 242–263. [CrossRef]

24.   Antonanzas, J.; Osorio, N.; Escobar, R.; Urraca, R.; Martinez-de_Pison, F.; Antonanzas-Torres, F. Review of photovoltaic power forecasting. *Solar Energy* **2016**, *136*, 78–111. [CrossRef]

25.   Marquez, R.; Coimbra, C. Intra-hour DNI forecasting based on cloud tracking image analysis. *Sol. Energy* **2013**, *91*, 327–336. [CrossRef]

26.   Marquez, R.; Coimbra, C. Proposed Metric for Evaluation of Solar Forecasting Models. *J. Sol. Energy Eng.* **2013**, *135*, 011016. [CrossRef]

27.   Gohari, S.; Urquhart, B.; Yang, H.; Kurtz, B.; Nguyen, D.; Chow, M.; Kleissl, J. Comparison of solar power output forecasting performance of the Total Sky Imager and the University of California, San Diego Sky Imager. *Energy Procedia* **2014**, *49*, 2340–2350. [CrossRef]

28.   Urquhart, B.; Kurtz, B.; Dahlin, E.; Ghonima, M.; Shields, J.; Kleissl, J. Development of a sky imaging system for short-term solar power forecasting. *Atmos. Meas. Tech. Discuss.* **2014**, *7*, 4859–4907. [CrossRef]

29.   Chow, C.W.; Urquhart, B.; Lave, M.; Dominguez, A.; Kleissl, J.; Shields, J.; Washom, B. Intro-hour forecasting with a total sky imager at the UC San Diego solar energy testbed. *Sol. Energy* **2011**, *85*, 2881–2893. [CrossRef]

30.   Chu, Y.; Pedro, H.; Coimbra, C. Hybrid intra-hour DNI forecasts with sky image processing enhanced by stochastic learning. *Solar Energy* **2013**, *98*, 592–603. [CrossRef]

31. Yang, H.; Kurtz, B.; Nguyen, D.; Urquhart, B.; Chow, C.; Ghonima, M.; Kleissl, J. Solar irradiance forecasting using a ground-based sky imager developed at UC San Diego. *Sol. Energy* **2014**, *103*, 502–524. [CrossRef]
32. Bernecker, D.; Riess, C.; Angelopoulou, E.; Hornegger, J. Continuous short-term irradiance forecasts using sky images. *Sol. Energy* **2014**, *110*, 303–315. [CrossRef]
33. West, S.; Rowe, D.; Sayeef, S.; Berry, A. Short-term irradiance forecasting using skycams: Motivation and development. *Sol. Energy* **2014**, *110*, 188–207. [CrossRef]
34. Wood-Bradley, P.; Zapata, J.; Pye, J. Cloud tracking with optical flow for short-term solar forecasting. In Proceedings of the Conference of the Australian Solar Energy Society, Melbourne, Australia, 21 August 2012.
35. Wang, F.; Zhen, Z.; Mi, Z.; Sun, H.; Su, S.; Yang, G. Solar irradiance feature extraction and support vector machines based weather status pattern recognition model for short-term photovoltaic power forecasting. *Energy Build.* **2014**, *86*, 427–438. [CrossRef]
36. Zhu, T.; Wei, H.; Zhang, C.; Zhang, K.; Liu, T. A Local Threshold Algorithm for Cloud Detection on Ground-based Cloud Images. In Proceedings of the 34th Chinese Control Conference, Hangzhou, China, 28–30 July 2015.
37. Peng, Z.; Yoo, S.; Yu, D.; Huang, D. Solar Irradiance Forecast System Based on Geostationary Satellite. In Proceedings of the IEEE International Conference on Smart Grid Communications (SmartGridComm), Vancouver, BC, Canada, 21–24 October 2013.
38. Yang, D.; Kleissl, J.; Gueymard, C.; Pedro, H.; Coimbra, C. History and trends in solar irradiance and PV power forecasting: A preliminary assessment and review using text mining. *Sol. Energy* **2018**, *168*, 60–101. [CrossRef]
39. Uriate, F.; Smith, C.; VanBroekhoven, S.; Hebner, R. Microgrid Ramp Rates and the Inertial Stability Margin. *IEEE Trans. Power Syst.* **2015**, *10*, 3209–3216. [CrossRef]
40. Bullich-Massagué, E.; Aragüés-Peñalba, M.; Sumper, A.; Boix-Aragones, O. Active power control in a hybrid PV-storage power plant for frequency support. *Solar Energy* **2017**, *144*, 49–62. [CrossRef]
41. Pourmousavia, T.K.S.S.A. Evaluation of the battery operation in ramp-rate control mode within a PV plant: A case study. *Sol. Energy* **2018**, *166*, 242–254. [CrossRef]
42. Parra, I.; Marcos, J.; García, M.; Marroyo, L. Dealing with the implementation of ramp-rate control strategies—Challenges and solutions to enable PV plants with energy storage systems to operate correctly. *Solar Energy* **2018**, *169*, 242–248.
43. Available online: https://www.nrel.gov/esif/assets/pdfs/omnetric-industry-day.pdf (accessed on 16 February 2019).
44. Available online: https://apps.dtic.mil/dtic/tr/fulltext/u2/688845.pdf (accessed on 16 February 2019).
45. Bendele, B. UTSA Solar Project Investigation—A Study to Measure the Current State of the SECO Project to Guide Preparation for the Next Stage of Research and Development. Texas Sustainable Energy Research Institute Internal Report. 2016.
46. Peng, Z.; Yu, D.; Huang, D.; Heiser, J.; Yoo, S.; Kalb, P. 3D cloud detection and tracking system for solar forecast using multiple sky imagers. *Solar Energy* **2015**, *118*, 496–519. [CrossRef]
47. Lucas, B.; Kanade, T. An Iterative Image Registration Technique with an Application to Stereo Vision. In Proceedings of the 7th International Joint Conference on Artificial Intelligence (IJCAI), Vancouver, BC, Canada, 24–28 August 1981.
48. Kuhn, P.; Wirtz, M.; Killius, N.; Wilbert, S.; Bosch, J.L.; Hanrieder, N.; Nouri, B.; Kleissl, J.; Ramírez, L.; Schroedter-Homscheidt, M.; et al. Benchmarking three low-cost, low-maintenance cloud height measurement systems and ECMWF cloud heights against a ceilometer. *Solar Energy* **2018**, *168*, 140–152. [CrossRef]
49. King, D.L.; Boyson, W.E.; Hansen, B.R.; Bower, W.I. *Improved Accuracy for Low-cost Irradiance Sensors*; Photovoltaic Systems Department: Albuquerque, NM, USA, 1998.
50. Farnebäck, G. Two-frame motion estimation based on polynomial expansion. *Image Anal.* **2003**, *2749*, 363–370.
51. Wolff, B.; Kühnert, J.; Lorenz, E.; Kramer, O.; Heinemann, D. Comparing support vector regression for PV power forecasting to a physical modeling approach using measurement, numerical weather prediction, and cloud motion data. *Sol. Energy* **2016**, *135*, 197–208. [CrossRef]
52. Voyant, C.; Notton, G.; Kalogirou, S.; Nivet, M.L.; Paoli, C.; Motte, F.; Fouilloy, A. Machine learning methods for solar radiation forecasting: A review. *Renew. Energy* **2017**, *105*, 569–582. [CrossRef]

53. Dong, B.; Li, Z.; Rahman, S.; Vega, R. A hybrid model approach for forecasting future residential electricity consumption. *Energy Build.* **2016**, *117*, 341–351. [CrossRef]
54. Li, Z.; Rahman, S.; Vega, R.; Dong, B. A hierarchical approach using machine learning methods in solar photovoltaic energy production forecasting. *Energies* **2016**, *9*, 55. [CrossRef]
55. Salakhutdinov, R.; Hinton, G. Deep Boltzmann Machines. In Proceedings of the Twelfth International Conference on Artificial Intelligence and Statistics (AISTATS'09), Clearwater Beach, FL, USA, 16 April 2009.
56. Bengio, Y.; LeCun, Y.; Hinton, G. Deep Learning. *Nature* **2015**, *521*, 436–444.
57. Schmidhuber, J. Deep Learning in Neural Networks: An Overview. *Neural Netw.* **2015**, *61*, 85–117. [CrossRef]
58. Mallat, S. Understanding deep convolutional networks. *Phil. Trans. R. Soc. A* **2016**, *374*, 20150203. [CrossRef]
59. Raschka, S. *Python Machine Learning, Birmingham*; PACKT Publishing: London, UK, 2016.
60. Hofmann, M.; Klinkenberg, R. (Eds.) *RapidMiner—Data Mining Use Cases and Business Analytics Applications*; Chapman and Hall/CRC: New York, NY, USA, 2013.
61. Available online: http://www.h2o.ai/wp-content/themes/h2o2016/images/resources/DeepLearningBooklet.pdf (accessed on 16 February 2019).
62. Available online: http://h2o-release.s3.amazonaws.com/h2o/rel-turchin/3/docs-website/h2o-docs/booklets/GBM_Vignette.pdf (accessed on 16 February 2019).
63. Pedregosa, F.; Varoquaux, G.; Gramfort, A.; Michel, V.; Thirion, B.; Grisel, O.; Blondel, M.; Prettenhofer, P.; Weiss, R.; Dubourg, V.; et al. Scikit-learn: Machine learning in Python. *J. Mach.Learn. Res.* **2011**, *12*, 2825–2830.
64. Reno, M.; Hansen, C.; Stein, J. *Global Horizontal Irradiance Clear Sky Models: Implementation and Analysis*; SANDIA: Albuquerque, NM, USA, 2012.
65. Hinton, G.; Salakhutdinov, R. Reducing the dimensionality of data with neural networks. *Science* **2006**, *313*, 504–507. [CrossRef]
66. Nguyen, D.; Kleissl, J. Stereographic methods for cloud base height determination using two sky imagers. *Solar Energy* **2014**, *107*, 495–509. [CrossRef]
67. Lowe, D.G. Distinctive Image Features from Scale-Invariant Keypoints. *Int. J. Comput. Vis.* **2004**, *60*, 91–110. [CrossRef]
68. Li, Q.; Lu, W.; Yang, J.; Wang, J. Thin Cloud Detection of All-Sky Images Using Markov Random Fields. *IEEE Geosci. Remote Sens. Lett.* **2012**, *9*, 1545–1558. [CrossRef]
69. Mertens, T.; Kautz, J.; van Reeth, F. Exposure fusion. In Proceedings of the 15th Pacific Conference on Computer Graphics and Applications (PG'07), Maui, HI, USA, 29 October–2 November 2007; pp. 382–390.
70. Traonmilin, Y.; Aguerrebere, C. Simultaneous High Dynamic Range and Superresolution Imaging without Regularization. *SIAM J. Imaging Sci.* **2014**, *7*, 1624–1644. [CrossRef]
71. Available online: https://opencv.org/ (accessed on 16 February 2019).
72. Available online: https://docs.opencv.org/3.0-beta/opencv2refman.pdf (accessed on 16 February 2019).
73. Nagothu, K.; Kelley, B.; Jamshidi, M.; Rajaee, A. Persistent Net-AMI for Microgrid Infrastructure Using Cognitive Radio on Cloud Data Centers. *IEEE Syst. J.* **2012**, *6*, 4–15. [CrossRef]

 *applied sciences*

Article

Evaluation Model of Demand-Side Energy Resources in Urban Power Grid Based on Geographic Information

Feifan Chen [1,*], Shixiao Guo [1], Yajing Gao [2,*], Wenhai Yang [3], Yongchun Yang [1], Zheng Zhao [4] and Ali Ehsan [5]

[1] State Key Laboratory of Alternate Electrical Power System with Renewable Energy Sources, North China Electric Power University, Baoding 071003, China; 2162213087@ncepu.edu.cn (S.G.); 51351574@ncepu.edu.cn (Y.Y.)
[2] China Electricity Council Electric Power Development Research Institute, Beijing 100031, China
[3] Huaneng Power International, Inc., Beijing 100031, China; hddahai@163.com
[4] Department of automation, North China Electric Power University, Baoding 071003, China; zheng_zhao@ncepu.edu.cn
[5] College of Electrical Engineering, Zhejiang University, Hangzhou 310027, China; aliehsan369@hotmail.com
* Correspondence: xzllcff@163.com (F.C.); 51351706@ncepu.edu.cn (Y.G.)

Received: 23 July 2018; Accepted: 27 August 2018; Published: 29 August 2018

Featured Application: This paper contributes to guiding government planners and DSER investors to compare and select various DSER, furthermore, they can make a relatively reasonable planning scheme or investment scheme. Meanwhile, it provides a strong support for the long-term optimization planning and medium-term optimization aggregation of DSER, which is conducive to the development and utilization of DSER.

Abstract: In the context of current energy shortage and environmental degradation, the penetration rate of demand-side energy resources (DSER) in the power grid is constantly increasing. To alleviate the problems concerning the energy and environment, it is of tremendous urgency to develop and make effective use of them. Therefore, this paper proposes the evaluation model of DSER in urban power grid based on geographic information, and a variety of demand-side energy resources in a region is evaluated. Firstly, as for five kinds of DSER, revolving wind power generation (WG), photovoltaic power generation (PV), electric vehicle (EV), energy storage (ES), and flexible load, the commonality indexes and individuality indexes of all kinds of resources are selected based on geographic information. The commonality indexes are common indexes of all DSER, and the individuality indexes are unique indexes of all DSER. Then the weight of each subindex under the commonality and individuality indexes are determined by analytic hierarchy process (AHP) and entropy weight method, respectively. Finally, weighted overlay are made according to the weights and quantized values of each index, and a comprehensive score is obtained from the commonality indexes and individuality indexes upon various demand-side energy resources in the region. The result depicts that the proposed evaluation model of demand-side energy resources is of well practicability and effectiveness, which is beneficial to the planning of the city and the power grid. Most of all, such model provides a strong support for the long-term optimization planning and the medium-term optimization aggregation of DSER.

Keywords: DSER; evaluation model; geographic information; commonality indexes; individuality indexes

---

1. Introduction

In the context of current energy shortage and environmental degradation, the proportion of energy resources, such as distributed generation, ES and flexible load, is increasing in the power grid; these resources are widespread existent on the demand side, thus are called demand-side energy resources (DSER). In China, the National Development and Reform Commission and other departments have issued a series of policies on distributed photovoltaic (PV) and decentralized wind power generation (WG) since 2007. DSER plays an increasingly important role in the transformation of new energy systems in China. By the end of 2017, the newly installed capacity of PV in China is 53.06 million kilowatts, and the cumulative installed capacity is 130.25 million kilowatts. The cumulative installed capacity of ES is 28.9 million kilowatts, an increase of 19%. In addition, according to the global distributed energy technology report, the installed capacity of global distributed energy in 2017 is about 132.4 million kilowatts. At the same time, to support the development of energy resources such as WG and distributed PV, many countries have formulated relevant policies. Brazil, Canada, Spain, the United States and other countries have formulated a localization rate policy for WG equipment. The European Union (EU) has set a target that renewable energy will account for 27% of the energy demand structure in 2030, promoting PV development. Nowadays, urban power grids are facing severe challenges including increasing load density, increasing peak and valley difference, shortage of power grid construction land and difficulties in raising funds for construction. At the same time, DSER receives attention from all sectors of society and policy support from the government, and its penetration rate in the power grid, especially in the urban power grid, is constantly increasing, providing a huge potential for the implementation of various demand-side projects. At this time, how to fully tap and exploit the potential of DSER, furthermore, ensuring their participation in the planning, construction, and operation of urban power grids, has become a key issue in solving the current difficulties existing in urban power grids [1–5].

It can be seen that the rational development and the effective use of DSER will have a far-reaching impact on China's energy power industry. The evaluation model of demand-side energy resources can be used to explore the status, characteristics, and development prospects of DSER. It is an important tool to develop and use DSER rationally. Therefore, the establishment of an evaluation model of demand-side energy resources is very important for the development and utilization of DSER as well as the energy power industry.

Therefore, many literature works discuss the evaluation of DSER. In literature [6], the meteorological data of Kahnuj in Iran have been measured at an altitude of 10 m, over a four-year period. Also, the monthly, annual and seasonal wind speed variations are investigated, and the economic feasibility is determined of installing wind turbine at the site. Literature [7] proposes a method to simulate a large wind farm and determine its capacity value, taking into account the mechanical failure of the wind turbine and the influence of the transmission system, and evaluate and compare the wind capacity values under different conditions. Literature [8] presents a study of the installation of a hybrid PV-WG system for social interest houses in the city of San Luis Potosi, Mexico. To assess the benefits of the implementation of this type of system, a technological, economic, and environmental evaluation is carried out based on the available renewable energy resources and considering a typical load profile of consumers. Literature [9] obtained a probability model for system power output by analyzing the structural characteristics of the PV power system and by examining its component failure mode and the effect of the component failure mode on the output power. In literature [10], through the analysis of the results of the various operating parameters of large-scale PV, the operation of large-scale grid connected PV grid is mastered. The comprehensive evaluation system is established by using the analytic hierarchy process. In literature [11], the electric vehicle (EV) charging load in a day is calculated by the traffic simulation and evaluated in order to assess the influence of EV charging on the grid system and the applicability of EV in the smart-grid system. Literature [12] proposes a reliability and economic evaluation model of distribution network under large-scale EV's access. The Monte Carlo method was utilized to evaluate the reliability of distribution network. In literature [13], on the basis of analyzing the traditional site selection program evaluation method, fuzzy analytic hierarchy process (AHP) is proposed to evaluate the EV charging station sitting programs, also the evaluation model algorithms are given, wishing to achieve the optimal program. Literature [14] evaluates the benefits of ES systems applied to renewable intermittent sources like wind. Literature [15] presents a real-time evaluation and simulation approach of ES system based on large renewable-based electricity generation, which can be used for grid support. In literature [16], a AHPPROMETHEE-GAIA method based on the analysis of the characteristics of the existing ES ways is presented. Taking the factors, such as technology maturity, cost, life, efficiency, response speed, and environment as assessment criterion, a comprehensive evaluation for the existing ES ways is made.

All of the above literatures have evaluated DSER, but the evaluation model constructed has only evaluated single DSER without taking a variety of DSER into comprehensive consideration. Therefore, this paper evaluates five kinds of DSER, including WG, distributed PV, EV, ES, and flexible loads. There are many factors to be considered when evaluating the DSER. Firstly, the development of various types of DSER will bring more or less benefits, which reflects the significance of these resources. Also, benefits are the direct purpose of digging deeply into DSER. At the same time, due to the difference in geographic location and development conditions, the development of different DSER varies from region to region. Therefore, resource development, which means proportion of resources currently developed in a certain region to the total resources of the region, is an important index for evaluating the development status of DSER. In addition, the future development potential of DSER is the focus of government and investment institutions, which lays fundamentals for the government to formulate relevant policies and incentives for various types of DSER. The above indexes need to be considered when all kinds of DSER are evaluated, which are defined as commonality indexes. In addition, various DSER have different characteristics. To facilitate the evaluation of the characteristics of DSER, DSER is divided into three groups and the key characteristics of evaluation concern are sorted out, defined as individuality indexes.

The advantages of WG and PV lie in the large range of distribution areas, high development value, almost zero pollution and inexhaustible [17,18]. However, the output of WG and PV are affected by the meteorological conditions such as wind speed and light intensity. The output has characteristics of volatility and intermittency, which increase the difficulty in regulating the peak of conventional power supply in power grid. Therefore, it is necessary to evaluate the output characteristics of WG and photoelectric, and the reliability of WG and photoelectric is also an important role in affecting the stability of the power system.

EV and ES can effectively fulfill the demand-side response and behave interactively with the grid. They can be used to cut peaks and valleys, smooth load fluctuations, and promote the use of intermittent energy resources. Therefore, the peak shaving capacities of EV and ES are particularly important, concerning that they will be connected to the grid in large scale, the reliability of EV and ES resources cannot be ignored as well [19,20].

Flexible loads resources refer to flexible loads that have a power demand response. Demand response technology is one of the core technologies in the smart grid. The application of demand response technology can fully exploit the load-side resources and realize the comprehensive optimization of resources configuration [21]. Different users' response willingness and responsive device capacities are different, such as supermarkets, shopping malls, and hotels and other commercial users with less response willingness and larger response capacity, the residents are the opposite, with greater willingness to respond and smaller response capacity. The response capacities of various users are very different. Therefore, as for flexible loads, features related to response are evaluation-focused.

It can be seen that the current evaluation systems focus on the evaluation of single DSER, and the selected indexes are often individuality indexes of DSER. So, a set of general evaluation system should be set up to evaluate the commonality indexes and individuality indexes of all kinds of DSER. At the same time, these DSER are widespread distributed in the urban power grid, their distribution has natural geographical features and is inseparable from the way of urban planning and partition.

According to the above situation, this paper proposes an evaluation model of DSER based on geographic information. Based on the functional partition of the city, the commonality indexes system and the individuality system are established by analyzing the commonality and individuality of various DSER, and the commonality indexes and individuality indexes are selected separately, that is, each index is defined. Then AHP is used to determine the weight of the commonality indexes. The entropy weight method is used to determine the weight of the individuality indexes. Finally, the comprehensive scores of the commonality indexes and individuality indexes are obtained respectively. The specific process is shown in Figure 1. The comprehensive scores of various DSERs in various regions are favorable for urban and power grid planning, guiding DSER investors and government planners, and providing a strong support for the long-term optimization planning and medium-term optimization aggregation of DSER.

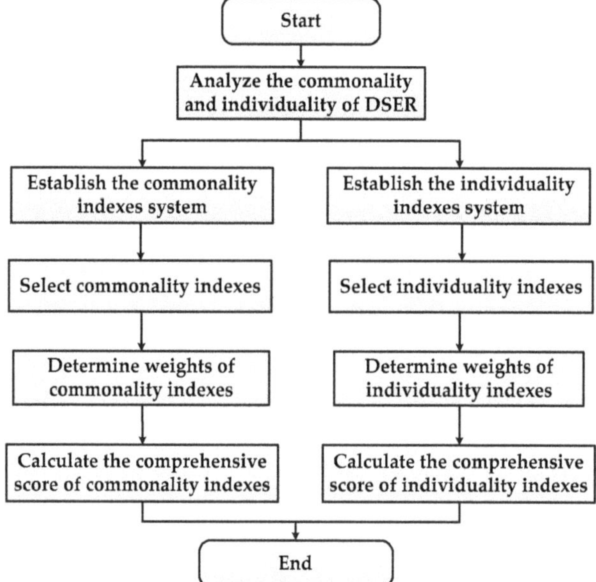

Figure 1. Flow chart for calculating the composite score. The Establishment of Commonality Indexes System for demand-side energy resources (DSER).

1.1. The Commonality Indexes System of Evaluation Model of DSER

According to the above analysis, benefits as well as resource development and development potential are selected as commonality indexes of DSER, which are the indexes common to all types of DSER, and the abovementioned first-level indexes are further refined to obtain a commonality indexes system of evaluation model of DSER, as shown in Figure 2.

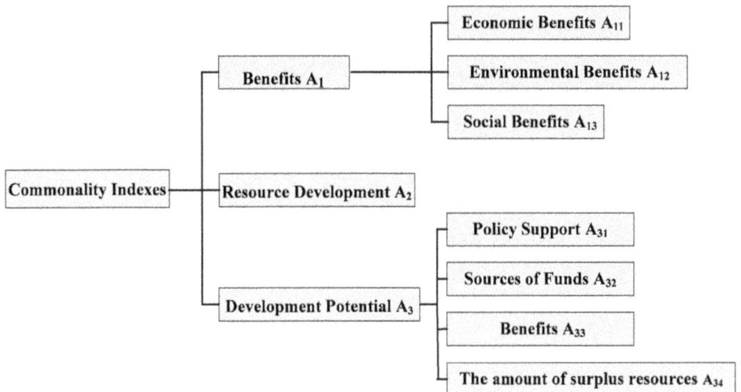

Figure 2. The framework of the commonality indexes.

1.2. Selection of Commonality Indexes

(1) Benefits

Benefits are indexes that measure the effects and profits of a project. The benefits of various DSER are evaluated in terms of three aspects of economic benefits, environmental benefits, and social benefits. Taking WG as an example, the specific definitions are shown in Table 1.

Table 1. Definition of benefits.

Subindex	Definition
Economic Benefits	$A_{11} = G_{wind} / C_{wind}$
Environmental Benefits	$A_{12} = E_{wind} / S_q$
Social Benefits	$A_{13} = M_{wind}$

- G_{wind}—The production value of WG in a certain region
- C_{wind}—The production cost of WG in the region
- E_{wind}—The amount of carbon dioxide emission reduction due to WG in a certain region
- S_q—The area of the region
- M_{wind}—The average satisfaction of residents in a certain area to WG, which is quantified by percentile system.

(2) Resource Development

The development of resources is another important index to evaluate DSER. The development of various DSER is measured by Equation (1).

$$A_2 = \frac{R_{use}}{R_{sum}} \tag{1}$$

A_2—Resource development

R_{use}—The amount of development of a DSER in a region

R_{sum}—The total amount of the DSER in the region

(3) Development Potential

All kinds of DSER must be taken into consideration to settle the long-term development problem, so the development potential is also an essential index for evaluating all kinds of DSER.

The development potential of various DSER is evaluated in terms of four aspects upon policy support, sources of funds, benefits, and the amount of surplus resources. The specific definitions are shown in Table 2.

Table 2. Definition of development potential.

Subindex	Definition
Policy support	According to the relevant documents issued by the government, the support is divided into 1–5 levels from small to large
Sources of funds	The amount of funds provided by country for a certain DSER
Benefits	$w_1 A_{11} + w_2 A_{12} + w_3 A_{13}$
The amount of surplus resources	$1 - A_2$

- A_{11}—The economic benefits of a certain DSER
- A_{12}—The environmental benefits of a certain DSER
- A_{13}—The social benefits of a certain DSER
- A_2—Resource development, which means proportion of resources currently developed in a certain region to the total resources of the region
- w_1—The weight of the economic benefits of the DSER
- w_2—The weight of the environmental benefits of the DSER
- w_3—The weight of the social benefits of the DSER

2. The Establishment of Individuality Indexes System for DSER

2.1. The Individuality Indexes System of Evaluation Model of DSER

Based on the above analysis, indexes that can represent the characteristics of various types of DSER are selected as individuality indexes. Among them, WG and PV resources share a set of individuality indexes, EV and ES resources share a set of individuality indexes, and flexible load resources use a set of individuality indexes. Output characteristics and reliability are selected as first-level indexes of WG and PV, reliability, peak shaving capacity and discharge performance are selected as first-level indexes of EV and ES, response capacity is selected as first-level indexes of flexible loads. The framework is shown in Figure 3.

2.2. Selection of Individuality Indexes

2.2.1. The Individuality Indexes of WG and PV

(1) Output characteristics

The output characteristics of WG and PV resources are of great importance for the consumption of WG and PV. The day with the maximum load is selected as the typical day. The output characteristics of WG and PV resources are quantified and evaluated in terms of three aspects: daily load rate, peak valley difference, and daily load fluctuation rate [22]. The definitions of the indexes are shown in Table 3.

Table 3. Definition of output characteristics.

Subindex	Definition
Daily Load Rate	$B_{11} = P_{av}/P_{max}$
Peak Valley Difference	$B_{12} = (P_{max} - P_{min})/P_{max}$
Daily Load Fluctuation Rate	$B_{13} = S/P_{av}$

- P_{max}—The maximum load of the typical day
- P_{min}—The minimum load of the typical day
- P_{av}—The average load of the typical day
- S—Standard deviation of load.

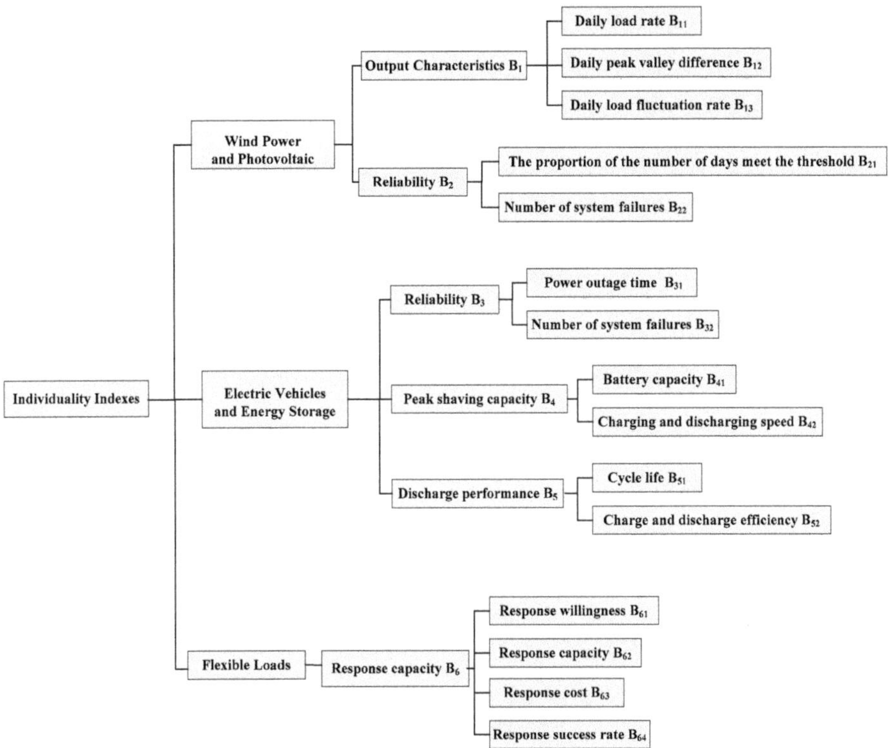

Figure 3. The framework of the individuality indexes.

(2) Reliability

The maximum daily average load of WG or PV is recorded as P, and $0.8P$ is defined as the threshold. The reliabilities of WG and PV in a certain region are quantified and evaluated in terms of two aspects: the proportion of days meet the threshold and number of system failures, correspondingly, the definitions of the indexes are shown in Table 4.

Table 4. Definition of reliability of WG and PV.

Subindex	Definition
The proportion of the days meet the threshold	$B_{21} = D_{th}/365$
Number of system failures	B_{22}

- D_{th}—The number of days in which the average daily load of a year meets the threshold
- B_{22}—The number of system failures in a year

2.2.2. The Individuality Indexes of EV and ES

(1) Reliability

The reliabilities of EV and ES in a certain region are quantified and evaluated in terms of two aspects: power outage time and number of system failures. Such two indexes are represented by the power outage time in one year in a region and the number of failures in the year.

(2) Peak shaving capacity

Realizing peak shaving and valley filling is one of the important goals for developing EV and ES. Therefore, Peak shaving capacity is an important index for evaluating EV and ES, and it is quantified and evaluated in terms of two aspects: battery capacity and charging and discharging speed, correspondingly, the definitions of the indexes are shown in Table 5.

Table 5. Definition of peak shaving capability.

Subindex	Definition
Battery capacity	B_{41}
Charging and discharging speed	B_{42}

- B_{41}—Total battery capacity in a certain region
- B_{42}—The average time required for a battery to be fully charged of battery once

(3) Discharge performance

A cycle means the ES system undergoes a charge–discharge process. Discharge performances of EV and ES in a certain region are quantified and evaluated in terms of two aspects: cycle life and charge and discharge efficiency, correspondingly, the definitions of the indexes are shown in Table 6.

Table 6. Definition of discharge performance.

Subindex	Definition
Cycle life	B_{51}
Charge and discharge efficiency	$B_{52} = E_r / E_i$

- B_{51}—The number of cycles the ES system can withstand before retired
- E_r—Energy released by an ES system after charging
- E_i—Initial storage energy of ES system

2.2.3. The Individuality Indexes of Flexible Loads

The response capacity of flexible loads is quantified and evaluated in terms of four aspects: users' willingness, responsive device capacity, response cost, and response success rate. Taking the users' willingness as an example, users' willingness in a certain region are quantified from low to high as 1–5, according to different types of users (public, commercial, industrial, and residents). Then, the quantized value upon the user's willingness of each type of user is multiplied by the proportion of each type of user in the region, and the comprehensive quantized value upon the user's willingness in the region is obtained. Specific quantization rule is shown in Table 7.

Table 7. The rule of quantifying users' response capacity.

User Type	User Representative	Quantized Value			
		Response Willingness	Response Capacity	Response Success Rate	Response Cost
Public user	office, school, hospital	1–2	1	1–2	1
Commercial user	supermarket, shopping mall, hotel	1–2	3	1–2	3–4
Industrial user	spinning	4	4	3–4	2–3
	cement	3–4	4	3–4	2–3
	steel	1–2	2–3	1–2	5
	mechanical	1–2	3–4	1–2	4–5
Resident user	Residential area	4–5	1	4–5	1

3. Determination of Commonality Indexes Weights and Calculation of Comprehensive Score

3.1. Standardization of Evaluation Indexes

The dimensionlessness in the evaluation index system is the prerequisite for the integration of indexes. If the nondimensionalized value of the index is called the index evaluation value, then the dimensionlessness process is the process of converting the actual value of the index into the evaluation value of the index, and the dimensionlessness method is to eliminate the influence of the primitive variable (index) dimension by the mathematical transformation. When the indexes are nondimensionalized, it is necessary to note that the positive and negative indexes have different effects on the overall target. For example, indexes such as economic benefits, policy support, and response capacity are positive indexes. The higher the index value, the better; the indexes such as power outage time, number of failures, and response cost are negative indexes, the lower the index value, the better.

In summary, the threshold method in the linear nondimensionalization method is used to nondimensionalize the index; threshold method is a dimensionless method to get the index evaluation value through the ratio of the actual value to the threshold of the index. The corresponding formulas are as follows.

Assume that there are m regions, n evaluation indexes, and x_{ij} represents the index value of the ith region under the jth index [23].

Positive indexes

$$x'_{ij} = \frac{x_{ij} - \min\{x_{1j}, \cdots, x_{mj}\}}{\max\{x_{1j}, \cdots, x_{mj}\} - \min\{x_{1j}, \cdots, x_{mj}\}} \tag{2}$$

Negative indexes

$$x'_{ij} = \frac{\max\{x_{1j}, \cdots, x_{mj}\} - x_{ij}}{\max\{x_{1j}, \cdots, x_{mj}\} - \min\{x_{1j}, \cdots, x_{mj}\}} \tag{3}$$

3.2. The Calculation of Weights for Commonality Indexes

Since the commonality indexes such as benefits, resource development, and development potential are indexes shared by the five DSER, and the preference of the decision makers for these indexes is more obvious, the Analytic Hierarchy Process (AHP) is used to evaluate the commonality indexes.

On one hand, the AHP takes the subjective experience judgement of expert scoring into account. On the other hand, the expert judgement is transformed into a mathematical model for quantitative calculation, so that the proportion of each index in the company evaluation index can be calculated. The combination of analysis and calculation is extremely useful for highlighting corporate evaluation in different periods.

The basic idea of the AHP is to build the problem hierarchically based on the decision goal (as is shown in Figure 4). The highest level is the target level, several intermediate levels are the criterion level, and the bottom level is the various options selected for solving the problem, which is called the plan level [24,25].

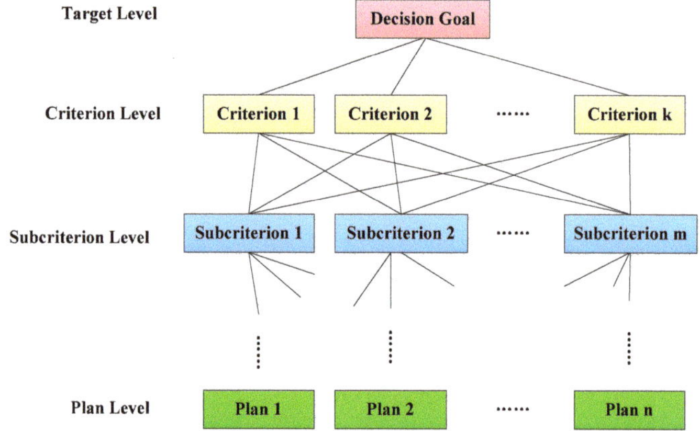

Figure 4. Hierarchical structure.

The weights of indexes are determined by AHP as follows:

(1) Compare pairs of indexes in the same level, and refer to the number 1–9 and its reciprocal as a scale to define the judgment matrix A, as shown in Table 8.

Table 8. The definition of the scale in the matrix.

Scale	Meaning
1	two factors of equal importance
3	two factors, the former is slightly more important than the latter
5	two factors, the former is obviously more important than the latter
7	two factors, the former is mightily more important than the latter
9	two factors, the former is extremely more important than the latter
2, 4, 6, 8	the middle value of the above adjacent judgment
reciprocal	If the ratio of the importance of factor i to factor j is a_{ij}, the ratio of the importance of factor j to the factor i is $a_{ji} = 1/a_{ij}$

(2) Calculate consistency ratios and test consistency

$$CR = \frac{CI}{RI} \tag{4}$$

$$CI = \frac{\lambda_{\max} - n}{n - 1} \tag{5}$$

where *CI*—Consistency indicator; *RI*—Random consistency indicator; *CR*—Test coefficient; $\lambda_{\max}$—The maximum eigenvalue of the judgment matrix A; n—The order of the judgment matrix A.

The value of *RI* is shown in Table 9

Table 9. Average random consistency.

n	1	2	3	4	5	6	7	8	9
RI	0	0	0.58	0.9	1.12	1.24	1.32	1.41	1.45

When CR < 0.10, the consistency of the judgment matrix is considered acceptable, otherwise the judgment matrix should be properly modified.

(3) The calculation of weight vector W

The weight vector W in AHP is calculated by the eigenvector method, and the weight vector W is multiplied by the judgement matrix A

$$AW = \lambda_{\max}W \tag{6}$$

3.3. The Calculation of the Comprehensive Score upon Commonality Indexes

The comprehensive score upon the commonality indexes of all kinds of DSER can be obtained by weighted overlaying the quantized values of each index and their corresponding weights in a certain region, as shown in formula (7). Suppose there are n commonality indexes,

$$f_A = \sum W_{Aj} \cdot x_{Aj} \tag{7}$$

Taking WG as an example, W_{Aj} is the weight of the jth index in the general index of WG, x_{Aj} is the quantified value of WG in jth index, and f_A is the comprehensive score of the commonality index of WG.

4. Determination of Individuality Indexes Weights and Calculation of Comprehensive Score

4.1. Standardization of Evaluation Indexes

The evaluation indexes are nondimensionalized, same as Section 3.1.

4.2. The Calculation of Weights for Individuality Indexes

Since the individuality indexes upon various DSER are different, and the relative importance of each index is difficult to divide artificially, therefore, the entropy weight method is used to evaluate the individuality index.

Entropy, one of the parameters that characterize the state of matter in thermodynamics, is a measure of the degree of chaos in the system. Entropy method is to use the degree of variation of information entropy to calculate the weight of each index, to evaluate the importance of each index [26].

Assuming that there are m regions, n evaluation indexes, the weights of indexes are determined as follows:

(1) Calculate P_{ij}, the weight of the index values of each region under various indexes

$$P_{ij} = \frac{x_{ij}}{\sum_{i=1}^{m} x_{ij}} \tag{8}$$

where i represents the ith region, j represents the jth index, and x_{ij} represents the index value of the ith region under the jth index.

(2) Calculate e_j, the entropy of each index

$$e_j = -\frac{1}{\ln(m)} \sum_{i=1}^{m} p_{ij} \ln p_{ij} \tag{9}$$

(3) Calculate g_j, the difference coefficient of each index

$$g_j = 1 - e_j \tag{10}$$

(4) Calculate W_j, the weight of each index

$$W_{Bj} = \frac{g_j}{\sum_{j=1}^{n} g_j} \tag{11}$$

4.3. The Calculation of the Comprehensive Score upon Individuality Indexes

The comprehensive score upon the individuality indexes of all kinds of DSER can be obtained by weighted overlaying the quantized values of each index as well as their corresponding weights in a certain region, as shown in formula (12). Suppose there are n individuality indexes,

$$f_B = \sum W_{Bj} \cdot x_{Bj} \tag{12}$$

Taking WG as an example, W_{Bj} is the weight of the *j*th index in the general index of WG, x_{Bj} is the quantified value of WG in this index, and f_B is the comprehensive score upon the individuality index of WG.

5. Case Analysis

Twenty regions are selected as evaluation objects for example analysis. The schematic diagram is shown in Figure 5.

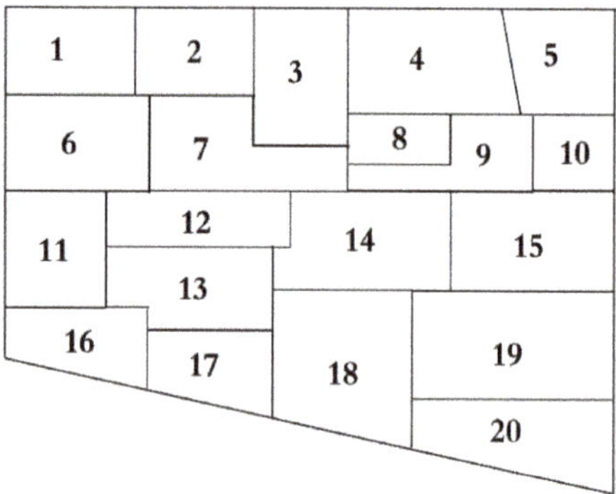

Figure 5. The schematic diagram of 20 regions.

5.1. Commonality Indexes

5.1.1. Standardization of Indexes

According to Equations (2) and (3), each index value can be processed to obtain the relative value of each index. Taking the commonality indexes of WG as an example, the processed index values are shown in Figure 6.

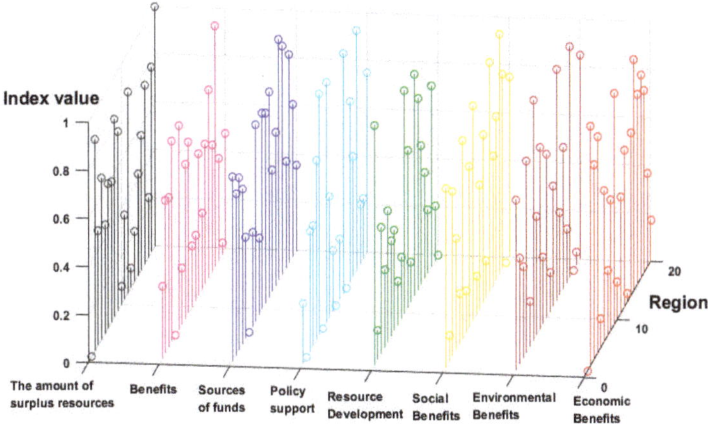

Figure 6. Value of commonality indexes of wind power generation (WG) in 20 regions.

As can be seen from Figure 6, among the 20 regions, WG in region 2 has the highest economic benefits, however, whose social benefits, resource development and policy support are almost behind those in other regions. The social benefits and funding sources of WG in region 18 are in a leading position, however, the environmental benefits are the lowest, moreover, policy support is not satisfactory.

5.1.2. The Calculation of Weights of Commonality Indexes

According to Equation (6), AHP is used to calculate the weights of benefits, resource development and development potential of the commonality indexes, so are the weights of their respective subindexes. The calculation results are shown in Table 10.

Table 10. Weight of commonality indexes.

Item	First Level Index	Weight	Subindex	Weight
	Benefit	0.5584	Economic Benefits	0.4567
			Environmental Benefits	0.3745
			Social Benefits	0.1688
Commonality indexes	Resource development	0.1220		
	Development potential	0.3196	Policy support	0.2956
			Sources of funds	0.3554
			Benefits	0.1026
			The amount of surplus resources	0.2464

As can be seen from Table 10, the weight of benefits in the commonality indexes is the largest, followed by the weight of development potential, and the weight of resource development is the least. For the subindexes, under the benefit index, the economic benefit accounts for the largest proportion, and under the development potential index, the sources of funds account for the largest proportion.

5.1.3. The Calculation of Comprehensive Score

After calculating the weights of the subindexes under the commonality indexes, the comprehensive scores of the commonality indexes of various DSER in various regions are calculated according to Equation (7). Analyzed results are as follows:

As shown in Figure 7, the scores of commonality indexes of the PV and EV in region 1 are relatively high, but that of WG, flexible load, and ES are in the middle level; the score of flexible loads in region 2 is relatively high, and the score of PV is relatively low. As is shown in Figure 8, the scores of commonality indexes of WG in regions 15 and 17 are relatively high, the scores of WG in regions 5, 7, and 19 are relatively low; the scores of commonality indexes of EV in regions 1, 3, 8, and 15 are relatively high, the scores of EV in regions 6, 12, 16, and 18 are relatively low.

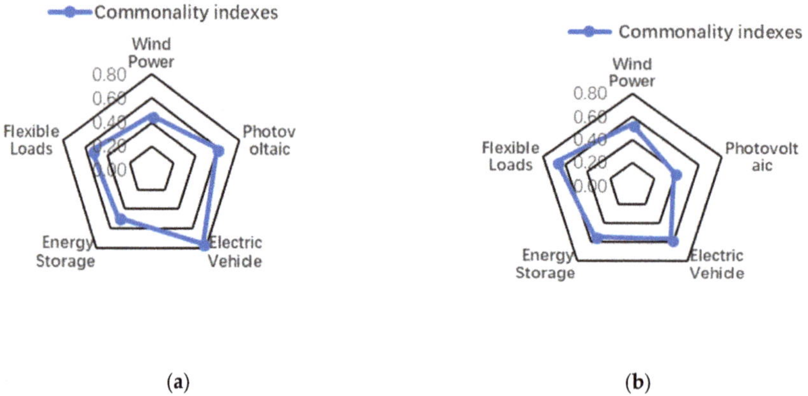

(a)

(b)

Figure 7. The comprehensive scores of commonality indexes of various DSER in a certain region. (a) The score of various DSER in region 1; (b) The scores of various DSER in region 2.

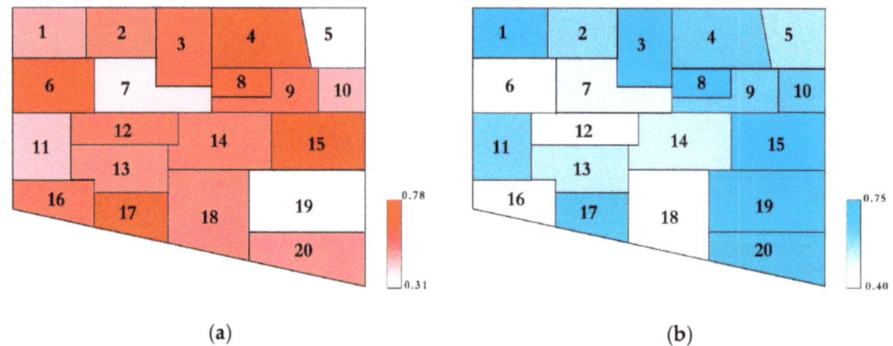

(a)

(b)

Figure 8. The comprehensive scores of commonality indexes in various regions. (a) The scores of WG in various regions; (b) The scores of electric vehicle (EV) in various regions.

5.2. Individuality Indexes

5.2.1. Standardization of Indexes

According to Equations (2) and (3), each index value can be processed to obtain the relative value of each index. Taking the individuality indexes of WG as an example, the processed index values are shown in Figure 9.

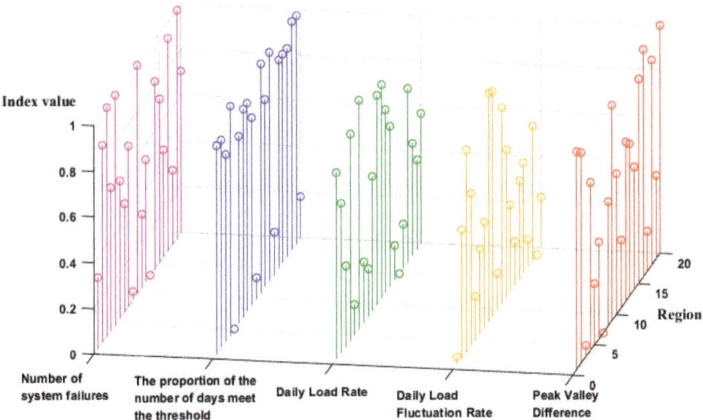

Figure 9. Value of individuality indexes of WG in 20 regions.

It can be seen from Figure 9 that among the 20 regions, the number of system failures of WG in region 5 is the lowest, but the remaining indexes are in a backward position in the ranking. The daily peak rate and daily load fluctuation rate of WG in region 9 are relatively good, but the system has more failures.

5.2.2. The Calculation of Weights of Individuality Indexes

The entropy weight method is used to calculate the weights of the individuality indexes upon various DSER according to Equations (8) to (11), as shown in Figure 10.

As can be seen in Figure 10, among the individual indexes of WG and PV, the weight of the daily load fluctuation rate is the largest, and the proportion of the number of days meet the threshold accounts for the smallest part. Among the individual indexes of EV and ES, the weight of the power outage time, number of system failures and the battery capacity index is relatively large, and the weight of charge and discharge speed is relatively small. Among the individual indexes of flexible loads, the weight of the response success rate index is the largest, and the response cost index has the smallest weight value.

5.2.3. The Calculation of Comprehensive Score

After calculating the weights of the subindexes under the individuality indexes, the comprehensive scores of the individuality indexes upon various DSER in various regions are calculated according to Equation (12). Analyzed results are as follows:

As shown in Figure 11, the score of individuality indexes of the PV in region 1 is relatively low, whereas that of the other DSER are relatively high. The scores of WG, PV, ES, and flexible loads in region 2 are higher than the score of EV. As is shown in Figure 12, the scores of individuality indexes of WG in regions 2 and 4 are relatively high, the scores of WG in regions 5 and 14 are relatively low; the scores of individuality indexes of EV in regions 4, 6, and 13 are relatively high, the scores of EV in regions 16 and 17 are relatively low.

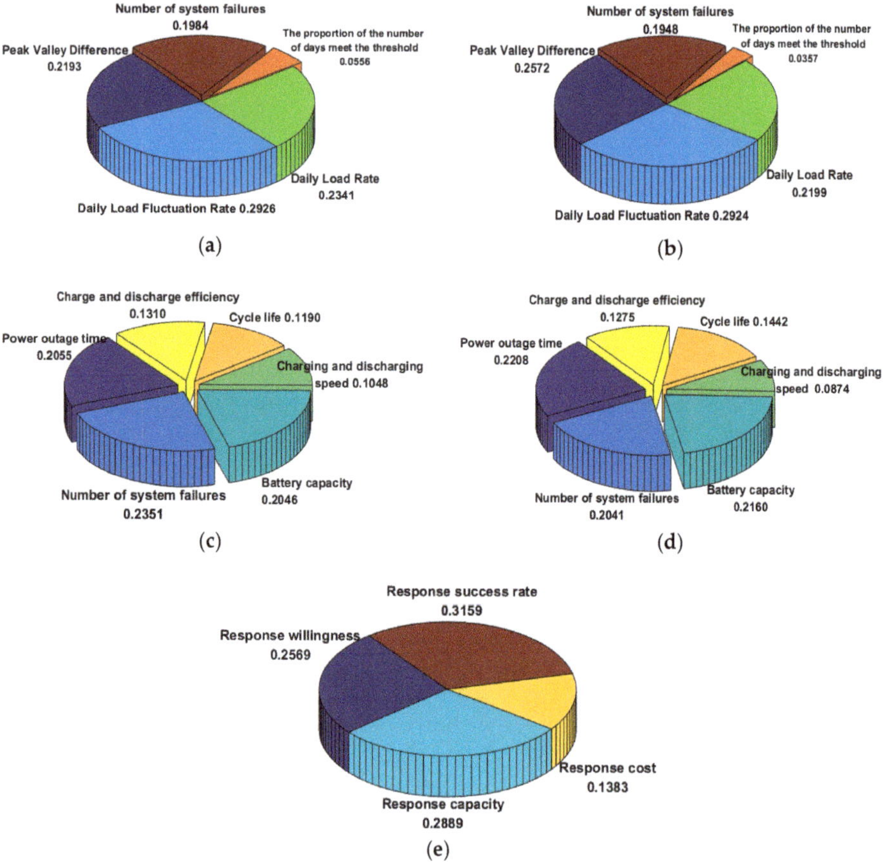

Figure 10. The weight of individuality indexes of various DSER. (**a**) The weight of indexes of WG; (**b**) The weight of indexes of photovoltaic (PV); (**c**) The weight of indexes of EV; (**d**) The weight of indexes of energy storage (ES); (**e**) The weight of indexes of flexible loads.

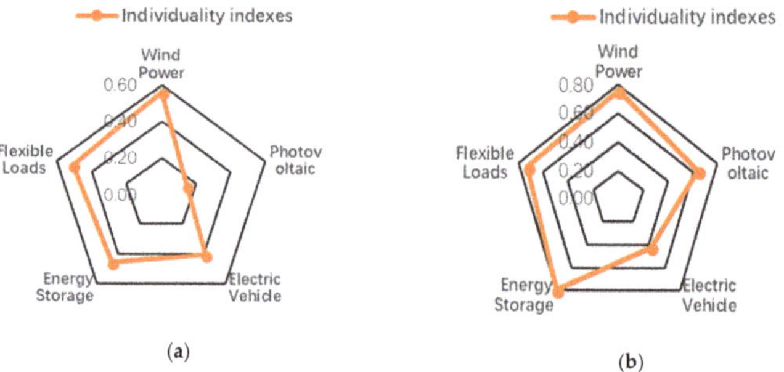

Figure 11. The comprehensive scores of individuality indexes of various DSER in a certain region. (**a**) The score of various DSER in region 1; (**b**) The scores of various DSER in region 2.

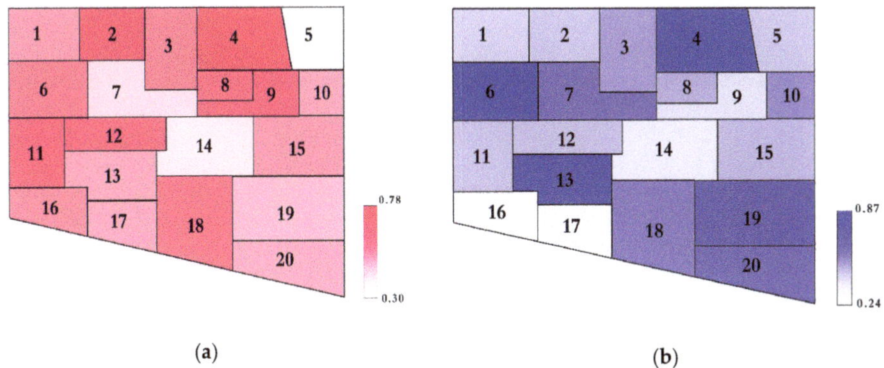

(a)   (b)

Figure 12. The comprehensive scores of individuality indexes in various regions. (**a**) The scores of WG in various regions; (**b**) The scores of EV in various regions.

6. Conclusions

To facilitate the decision-making of optimal planning and optimally aggregate utilization of DSER, this paper proposes an evaluation model of DSER based on geographic information, that is, various DSER in a region are evaluated based on geographic information. Firstly, for five kinds of demand-side energy resources of WG, PV, EV, ES, and flexible load, select the evaluation indexes of all kinds of resources and divide all indexes into commonality indexes and individuality indexes. Then, AHP is used to determine the weight of each subindex under the commonality indexes, and the entropy weight method is used to determine the weight of each subindex under the individuality indexes. Finally, weighted overlay is acquired according to the weights and quantized values of each index, and a comprehensive score is obtained for the commonality indexes and individuality indexes of various DSER in a region. The following conclusions are obtained through the cases analysis.

1. The evaluation results of commonality indexes are helpful to understand the development status and prospects of DSER, which is conducive to the future development and long-term optimization planning of DSER.
2. The evaluation results of individual indexes are helpful to understand the respective characteristics of DSER, which is conducive to achieving a good interaction between the DSER and the power grid, providing support for the medium-term optimization aggregation of the DSER.
3. The evaluation model of DSER based on geographic information proposed in this paper has good practicability and effectiveness, which matches the development of DSER with urban planning as well as power grid planning, providing a theoretical basis for the subsequent development of DSER.

Author Contributions: The author F.C. carried out the main research tasks and wrote the full manuscript, S.G. participated in the revision of the manuscript, and Y.G. proposed the original idea, analysed and double-checked the results and the whole manuscript. W.Y. contributed to data processing and to writing and summarizing the proposed ideas, Y.Y. and Z.Z. provided technical and financial support throughout, while A.E. modified the English grammar.

Funding: The Nature Science Foundation of China (51607068); The Fundamental Research Funds for the Central Universities (2018MS082); The Fundamental Research Funds for the Central Universities (2017MS090).

Acknowledgments: This work is supported in part by the Nature Science Foundation of China (51607068), the Fundamental Research Funds for the Central Universities (2018MS082) and the Fundamental Research Funds for the Central Universities (2017MS090).

Conflicts of Interest: The authors declare no conflict of interest.

References

1. Paliwal, N.K.; Mohanani, R.; Singh, N.K.; Singh, A.K. Demand side energy management in hybrid microgrid system using heuristic techniques. In Proceedings of the 2016 IEEE International Conference on Industrial Technology, Taipei, Taiwan, 14–17 March 2016; pp. 1910–1915.
2. Ehsan, A.; Yang, Q. Optimal integration and planning of renewable distributed generation in power distribution networks: A review of analytical techniques. *Appl. Energy* **2017**, *210*, 44–59. [CrossRef]
3. Gohla-Neudecker, B.; Wagner, U.; Hamacher, T. Analysis of renewable energy grid integration by range extension technologies of BEVs. In Proceedings of the 2013 International Conference on Clean Electrical Power (ICCEP), Alghero, Italy, 11–13 June 2013; pp. 155–162.
4. Amamra, S.; Meghriche, K.; Cherifi, A.; Francois, B. Multilevel Inverter Topology for Renewable Energy Grid Integration. *IEEE Trans. Ind. Electron.* **2017**, *64*, 8855–8866. [CrossRef]
5. Fang, X.; Yang, Q.; Wang, J.; Yan, W. Coordinated dispatch in multiple cooperative autonomous islanded microgrids. *Appl. Energy* **2017**, *162*, 40–48. [CrossRef]
6. Soltani, N.; Fazelpour, F. Evaluation of wind energy potential and economics for the city of Kahnuj in Kerman Province, Iran. In Proceedings of the 2016 IEEE 16th International Conference on Environment and Electrical Engineering (EEEIC), Florence, Italy, 6–8 June 2016; pp. 1–6.
7. Sulaeman, S.; Benidris, M.; Mitra, J. Evaluation of wind power capacity value including effects of transmission system. In Proceedings of the 2015 North American Power Symposium (NAPS), Charlotte, NC, USA, 4–6 October 2015; pp. 1–6.
8. Ospino-Castro, A.; Peña-Gallardo, R.; Rodríguez, A.H.; Segundo-Ramírez, J.; Muñoz-Maldonado, Y.A. Techno-economic evaluation of a grid-connected hybrid PV-WG system in San Luis Potosi, Mexico. In Proceedings of the 2017 IEEE International Autumn Meeting on Power, Electronics and Computing (ROPEC), Ixtapa, Mexico, 8–10 November 2017; pp. 1–6.
9. Li, H.; Ding, J.; Huang, J.; Dong, Y.; Li, X. Reliability evaluation of PV power systems with consideration of time-varying factors. *J. Eng.* **2017**, *13*, 1783–1787. [CrossRef]
10. Hong, F.; Si, C. A study of evaluation system based on large scale photovoltaic power generation. In Proceedings of the 2016 IEEE 11th Conference on Industrial Electronics and Applications (ICIEA), Hefei, China, 5–7 June 2016; pp. 2507–2510.
11. Ito, T.; Iwafune, Y.; Hiwatari, R. Evaluation of EV charging load contributing to stabilize PV output with consideration for rapid charge. In Proceedings of the 2012 IEEE PES Innovative Smart Grid Technologies (ISGT), Washington, DC, USA, 16–20 January 2012; pp. 1–6.
12. Chen, L.; Zhang, H.; Lin, X.; Yu, M.; Zhang, Z. Reliability and economic evaluation model of power distribution network under large scale EV's access. In Proceedings of the 2017 IEEE 3rd International Future Energy Electronics Conference and ECCE Asia (IFEEC 2017-ECCE Asia), Kaohsiung, Taiwan, 3–7 June 2017; pp. 2042–2047.
13. Meng, W.; Kai, L.; Songhui, Z. Evaluation of Electric Vehicle Charging Station Sitting Based on Fuzzy Analytic Hierarchy Process. In Proceedings of the 2013 Fourth International Conference on Digital Manufacturing & Automation, Qingdao, China, 29–30 June 2013; pp. 568–571.
14. Faias, S.; Santos, P.; Matos, F.; Sousa, J.; Castro, R. Evaluation of energy storage devices for renewable energies integration: Application to a Portuguese wind farm. In Proceedings of the 2008 5th International Conference on the European Electricity Market, Lisboa, Portugal, 28–30 May 2008; pp. 1–7.
15. Cha, S.T.; Zhao, H.; Wu, Q.; Østergaard, J.; Nielsen, T.S.; Madsen, H. Evaluation of energy storage system to support Danish island of Bornholm power grid. In Proceedings of the 2012 10th International Power & Energy Conference (IPEC), Ho Chi Minh City, Vietnam, 12–14 December 2012; pp. 242–247.
16. Wei, L.; Hou, J.; Qin, T.; Yuan, Z.; Yan, Y. Evaluation of grid energy storage system based on AHP-PROMETHEE-GAIA. In Proceedings of the 2016 35th Chinese Control Conference (CCC), Chengdu, China, 27–29 July 2016; pp. 9787–9792.
17. Liu, J.; Wang, J.; Hu, Y.; Guo, J.; Zhang, B. Analysis of wind power characteristics of typical wind farm in Inner Mongolia area. In Proceedings of the 2017 29th Chinese Control and Decision Conference (CCDC), Chongqing, China, 28–30 May 2017; pp. 6721–6725.

18. Kang, S.W.; Yoon, H.S.; Kim, S.B.; Baatarbileg, A.; Sakong, J.; Lee, G.M. Regional generation characteristics of MW photovoltaic power plants in Jeju Island. In Proceedings of the 2018 5th International Conference on Renewable Energy: Generation and Applications (ICREGA), Al Ain, United Arab Emirates, 26–28 February 2018; pp. 52–55.

19. Sun, S.; Yang, Q.; Yan, W. A novel Markov-based temporal-SoC analysis for characterizing PEV charging demand. *IEEE Trans. Ind. Inf.* **2018**, *14*, 156–166. [CrossRef]

20. Psola, J.H.; Henke, M.; Pandya, K. Characteristics of energy storage in smart grids. In Proceedings of the 2014 Ninth International Conference on Ecological Vehicles and Renewable Energies (EVER), Monte-Carlo, Monaco, 25–27 March 2014; pp. 1–7.

21. Gao, C.; Li, Q.; Li, H.; Zhai, H.; Zhang, L. Methodology and Operation Mechanism of Demand Response Resources Integration Based on Load Aggregator. *Autom. Electr. Power Syst.* **2013**, *37*, 78–86.

22. Gao, Y.; Sun, Y.; Yang, W.; Xue, F.; Sun, Y.; Liang, H.; Li, P. Study on Load Curve's Classification Based on Nonparametric Kernel Density Estimation and Improved Spectral Multi-manifold Clustering. *Power Syst. Technol.* **2018**, *42*, 1605–1612.

23. Jiao, L. On Methods of Standardization Management of Index. *J. Anhui Agrotech. Teach. Coll.* **1999**, *3*, 9–12.

24. Goyal, R.K.; Kaushal, S. Deriving crisp and consistent priorities for fuzzy AHP-based multicriteria systems using non-linear constrained optimization. *Fuzzy Optim. Decis. Mak.* **2018**, *17*, 195–209. [CrossRef]

25. Naveed, Q.N.; Qureshi, M.R.N.; Alsayed, A.O.; Muhammad, A.; Sanober, S.; Shah, A. Prioritizing barriers of E-Learning for effective teaching-learning using fuzzy analytic hierarchy process (FAHP). In Proceedings of the 2017 4th IEEE International Conference on Engineering Technologies and Applied Sciences (ICETAS), Salmabad, Bahrain, 29 November–1 December 2017; pp. 1–8.

26. Hamid, T.; Al-Jumeily, D.; Hussain, A.; Mustafina, J. Cyber Security Risk Evaluation Research Based on Entropy Weight Method. In Proceedings of the 2016 9th International Conference on Developments in eSystems Engineering (DeSE), Liverpool, UK, 31 August–2 September 2016; pp. 98–104.

 *applied sciences*

Article

Analysis and Prioritization of the Floating Photovoltaic System Potential for Reservoirs in Korea

Sung-Min Kim [1], Myeongchan Oh [2] and Hyeong-Dong Park [2,3,*

1   Division of Graduate Education for Sustainability of Foundation Energy, Seoul National University, Seoul 08826, Korea; snuhyrule@hanmail.net
2   Department of Energy Resources Engineering, Seoul National University, Seoul 08826, Korea; amir117@daum.net
3   Research Institute of Energy and Resources, Seoul National University, Seoul 08826, Korea
*   Correspondence: hpark@snu.ac.kr; Tel.: +82-2-880-8808

Received: 31 December 2018; Accepted: 21 January 2019; Published: 24 January 2019

Abstract: Photovoltaic (PV) energy is one of the most promising renewable energies in the world due to its ubiquity and sustainability. However, installation of solar panels on the ground can cause some problems, especially in countries where there is not enough space for installation. As an alternative, floating PV, with advantages in terms of efficiency and environment, has attracted attention, particularly with regard to installing large-scale floating PV for dam lakes and reservoirs in Korea. In this study, the potentiality of floating PV is evaluated, and the power production is estimated for 3401 reservoirs. To select a suitable reservoir for floating PV installation, we constructed and analyzed the water depth database using OpenAPI. We also used the typical meteorological year (TMY) data and topographical information to predict the irradiance distribution. As a result, the annual power production by all possible reservoirs was estimated to be 2932 GWh, and the annual GHG reduction amount was approximately 1,294,450 tons. In particular, Jeollanam-do has many reservoirs and was evaluated as suitable for floating PV installation because of its high solar irradiance. The results can be used to estimate priorities and potentiality as a preliminary analysis for floating PV installation.

Keywords: floating photovoltaic system; renewable energy; solar energy; photovoltaic energy; greenhouse gas

1. Introduction

To reduce greenhouse gas (GHG) and protect the environment, renewable energy sources are increasingly utilized worldwide. According to Renewable Capacity Statistics [1], the total renewable power capacity for power generation is 2179 GW in 2017, with an 8.3% increase over the previous year. In particular, the growth rate of photovoltaic (PV) energy is prominent because the price of solar panels is steadily declining, and various stimulus policies are being implemented. PV energy is considered to be one of the most promising energy alternatives due to its ubiquity and sustainability. Particularly in South Korea, PV energy is spreading very quickly after the introduction of the Renewable Portfolio Standards (RPS) program, which requires electricity providers to increase their renewable energy production. PV energy is the renewable energy that accounts for the largest portion in Korea today, except for waste energy and bio-energy. According to the Korea New and Renewable Energy Center (KNREC, Yongin, Korea), renewable energy supply ratio in Korea is about 19.5%, 61.7%, 4.3%, and 2.5% for solar, biomass, hydro, and wind, respectively in 2016. The supply of PV energy in Korea was only 7.6 toe (tonne of oil equivalent) until 2006 but increased to a value of 1092.8 toe in 2016. This is because government policies have been actively implemented, and the public has become actively engaged in PV projects to generate profits by selling the produced electricity. However, the excessive installation of PV facilities has caused various problems. In addition to the damage to the landscape caused by the installation of solar panels, the residents of areas where the panels have been installed have been inconvenienced by the reflected light and increase in the ambient temperature. In addition, solar panels that are swept away in the event of landslides or typhoons can cause safety accidents. In recent years, there has been an increase in complaints from local residents, as the solar power generation business has become a means of speculation in the area of PV power generation. These problems can be even more significant in countries such as South Korea, where available land is scarce, so that there is not enough area for the installation of solar panels. Therefore, floating PV systems, which can overcome or alleviate these problems, have attracted considerable attention in these countries.

In addition to being able to utilize the reflected light from the water surface, the floating PV is known to be about 11% more efficient than the terrestrial solar panel due to the temperature reduction effect in water [2]. Solar panels that cover the water can also reduce evaporation [3] and prevent algae [4] due to shadows by the panels. In other words, the floating PV is advantageous in terms of efficiency and environmental aspects, in addition to the lack of need for land. Because of these advantages, floating PV systems installed on water bodies, such as reservoirs or dam lakes, have increased worldwide and have already been deployed in several countries, including South Korea, Japan, China, and the US [5]. Although most technological growth occurred between 2007 and 2014, the global installed capacity for floating PV has increased significantly since 2015. Floating PV installations in the world were estimated at 94 MW by 2016, of which about 60% were known to be installed in Japan [6]. In 2017, however, the world's largest floating PV system was installed on the pit lake of a coal mine in China and is known to have capacity of 40 MW. The system is composed of 120,000 solar panels, supplying the electricity needs of 15,000 households [7]. Recently, South Korea has been rapidly improving its floating PV technology and increasing installation capacity. The Korea Water Resources Corporation (K-water, Daejeon, Korea) carried out the first demonstration project in South Korea at the Hapcheon reservoir in 2014 with a capacity of 100 kW. In addition, K-water operates South Korea's first commercial system by producing 500 kW of electricity for approximately 170 households. K-water plans to install 550-MW capacity dam lakes by 2022, and the Korea Rural Community Corporation (KRC, Naju, Korea) is planning 1900 MW of capacity for the reservoirs under management.

Various studies have been conducted to improve and utilize floating PV. There have been many studies on mooring systems to maintain the balance of panels on the water [8,9], and a solar tracking system has been actively studied to improve efficiency [10,11]. In recent years, it has become possible to transmit and receive data from a floating PV through the integration of ICT technology [12]. In addition, there have been several studies where Geographical Information System (GIS) techniques have been applied to determine a suitable location for field application and evaluation of the PV potential [13]. Song and Choi [14] evaluated the potential and economic feasibility of a pit lake on the abandoned mine site in Gangwon-do, South Korea, by assessing shadow effects based on GIS techniques and a fish-eye lens camera. Lee and Lee [15] evaluated the applicability of floating PV by region based on the analytic hierarchy process (AHP), considering geographical conditions and weather conditions. Lee et al. [16] evaluated additional parameters to be considered because of the difference between conventional and floating PV systems and evaluated the suitability of the Hapcheon dam area.

There have been many studies regarding individual reservoirs, lakes, and dams, but little research has been conducted on a national scale. Particularly in countries like Korea where large-capacity floating PV systems will be installed in the long term, the PV potential of each available area should be assessed. The purpose of this study is to evaluate the applicability and potential of floating PV for the approximately 3400 reservoirs registered and managed in Korea.

2. Study Area and Data

In this study, we investigated whether floating PV installation is suitable for reservoirs distributed in South Korea. There are 3401 reservoirs managed by the KRC as a database that contains information such as the area of the reservoir, dead water level, full water level, and water storage capacity. The database was downloaded from the Public Data Portal (PDP, data.go.kr). Spatial data such as location and boundary information for each reservoir are also managed and provided to the public. They were downloaded from the National Spatial Data Infrastructure Portal (NSDI, data.nsdi.go.kr). Figure 1a shows reservoir locations, provinces, and some major cities, and Figure 1b shows reservoir density across the country. It can be seen that the most reservoirs are concentrated in Jeollanam-do. Using this location data, each reservoir is constructed as point data, and each point object has the name and standard code of the reservoir. Since these standard codes are used in the same way in all the databases, they are used to perform joint work between different data sources. To calculate the area for each reservoir and perform geospatial analysis, information on the boundary of the reservoir was used. Most of the reservoir boundaries and locations are very accurately constructed, except some of the very small reservoirs, when compared with satellite images. To operate floating PV properly, the reservoir needs to have sufficient water depth. Therefore, information on reservoir depth is required for this study, and the PDP provides the level of the reservoir managed by the KRC on a daily basis. This information is provided through OpenAPI, which is a publicly available application programming interface that provides developers with programmatic access to web service. In this study, we constructed a database of the daily reservoir depth over one year in 2017 for all reservoirs through Python coding for OpenAPI. To evaluate the potential of floating PV installed in the reservoir, it is necessary to use the solar irradiance for each site. In this study, we use the typical meteorological year (TMY) data constructed by the Korea Institute of Energy Research (KIER, Daejeon, Korea) to predict the amount of solar irradiance at each site. Global horizontal irradiance (GHI) and direct normal irradiance (DNI) derived from TMY data are used in this study. TMY datasets in 16 cities are created by KIER and distributed by the National Center for Standard Reference Data (NCSRD, Daejeon, Korea). In addition, it is necessary to analyze the shadowing effect to evaluate the power production of floating PV. For this purpose, information about the terrain is needed. In this study, the digital elevation model (DEM) of 90-m resolution, downloaded from the National Geographic Information Institute (NGII, Suwon, Korea), is used. The DEM and TMY observation points for South Korea are shown in Figure 2.

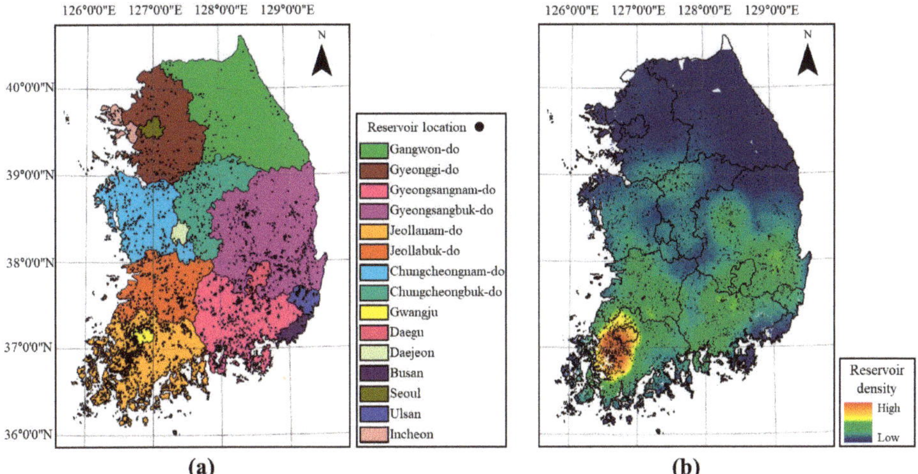

(a)　　　　　(b)

Figure 1. Reservoirs in Korea: (**a**) reservoir locations, provinces and some major cities; (**b**) reservoir density across the country.

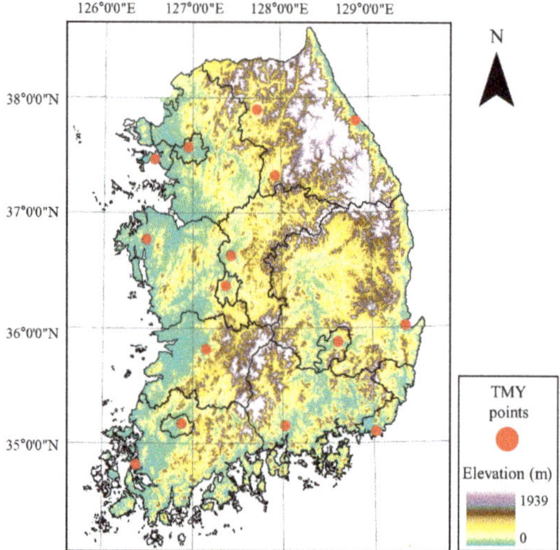

Figure 2. Digital Elevation Model (DEM) and Typical Meteorological Year (TMY) data observation points in Korea.

3. Methods

Figure 3 shows the study process used to select suitable reservoirs for installing floating PV from approximately 3400 reservoirs in South Korea and to evaluate the expected effect. This study process can be summarized as follows. In the first step, TMY data were used to predict the daily average solar irradiance, considering the annual irradiance of South Korea. In this process, an interpolation method was applied. However, this solar irradiance map does not take into account the effects of the terrain. In the second step, we performed topographic analysis using the DEM to account for the influence of the shadow caused by the terrain. An average daily irradiance map was created based on topography, taking into account the shadowing ratio and solar irradiance map created in the first step. In the third step, reservoirs that do not have sufficient water depth were excluded from the analysis. In the fourth step, the possible power production of the floating PV system of each reservoir was estimated by considering the meteorological data and system design parameters. In the fifth step, the economic feasibility and reduction in GHG emissions by administrative districts were evaluated based on the expected power production.

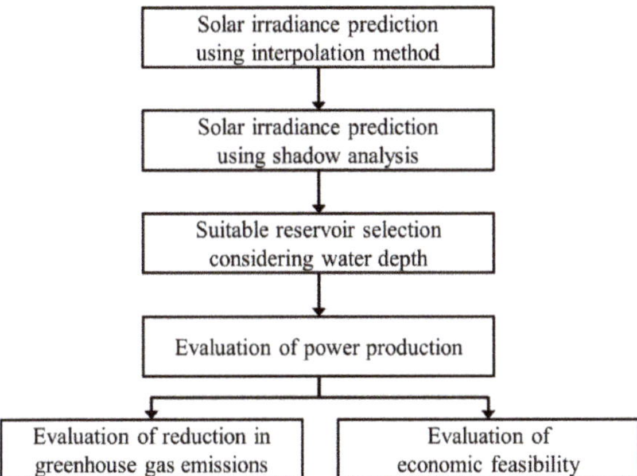

Figure 3. Overall study process to select suitable reservoirs and assess the potential of floating photovoltaic systems in Korea.

3.1. Point Irradiance Calculation and Interpolation

The solar irradiance was calculated based on an equal-area-equal-angle sky division grid [17] with the TMY dataset of 16 cities. Beam and diffuse irradiances for each position in the celestial sphere were calculated by summing all the irradiance over the entire year for a time interval. This study refers to this irradiance intensity map of the sky as Sunmap. Figure 4a shows the average of Sunmap for a tilt angle of 20° from all 16 cities, and Figure 4b shows the average of total irradiance as a function of solar panel direction.

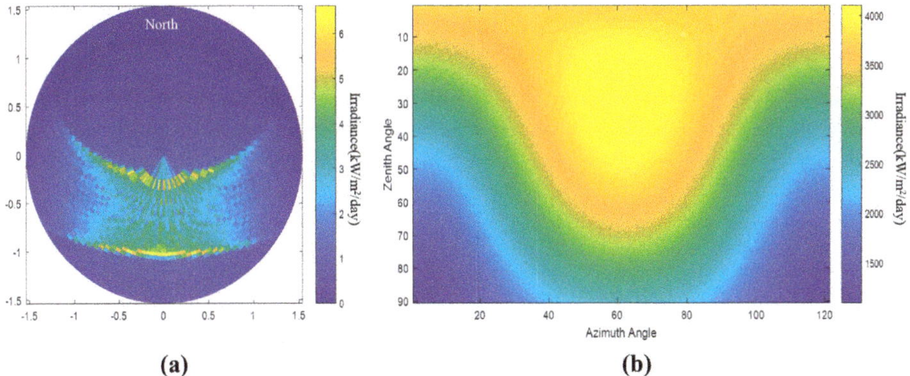

Figure 4. (a) Mean Sunmap of TMY dataset in Korea with a tilt angle 20°; (b) Mean total irradiance with respect to solar panel direction.

The top is north, and the center point represents the zenith. The sky diffuse model [18] is used when calculating the diffuse irradiance in the sky. The total incident irradiance of the tilted surface is the summation of all positive normal intensities from the entire sky. Considering the irradiance of each city in the TMY dataset, it was assessed that the optimal panel direction of Korea is southward, and the optimal tilt angle is about 30° for most cities. However, many solar panels in the floating PV system have been installed with a tilt angle close to horizontal. National Renewable Energy Laboratory (NREL, Golden, Colorado, USA) research has shown that a tilt angle of 11° is the typical mounting angle for floating solar systems in Tokushima, Japan [19]. Other installed systems in Italy and Singapore also have a tilt angle close to the horizontal [20]. This preference is derived for reasons such as maximizing power density or structural stability. Therefore, this study sets a tilt angle of 20 °, which has a difference in irradiance of less than 1%, compared to a tilt of 30 °.

Point irradiance data was interpolated to generate an irradiance map to estimate the irradiance in all areas in Korea. Interpolation is a technique to estimate values between known data. Kriging is a widely used geostatistical interpolation method, which calculates weighting from the correlation of known points. Although it is statistically powerful and is preferred in many cases, a sufficient number of points is necessary to estimate the statistical relationship of each point. Therefore, kriging is not applicable in this study, so the inverse distance weighting (IDW) method was applied. IDW is a deterministic method, which is frequently used in many studies [21]. As shown in Equation (1), the IDW method interpolates unknown values based on distances and values of known points.

$$I_p = \frac{\sum_{i=1}^{n} \frac{I_i}{d_i^P}}{\sum_{i=1}^{n} \frac{1}{d_i^P}}, \tag{1}$$

where I_p is the irradiance of an unknown point, n is number of points, i is an index of a point, I_i is irradiance of the known ith point, d_i is the distance between the unknown point and the known ith point, and P is power parameter. This method was used to generate a renewable energy resource atlas in Korea [22]. In addition, Evrendilek et al. [23] showed that the accuracies of IDW and co-kriging are similar for Turkey.

3.2. Solar Irradiance Map based on Shadow Analysis

As the study area has mountainous terrain, calculating shadowing is an essential process. Whether the sky is covered by terrain or not is determined using the DEM for each position in the Sunmap grid. This sky coverage map is referred to as Viewmap and is used to calculate the shadowed irradiance or sunshine duration. Details of the process are described in Oh et al. [17]. The annual sunshine duration is calculated based on Sunmap and Viewmap.

The annual sunshine duration is calculated considering the effect of shadows, but the difference in irradiance at different times is not considered. For example, even though the sunshine duration in the morning or evening is the same as the sunshine duration at noon, the irradiance in the morning or evening is smaller. To solve this problem, this study used an hourly weight. Figure 5 shows the hourly distribution of irradiance, which is the hourly irradiance divided by the total irradiance, for each city in the TMY dataset. This shows that the hourly distribution for all cities is similar. Therefore, this study assumes that the pattern of hourly distribution of irradiance is the same for every location. Based on this assumption, the sunshine duration for each hour is calibrated as the shadowless ratio by considering the weighting according to the hourly distribution of irradiance.

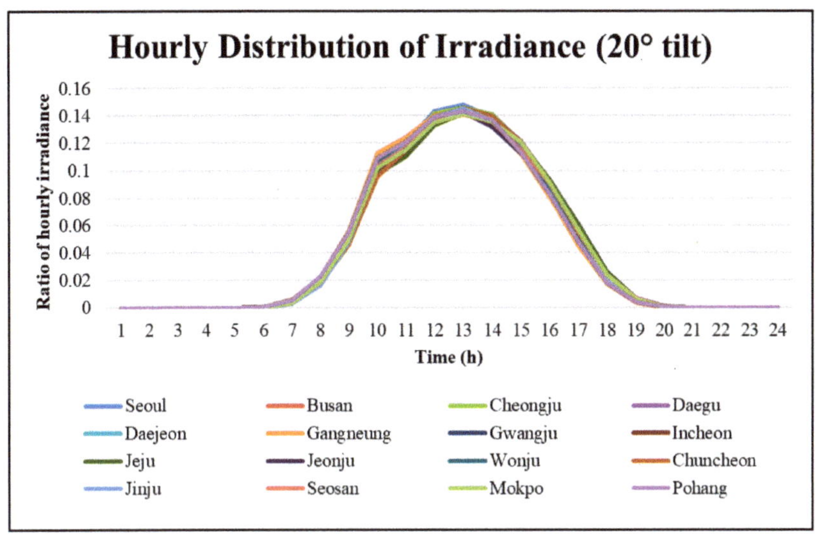

Figure 5. Graph of the ratio of hourly irradiance of each city in TMY dataset.

Equation (2) shows the equation for the hourly weight calculation, and Equation (3) shows the equation for irradiance considering shadows,

$$W_h = \frac{S_h}{S_{maxh}} \times \frac{I_{th}}{\sum_{i=1}^{24} I_{ti}} \qquad (2)$$

$$I_{sp} = \left(\sum_{h=1}^{24} W_h \right) I_p \qquad (3)$$

where h is the time index W_h is the weight at each time h, I_{th} is annual irradiance of the TMY dataset at time h, S_h is sunshine duration at time h, S_{maxh} is maximum sunshine duration at time h, and I_{sp} is irradiance considering shadows of the unknown point. The ratio of sunshine duration and the hourly distribution of irradiance of the TMY dataset is calculated in Equation (2). Mean values for the entire TMY dataset are used in the calculation. The weight at each time is the multiplication of these two ratios. The summation of all hourly weights is used as the total weight, equal to 1 when there are no shadows and equal to 0 when that position is fully shadowed.

The irradiance map assuming shadows (Figure 6c) is generated by the multiplication of the interpolated irradiance map (Figure 6a) and total weight map (Figure 6b). This map can represent both the influence of weather and terrain. The spatial resolution of the map is 90 m, which is equal to that of the DEM.

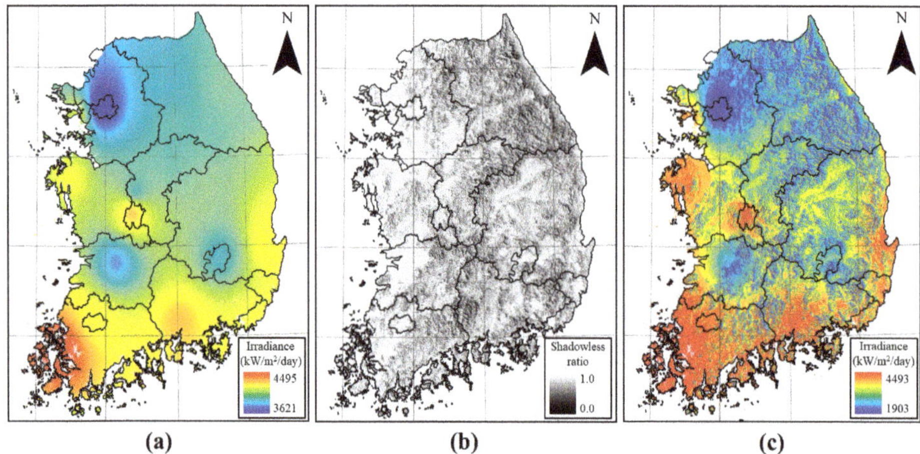

Figure 6. Irradiance prediction: (**a**) interpolated irradiance map; (**b**) shadowless ratio map; (**c**) irradiance map considering shadow by terrain.

3.3. Constraint Analysis on Reservoir Depth

In general, for smooth installation and operation of the floating PV, a reservoir with a water depth greater than 5 m is recommended, maintaining a water depth of at least 1 m [24]. Therefore, reservoirs that do not meet these conditions were excluded in this study. In fact, a floating PV of 465 kW installed in the Geumgwang reservoir in Korea had difficulty in operation because of a summer drought that caused the bottom of the reservoir to be revealed [25]. The KRC provides the daily water level for each reservoir using OpenAPI. The data format is XML, and the API service type is REST type. When the standard code and search period are requested as parameters, the server provides the daily level for that period. Since only one standard code for a particular reservoir can be input in one request, Python-based coding was performed to obtain data for more than 3000 reservoirs. The Requests module from Python was used for OpenAPI requests, and the BeautifulSoup module was used for parsing XML data. Since the water level is provided as an elevation, the water depth was calculated by subtracting the dead water level of each reservoir from the water level. This study constructed a reservoir depth database of 2017 and used it for analysis because for some reservoirs there is no data before 2017.

3.4. Design and Evaluation of the PV System

In this study, it is assumed that the solar PV panels are installed in a fixed-tilt array with a 20° slope facing south. The annual power production of the PV system was calculated using the formula of RETScreen software developed by Natural Resources Canada (NRC, Ottawa, Canada). RETScreen evaluates the power of the PV system via Equation (4),

$$E_A = H_t \times S \times \eta_r \times \eta_{inv} \times \left[1 - \beta_p \times (T_c - 25) \right] \times (1 - \lambda_p) \times (1 - \lambda_c) \tag{4}$$

where E_A is the amount of power (kWh/h) produced by a PV system, H_t is solar irradiation per unit area per unit time (kWh/m²/h), S is a surface area (m²) of the solar array, η_r is conversion efficiency (0–1) of the solar cell module, η_{inv} is conversion efficiency of the inverter, β_p is temperature coefficient related to the efficiency of the solar cell module, T_c (°C) is average temperature of the solar cell module, λ_p is loss coefficient of the solar cell module, and λ_c is loss coefficient of the inverter [26]. Table 1 shows the parameters for the solar panel and inverter used in this study. In this study, the number and surface area of the panels that can be installed are calculated, considering the area of each reservoir, the area of one panel, and array spacing. Then, this result is substituted into equation (4), and finally, the power production is estimated.

Table 1. Design parameters for the photovoltaic systems considered in this study.

Type	Parameter	Value
	Model	SPR-210-BLK
	Length (m)	1.56
	Width (m)	0.80
Solar cell module	Power capacity (kW/unit)	0.21
	Efficiency (%)	16.9
	Nominal Operating cell temperature (°C)	46
	Temperature coefficient (%/°C)	−0.4
	Losses (%)	1
	Model	SPR-12000f
Inverter	Efficiency (%)	95.5
	Capacity (kW/unit)	12.5
	Losses (%)	0

The parameters of solar panel and inverter in this study are the values of the products that are commercially available and are produced on an industrial scale. If more efficient equipment is installed in the reservoirs than Table 1, then the total power production and environmental benefit will increase further. In that case, however, the overall cost will increase, and the economic burden will also increase. Therefore, appropriate selection should be made according to technology level and economic conditions at installation. According to Battaglia et al. [27], current silicon solar cells technology achieved over 25% of efficiency, and several advanced concepts have been proposed to overcome single-junction solar cells. In addition, silicon hetero-junction solar cell achieved over 26% of efficiency [28]. Therefore, more efficient equipment is expected to be installed at a lower cost in the future.

3.5. Economic Assessment and GHG Reduction

In this study, an economic assessment is performed in terms of the cost of the PV system and the profit from the power production. The cost of a PV system can be separated into the initial installation costs and operating costs. According to the Korea Energy Economics Institute (KEEI, Ulsan, Korea) [29], the installation cost is about 1.43 USD/W, and the annual operating cost is about 10.38 USD/kW. The RPS system was used to estimate the revenue generated by electricity production and sales. The RPS system calculates the electricity sales revenue by summing the system marginal price (SMP) and renewable energy certificates (REC). Here, SMP refers to the revenue from the sale of electricity through the Korea Power Exchange (KPX, Naju, Korea). REC is a tradable, non-tangible energy commodity that represents 1 MWh of electricity generated from an eligible renewable energy resource. Since the prices of SMP and REC are constantly changing, this study used the average value of the amount traded in Korea since 2018. SMP and REC prices are 82.93 USD/MWh and 89.06 USD/MWh, respectively. In addition, in Korea, the power generated by a floating PV is multiplied by a REC weight of 1.5 308.

The net present value (NPV) of the PV system was also calculated by Equation (5).

$$\text{NPV} = \sum_{t=1}^{N} \frac{E_t - C_t}{(1+r)^t} - C_0 \tag{5}$$

where N is the system operating period (20 y in this study), E_t is annual electricity sales revenue (USD), C_t the annual operating cost (USD), r is the discount rate, and C_0 is initial cost (USD). A discount rate of 5.5% was applied in this study, according to KEEI [30]. The payback year can be determined by calculating N, satisfying the condition that NPV becomes zero in Equation (5).

GHG reduction was also calculated in this study by multiplying the GHG emission reference value (0.4415 tCO$_2$/MWh) and total amount of power (MWh/year). GHG reduction refers to the amount of GHG generated when the same amount of power produced through renewable energy is generated through the fossil energy system.

4. Results

Figure 7 shows the variation in the mean water depth over one year by distinguishing reservoirs for each province. In Gangwon-do, the number of reservoirs is small, but the reservoirs have a deep water depth. On the other hand, Jeollanam-do has the highest number of water reservoirs but the lowest average water depth. Commonly, the water depth falls rapidly from May to July and tends to recover after July. Figure 8 shows the map of the average water depth for each reservoir over the course of a year. There are a considerable number of reservoirs that are not suitable for floating PV installations, in which the average water depth is less than 5 m. There are many reservoirs in Jeolla-do and Gyeongsang-do, but many reservoirs are low in water depth.

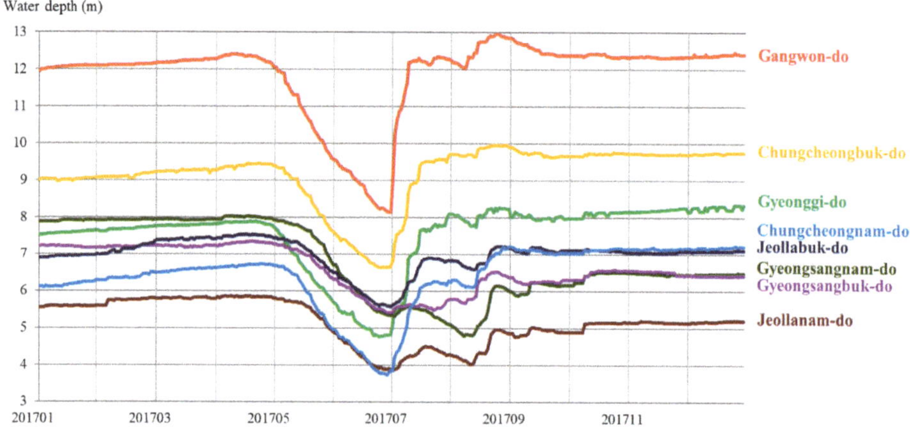

Figure 7. The variation of mean water depth over one year for reservoirs of each province.

The floating PV capacity and power production of each reservoir was calculated considering the solar irradiance, characteristics of the panel, and installation area of the panel. In this study, an additional 11% efficiency was considered for the use of the floating PV. The results for all reservoirs are shown in Table 2, based on the ratio of the installation area for the reservoirs satisfying the water depth requirements assumed above. Only reservoirs with a floating PV capacity of 100 kW or greater were considered. Generally, it is common to install floating PV panels in 10% of the reservoir area. In this case, 1134 reservoirs satisfy the condition, and the total installed capacity of the panels is about 2103 MW, with an annual power production of 2932 GWh. In recent years, a floating PV has been installed for a high percentage of the reservoir area in several cases [31]. If a floating PV is installed over the entire area of the reservoir under extreme assumptions, the total installed capacity of the panel is estimated to be 21,093 MW, and the annual power production is estimated to be 29,409 GWh.

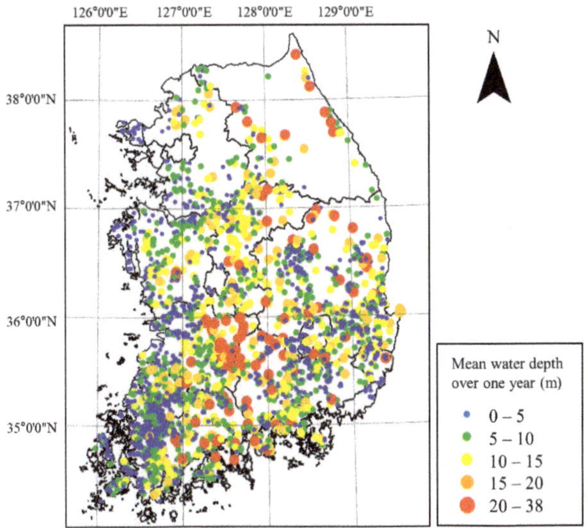

Figure 8. The map of the average water depth for each reservoir over a year.

Figure 9a is a map showing the annual electricity usage for the administrative units of Korea in 2017. Metropolitan cities or industrial cities such as Seoul, Ulsan, Incheon, and Busan show relatively high electricity usage, and their annual electricity consumption is about 46,493 GWh, 32,095 GWh, 23,876 GWh, and 20,467 GWh, respectively. Figure 9b shows what percentage of electricity usage can be satisfied when a floating PV is installed in each administrative unit. It is assumed that solar panels are installed in 10% of the reservoir area. It is estimated that Jeolla-do, which has a relatively low power consumption and is favorable for floating PV installation, can supply a large portion of the electric power demand in comparison with other provinces. For example, it is estimated that about 41% of the power demand of Imsil in Jeollabuk-do and 16% of the power demand of Gangjin in Jeollanam-do can be covered by floating PV power generation.

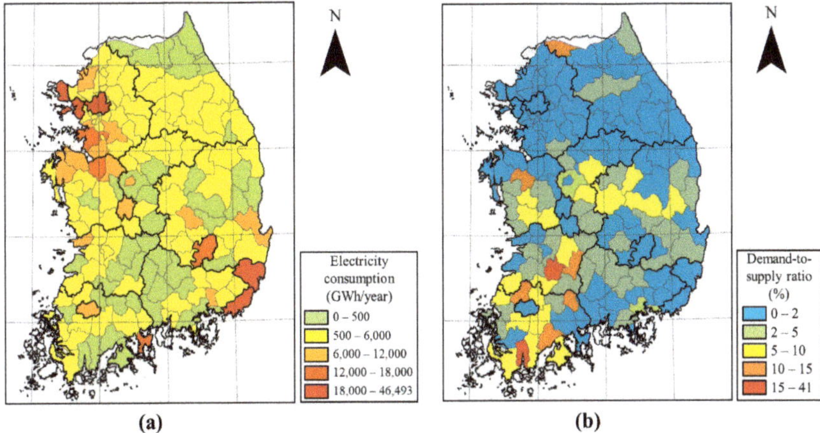

Figure 9. (**a**) Map of the annual electricity usage for the administrative units and (**b**) map of the demand-to-supply ratio for electricity usage.

Table 3 shows the overall results of reservoir condition, installed floating PV capacity, annual power production, and GHG reduction of each province. It can be seen that Jeollanamdo, Gyeongsangbuk-do, and Chungcheongnam-do, which have high solar irradiance and a large number of reservoirs, have high power production and high GHG reduction. If a floating PV is installed in 10% of the area for all reservoirs, the expected annual power production is 2932 GWh, and the annual GHG reduction amount is approximately 1,294,450 ton.

Assuming that a floating PV is installed in 10% of the reservoir area, the analysis results of the top 10 reservoirs (Figure 10) with the highest power production are shown in Table 4. In this case, a larger reservoir area corresponds to more power generation. Therefore, the top 10 reservoirs with high economic efficiency were also analyzed for an installed 100-kW floating PV (476 units of PV panel with 0.21-kW capacity). For a floating PV installed in proportion to the area of the reservoir, the analysis shows that maximum power production corresponds to a PV installed in Yedang reservoir, located in Chungcheongnam-do. In this case, reservoirs suitable for floating PV installations are distributed nationwide. On the other hand, in terms of economic efficiency, reservoirs of Jeollanam-do with high solar irradiance were found to be advantageous.

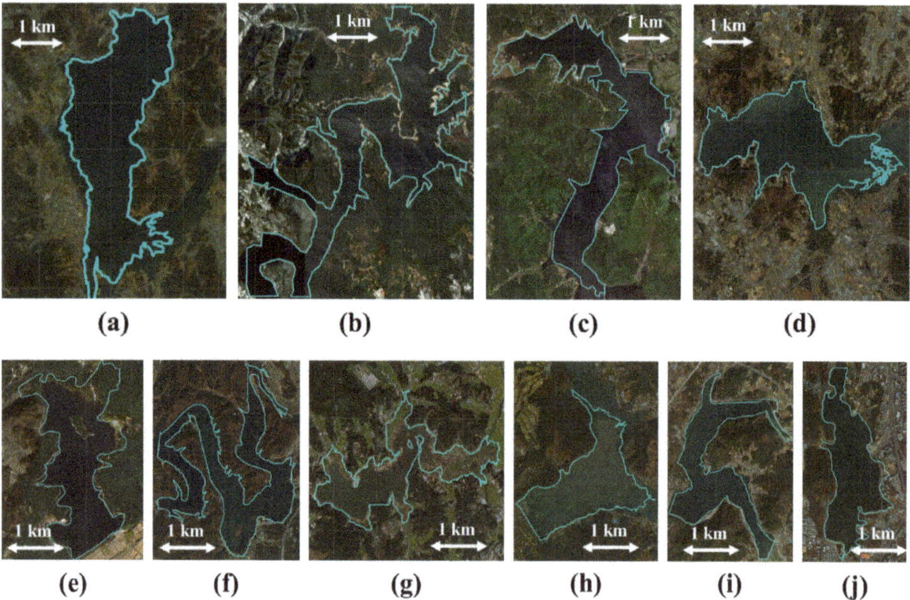

Figure 10. Boundary data and satellite images of top 10 reservoirs with the highest power production: (**a**) Yedang; (**b**) Seomjin; (**c**) Najuho; (**d**) Topjeong; (**e**) Togyo; (**f**) Miho; (**g**) Gosam; (**h**)Bulgop; (**i**) Yidong; (**j**) Deokdong.

Table 2. The results of floating photovoltaic system installations according to the ratio of the installation area for all reservoirs satisfying the water depth requirements.

Installation area to reservoir area ratio (%)	10	20	30	40	50	60	70	80	90	100
Number of reservoirs satisfying 100 kW	1134	1201	1218	1225	1230	1234	1236	1236	1236	1236
Maximum capacity (MW)	75.85	151.70	227.55	303.40	379.25	455.10	530.95	606.80	682.65	758.50
Total capacity (MW)	2102.89	4216.33	6326.63	8436.33	10,545.97	12,655.63	14,765.12	16,874.42	18,983.73	21,093.03
Total power production (GWh/year)	2931.94	5878.65	8820.95	11,762.40	14,703.79	17,645.20	20,586.38	23,527.30	26,468.21	29,409.12

Table 3. The results of reservoir condition, installed floating photovoltaic (PV) capacity, annual power production, and greenhouse gas (GHG) reduction of each province.

	Number of Reservoirs	Mean Water Depth (m)	Mean Reservoir Area (km^2)	Mean Irradiance (kW/m^2 /day)	Mean PV Capacity (MW)	Maximum PV Capacity (MW)	Total PV Capacity (MW)	Mean Power Production (GWh/year)	Maximum Power Production (GWh/year)	Total Power Production (GWh/year)	Total GHG Reduction (1000 tCO^2/year)
Gangwon-do	48	12.09	0.50	3714.40	2.43	30.28	116.81	3.28	41.21	157.36	69.47
Gyeonggi-do	50	9.78	0.64	3704.45	3.61	24.60	180.60	4.88	34.20	244.19	107.81
Gyeongsangnam-do	167	12.23	0.22	3864.00	0.81	6.22	135.27	1.11	8.73	184.85	81.61
Gyeongsangbuk-do	227	11.26	0.38	3814.96	1.75	20.48	397.25	2.45	29.07	556.32	245.62
Jeollanam-do	285	9.96	0.32	3987.81	1.50	54.94	427.50	2.17	80.51	617.29	272.53
Jeollabuk-do	149	12.37	0.38	3798.40	1.95	75.25	290.11	2.67	102.39	397.32	175.42
Chungcheongnam-do	98	9.78	0.61	3877.83	3.65	75.85	357.75	5.16	107.58	505.29	223.09
Chungcheongbuk-do	110	12.43	0.40	3781.62	1.79	26.56	197.18	2.45	36.71	269.32	118.90
Sum	1134						2068.53			2931.94	1294.45

Table 4. The analysis results of the top 10 reservoirs with the highest power production and with the highest efficiency.

	Rank	Reservoir Name	Province	City	Reservoir Area (km²)	Capacity (MW)	Power Production (GWh/year)	GHG Reduction (ton)	NPV (10,000 USD)	Pay Back (year)
Floating PV installed in proportion to the area of the reservoir (High power production)	1	Yedang	Chungcheongnam-do	Yesan	8.83	75.85	107.58	47,494.86	17,513.99	4.42
	2	Seomjin	Jeollabuk-do	Imsil	10.11	75.25	102.39	45,205.39	16,191.54	4.70
	3	Najuho	Jeollanam-do	Naju	7.19	54.94	80.51	35,544.71	13,392.70	4.21
	4	Topjeong	Chungcheongnam-do	Nonsan	5.82	47.20	67.77	29,921.57	11,125.65	4.34
	5	Togyo	Gangwon-do	Cheorwon	3.92	30.28	41.21	18,195.17	6518.30	4.69
	6	Miho	Chungcheongbuk-do	Jincheon	4.42	26.56	36.71	16,206.52	5869.99	4.59
	7	Gosam	Gyeonggi-do	Anseong	3.56	24.60	34.20	15,101.31	5493.85	4.55
	8	Bulgop	Jeollanam-do	Yeonggwang	3.20	21.44	31.70	13,996.68	5304.63	4.15
	9	Yidong	Gyeonggi-do	Yongin	2.87	22.00	30.31	13,382.42	4836.23	4.61
	10	Deokdong	Gyeongsangbuk-do	Gyeongju	3.11	20.48	29.08	12,836.71	4736.96	4.41
Floating PV with 100 kW capacity (High efficiency)	1	Naesan	Jeollanam-do	Haenam	0.11	0.10	0.16	69.08	27.08	3.82
	2	Jangsan	Jeollanam-do	Shinan	0.14	0.10	0.16	69.04	27.06	3.82
	3	Gopyeong	Jeollanam-do	Haenam	0.08	0.10	0.16	68.86	26.95	3.84
	4	Cheongsu	Jeollanam-do	Muan	0.15	0.10	0.16	68.54	26.75	3.86
	5	Illo	Jeollanam-do	Muan	0.75	0.10	0.15	68.39	26.66	3.88
	6	Gwangdae	Jeollanam-do	Shinan	0.41	0.10	0.15	68.21	26.55	3.89
	7	Sindeok	Jeollanam-do	Yeongam	0.13	0.10	0.15	68.20	26.54	3.89
	8	Dopyeong	Jeollanam-do	Jindo	0.11	0.10	0.15	68.09	26.48	3.90
	9	Sungdong	Jeollanam-do	Muan	0.17	0.10	0.15	68.02	26.43	3.91
	10	Sungyang	Jeollanam-do	Yeongam	0.40	0.10	0.15	67.77	26.28	3.93

5. Conclusions

In this study, we analyzed the water level data of 3401 reservoirs in Korea using OpenAPI and extracted reservoirs suitable for floating PV installation. In this analysis, the conditions of an average reservoir water depth greater than 5 m and minimum water depth greater than 1 m were considered. GIS spatial analysis was applied to the TMY dataset to predict the national distribution of solar irradiance for shadow conditions. Assuming an expected power production of 10% of the area of all reservoirs satisfying the conditions, the annual power production was estimated to be 2932 GWh, and the annual GHG reduction amount was estimated to be 1,294,450 ton. In particular, Jeollanam-do has many reservoirs, including many suitable for floating PV installation because of high solar irradiance. The results of this study can be used to estimate priorities and potentiality prior to actual floating PV installation and detailed analysis. This study did not take degradation into consideration because it is aimed at estimating priorities through relative comparisons. However, as floating PV is installed in the water, it is necessary to study the efficiency change and degradation related to the moisture in actual installation.

The results of this study are very useful for preliminary evaluation at the national level, but improvement is needed in the following points. First, there may be a limit to the prediction accuracy in that only 16 data points are used to predict the solar irradiance. If the irradiance is measured at many more points, accurate irradiance prediction can be made. However, because of budget limitations, it is necessary to continuously increase the number of solar irradiance observation points at the government level. It is also possible to improve the accuracy of solar irradiance prediction through complex analysis with satellite image data. Next, the water level data used in this study is not well managed by the government, and some data are missing or include errors. In this study, we manually identified and corrected these errors. To reflect water level data in real time during the operation of the floating PV in the future, more thorough management of water level data is needed. In addition, this study was carried out to evaluate reservoirs suitable for floating PV installation and to evaluate their overall potential. However, to actually install floating PV in individual reservoirs, a more accurate analysis of the area's solar irradiance, wind speed, accessibility, detailed topography, and other conditions needs to be implemented. Finally, since SMP and REC prices are constantly changing, it is necessary to periodically consider the economic feasibility. The results of this study have important implications in Korea for large-scale floating PV installation in the future. For the successful application of floating PV, a rigorous feasibility assessment and planning for the redistribution of profit is most important, to minimize environmental damage and repudiation from the local residents.

Author Contributions: H.-D.P. conceived and designed the analysis. S.-M.K. and M.O. performed the analysis. S.-M.K. and M.O. developed the code and analyzed the data. H.-D.P. contributed reagents/materials/analysis tools. S.-M.K. wrote the paper.

Funding: This research was funded by the National Research Foundation of Korea (NRF) (No. NRF-2017R1A2B4007623).

Acknowledgments: This research was supported by the Brain Korea 21 Project.

Conflicts of Interest: The authors declare no conflict of interest.

References

1. The International Renewable Energy Agency (IRENA). *Renewable Capacity Statistics 2018*; IRENA: Abu Dhabi, United Arab Emirates, 2018.
2. Choi, Y.K. A Study on Power Generation Analysis of Floating PV System Considering Environmental Impact. *Int. J. Softw. Eng. Appl.* **2014**, *8*, 75–84. [CrossRef]
3. Sahu, A.; Yadav, N.; Sudhakar, K. Floating photovoltaic power plant: A review. *Renew. Sust. Energ. Rev.* **2016**, *66*, 815–824. [CrossRef]

4. Galdino, M.; de Almeida Olivieri, M.M. Some Remarks about the Deployment of Floating PV Systems in Brazil. *J. Elec. Eng.* **2017**, *5*, 10–19. [CrossRef]

5. The World Bank. *Where Sun Meets Water: Floating Solar Market Report*; The World Bank: Washington, DC, USA, 2018.

6. Minamino, S. Floating Solar Plnats: Niche Rising to the Surface? Available online: https://www.solarplaza.com/channels/top-10s/11634/floating-solar-plants-niche-rising-surface/ (accessed on 28 December 2018).

7. Osborne, M. China Completes Largest Floating Solar Power Plant. Available online: https://www.pv-tech.org/news/china-completes-largest-floating-solar-power-plant (accessed on 28 December 2018).

8. Lee, Y.G.; Joo, H.J.; Yoon, S.J. Design and installation of floating type photovoltaic energy generation system using FRP members. *Sol. Energy* **2014**, *108*, 13–27. [CrossRef]

9. Kim, S.H.; Yoon, S.J.; Choi, W. Design and Construction of 1 MW Class Floating PV Generation Structural System Using FRP Members. *Energies* **2017**, *10*, 1142. [CrossRef]

10. Choi, Y.K.; Kim, I.S.; Hong, S.T.; Lee, H.H. A Study on Development of Azimuth Angle Tracking Algorithm for Tracking-type Floating Photovoltaic System. *Adv. Sci. Technol. Lett.* **2014**, *51*, 197–202. [CrossRef]

11. Choi, Y.K.; Lee, N.H.; Lee, A.K.; Kim, K.J. A Study on Major Design Elements of Tracking-Type Floating Photovoltaic Systems. *Int. J. Smart Grid Clean Energy* **2014**, *3*, 70–74. [CrossRef]

12. Lee, A.K.; Shin, G.W.; Hong, S.T.; Choi, Y.K. A Study on Development of ICT Convergence Technology for Tracking-Type Floating Photovoltaic Systems. *Int. J. Smart Grid Clean Energy* **2014**, *3*, 80–87. [CrossRef]

13. Song, J.; Choi, Y.; Yoon, S.H. Analysis of Photovoltaic Potential at Abandoned Mine Promotion Districts in Korea. *Geosyst. Eng.* **2015**, *18*, 168–172. [CrossRef]

14. Song, J.; Choi, Y. Analysis of the Potential for Use of Floating Photovoltaic Systems on Mine Pit Lakes: Case Study at the Ssangyong Open-Pit Limestone Mine in Korea. *Energies* **2016**, *9*, 102. [CrossRef]

15. Lee, K.R.; Lee, W.H. Floating Photovoltaic Plant Location Analysis using GIS. *J. Korean Soc. Geospat. Inf. Syst.* **2016**, *24*, 51–59. [CrossRef]

16. Lee, S.H.; Lee, N.H.; Choi, H.C.; Kim, J.O. Study on Analysis of Suitable Site for Development of Floating Photovoltaic System. *J. Korean Inst. Illum. Electr. Install. Eng.* **2012**, *26*, 30–38.

17. Oh, M.; Park, H.D. A new algorithm using a pyramid dataset for calculating shadowing in solar potential mapping. *Renew. Energy* **2018**, *126*, 465–474. [CrossRef]

18. Perez, R.; Seals, R.; Michalsky, J. All-Weather Model for Sky Luminance Distribution—Preliminary Configuration and Validation. *Sol. Energy* **1993**, *50*, 235–245. [CrossRef]

19. NREL. STAT FAQs Part 1: Floating Solar. Available online: https://www.nrel.gov/state-local-tribal/blog/posts/stat-faqs-part1-floating-solar.html (accessed on 28 December 2018).

20. Patil, S.S.; Wagh, M.M.; Shinde, N.N. A Review on Floating Solar Photovoltaic Power Plants. *Int. J. Sci. Eng. Res.* **2017**, *8*, 789–794.

21. Palmer, D.; Cole, I.; Betts, T.; Gottschalg, R. Interpolating and Estimating Horizontal Diffuse Solar Irradiation to Provide UK-Wide Coverage: Selection of the Best Performing Models. *Energies* **2017**, *10*, 181. [CrossRef]

22. Kang, Y.H. *KOREA Renewable Energy Resource Atlas*; Korea Institute of Energy Research: Daejeon, Korea, 2015.

23. Evrendilek, F.; Ertekin, C. Assessing solar radiation models using multiple variables over Turkey. *Clim. Dynam.* **2008**, *31*, 131–149. [CrossRef]

24. LSIS. *Floating Photovoltaic System*; LS Industrial Systems: Anyang, Korea, 2016.

25. Kim, G.H. Water Solar Power Plant That Sits Down on the Ground Due to Drought. Available online: https://www.yna.co.kr/view/AKR20170626109700061 (accessed on 28 December 2018).

26. RETScreen International Clean Energy Decision Support Centre. *Clean Energy Project Analysis, RETScreen Engineering & Cases Textbook*; CANMET Energy Technology Centre: Montreal, QC, Canada, 2005.

27. Battaglia, C.; Cuevas, A.; De Wolf, S. High-efficiency crystalline silicon solar cells: Status and perspectives. *Energy Environ. Sci.* **2016**, *9*, 1552–1576. [CrossRef]

28. Yoshikawa, K.; Kawasaki, H.; Yoshida, W.; Irie, T.; Konishi, K.; Nakano, K.; Uto, T.; Adachi, D.; Kanematsu, M.; Uzu, H.; et al. Silicon heterojunction solar cell with interdigitated back contacts for a photoconversion efficiency over 26%. *Nat. Energy* **2017**, *2*, 17032. [CrossRef]

29. Lee, C.Y. *Equalization Cost through Solar Cost Analysis International Comparative Analysis*; Korea Energy Economics Institute: Ulsan, Korea, 2017.
30. MOTIE. *Guidelines for Management and Operation of New and Renewable Energy Supply Mandatory System and Fuel Mixture Mandatory System*; Ministry of Trade, Industry and Energy: Sejong, Korea, 2018.
31. Europe's Largest Floating Solar Farm to Open. Available online: https://www.bbc.com/news/uk-england-london-35705345 (accessed on 28 December 2018).

Review

Photovoltaic Power Systems Optimization Research Status: A Review of Criteria, Constrains, Models, Techniques, and Software Tools

Samer Alsadi [1] and Tamer Khatib [2,*]

[1] Department of Electrical Engineering, Palestine Technical University-Khodori, Tulkarm 97300, Palestine; samer_sadi@yahoo.com

[2] Department of Energy Engineering, An-Najah National University, Nablus 97300, Palestine

* Correspondence: t.khatib@najah.edu; Tel.: +970-59-931-7172

Received: 9 September 2018; Accepted: 24 September 2018; Published: 29 September 2018

Abstract: The photovoltaic (PV) generating system has high potential, since the system is clean, environmental friendly and has secure energy sources. There are two types of PV system, which are grid connected and standalone systems. In the grid connected photovoltaic system (GCPV), PV generator supplies power to the grid, whether or not the whole or a portion of the generated energy will be used to supply load demands. Meanwhile, the standalone photovoltaic system (SAPV) is used to fulfil a load demand that close to its point of use. These days, many researchers study in term of optimization sizing of photovoltaic system, in order to select optimum number of PV modules, inverter, battery storage capacity, and tilt angle. Based on that, this review aims to give explanations on approaches done by previous researchers in order to find ultimate combinations for design parameters. Moreover, the paper discusses on modelling of PV system components, which includes PV panels' output power estimation and battery system. Finally, simulation softwares that used as sizing tools in previous studies are reviewed and studied.

Keywords: photovoltaic; optimum sizing; design; renewable energy; stand alone; grid connected

---

1. Introduction

Rapid reduction of fossil fuel resources and growing evidence of global warming phenomena cause the necessity of urgent search for alternative energy sources. Recent studies shows that renewable energy has great potential and can be used to fulfil world energy demand [1]. According to [2], the PV industry has grown more than 40% per year since last decades due to rapid decrease in PV technology cost. PV technology may become major alternative energy source in the future since it has several positive attributes, low maintenances, free and inexhaustible energy source and robust and long life time system's components [3]. However, solar energy is not always reliable, because solar radiation varies and frequently changes, due to unpredictable nature and dependence on weather and climate changes. Hence, generated energy does not match with load demand all the time. Energy generated in PV systems depends mainly on solar energy available at the site. Geographical location, ambient temperature, clearness index, tilt and orientation of PV panel are the main factors that affect solar energy collected by a PV panel. Hence, studying these meteorological data is very crucial in preliminary design of a PV system. Other factors may affect energy generated, such as shading effect, cable reduction factor, system elements and configurations losses.

There are two types of PV power systems namely grid connected and standalone systems. Grid connected photovoltaic (GCPV) systems engage PV technology with electricity grid network. In the GCPV system, an inverter converts DC electricity generated by PV modules into AC electricity. Then, output power from inverter is fed to the utility grid [4]. On the other hand, standalone photovoltaic (SAPV) systems are off grid systems, where PV technology is not connected to the grid, and power generated is not sold to the utility [5]. Before recommending and installing a PV system, it is crucial to ensure that the system is not over/under sized. In other words, the designer has to investigate the viability of system carefully. To efficiently and economically use solar energy, optimal sizing of the system is necessary so that the proposed system can operate in optimum condition in term of produced units costs and power reliability [6]. Electricity consumption is very important in when it comes to system sizing, as well as economical analysis, since overproduction can affect the feasibility of the system negatively. Thus, large self-consumption is desirable so as to ensure the lowest investment with full use of PV array and/or battery bank.

Some review papers for PV system optimization can be found in the literature. In [7], a review of grid connected PV system in terms of technical and economic aspects was done. Electrical performance of PV modules, energy analysis, potential technical problems, and inverter's role in GCPV was explained exquisitely. However, the authors did discuss the optimization of these systems. In the meanwhile, photovoltaic technology and its power generating capability are reviewed in [8]. The authors discussed different existing PV systems performance and reliability, evaluation models, sizing and control methodologies, grid connection and distribution configurations. However, the authors did not give the major focus to the optimization techniques and constrains. On the other hand, in [9], general discussion for optimization methods applied to renewable energy is presented. However, the PV systems were not given enough focus in this paper. In [10,11], the use of artificial intelligence techniques for PV system optimization is reviewed. The work done in these reviews mainly focuses on the implementation of a specific technique without focusing on the optimization problem itself. Furthermore, in [11], there is no mention of optimization methods for a grid connected photovoltaic system. In [12], a detailed review of the PV system optimization is presented for standalone, hybrid and grid connected systems. However, in this review there is a lack of discussion on available optimization software tools, recent used techniques optimization constrains and systems models. In [13], a review on solar photovoltaic software tools was done. Software's accessibility, prices, working platform, capacities, scopes, resultants and updatability were discussed in this paper. However, the authors did not evaluate the software's limitations, advantages, simulation scopes and what type of system configurations that can be simulated by the software. Meanwhile, in [14], the authors reviewed 19 software tools for hybrid system analysis, with further evaluation on HOMER and RETScreen. Yet, the authors did not mention the software's cost and availability.

Based on that, this review aims to give explanations on approaches done by previous researchers in order to find ultimate combinations for design parameters. Moreover, the paper discusses on modelling of PV system components, which includes PV panels' output power estimation and battery system. Finally, as none of the aforementioned reviews has discussed the available PV softwares, simulation softwares that used as sizing tools in previous studies are reviewed and studied in this article.

2. Modelling of PV System's Components

In a PV system, main and storage energy sources' performance is dependent on each other. Hence, it is important to investigate PV system with and without battery storage system. To predict system performance, each component needs to be modelled first and then the combination can be evaluated whether it meets the design objectives or not. As a fact, if power output prediction is accurate enough, the resultant combination will deliver power with least cost.

2.1. Modelling of Photovoltaic Panel

The accuracy of a PV model has a great significance on system design. To predict the energy output of the system, researchers have to investigate meteorological condition at system's location. It is because the performance of PV modules strongly depends on the sun light condition and cell temperature. Moreover, the energy generated also depends on components' rated characteristics, installation configuration, and surroundings' condition [7].

PV power generated from a PV panel is highly affected by total of solar radiation received. Solar radiation data provides information on how much energy strikes the specific earth location [15]. However, solar radiation value is different if the panel slanted at different angles and orientations. Solar radiation data often measured in horizontal plane without any obstacle or shading effect. However, in some situations, when PV panels are mounted in tilted position, such as in building integrated photovoltaic system application, the total radiation input for the tilted PV modules has to be calculated [3]. Following this, Khatib T. et al. reviewed solar energy modelling techniques in [15]. The authors presented linear, nonlinear and artificial intelligence approach models for both global and diffused solar energy model on horizontal PV panel. The direct solar radiation comes directly from the sun. Meanwhile, diffuse solar radiation scattered from dome of the sky without direction. However, for a tilted PV panel, solar radiation comprises direct portion, diffuse portion and reflected portion [15–17]. To model diffused radiation on a tilt surface, isotropic and anisotropic models can be used [15]. According to [18], the isotropic model assumes the intensity of sky-diffused radiation is uniform over the sky dome. Hence, the diffuse radiation incident on a tilted PV module depends on the fraction of the sky dome seen by it. The second model is the anisotropic model, where it assumes the anisotropy of sky diffused radiation in circumsolar region and isotropic distributed diffuse component from the rest of sky dome.

The output power of a PV system is accessed via voltage and current produced from PV module where the multiplication of voltage and current produces power [19]. However, it is important to analyze power estimation accuracy, since the actual power output is usually lower than modelled one [5]. In [20], the author reviewed methods to calculate annual photovoltaic generation, by sorting all of the proposed methods into three categories. In the first method, the authors construct an I–V curve by using atmospheric parameter value, and power generated was calculated from it. Meanwhile, in the second method the power is directly calculated using time series atmospheric parameter values. On the other hand, the last method employs some technical factors such as performance ratio and system efficiency in calculating the power generated from a PV panel/array. In [1], the authors conduct a simple review on classical and novel modelling techniques applied for various types of photovoltaic systems such as meteorological data forecasting using artificial intelligence techniques.

In general, two circuit diagrams can represent PV cell model namely single-diode and two-diode models. Single diode model is one of the most commonly used PV generator's models. Both models are based on the fact that the solar cell is an illuminated p-n junction in the reverse-bias, connected to a resistive load. From [21], a solar cell can be represented using an equivalent circuit diagram consisted of a dependent current source connected on parallel with diode in the reverse mode. These two components are also connected in parallel with a relatively large resistor to represent the dark current of this p-n junction. In addition to that a resistance is connected in series to the aforementioned components so as to represents the internal resistance of the solar cell (see Figure 1), where I_{PH} is light generated current (A), D is diode, I_D is diode current due to the p-n junction under forward bias (A), R_S is series resistor (Ω), and R_{SH} is shunt resistors (Ω). The general I–V characteristic of a PV panel based on the single exponential model is:

$$i = I_{PH} - I_O \left(e^{\frac{v + iR_s}{n_s V_t}} - 1 \right) - \frac{v + iR_s}{R_{SH}} \tag{1}$$

where I_O is dark saturation current in STC, A is diode quality (ideality) factor, k is Boltzmann's constant, q is electron charge, n_s is number of cells in series, and T_{sc} is temperature at STC. V_t is the thermal voltage and it can be given by,

$$V_t = \frac{AkT_{stc}}{q} \tag{2}$$

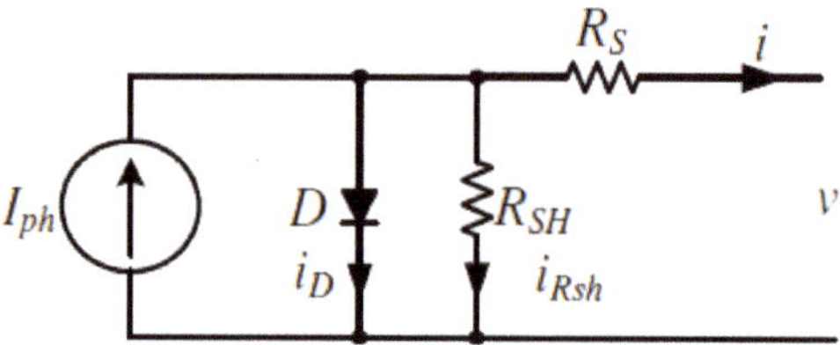

Figure 1. Electrical Equivalent of PV Cell on one-diode model [21].

On the other hand a diagram of a solar cell model using two diodes can be the same as Figure 1 but with additional diode in the parallel to the first diode. The general I–V characteristic of a PV panel based on the two diodes model can be described as,

$$i = I_{PH} - I_{O1}\left(e^{\frac{v+iR_s}{m_1 V_t}} - 1\right) - I_{O2}\left(e^{\frac{v+iR_s}{m_1 V_t}} - 1\right) - e^{\frac{v+iR_s}{R_P}} \tag{3}$$

However, one of the most important methods used to predict power generated from a photovoltaic panel is using current-voltage characteristic. Some researchers used maximum output power of the PV module to forecast PV system performance, calculated using datasheet's specification under standard test condition (STC). The calculation of maximum power output of a solar cell can be done by,

$$P_{mp\_stc} = V_{oc}I_{sc}FF \tag{4}$$

where P_{mp_stc} is maximum output power of solar cell (W), V_{oc} is solar cell open circuit voltage (V), I_{sc} is short circuit current of solar cell (A) and FF is fill factor [22]. FF is a term to show how much energy can be extracted from the module, and calculated as:

$$FF = \frac{P_{mp}}{V_{oc}I_{sc}} = \frac{I_{mp}V_{mp}}{V_{oc}I_{sc}} \tag{5}$$

Then, the mathematical model to estimate instantaneous PV module power output under real operating condition in term of P_{mp_stc} can be represented by following expression [5]:

$$P_{pv}(t) = \eta P_{mp\_stc} PSF \tag{6}$$

where PSF is peak sun factor (dimensionless) and η is de-rating factor. PSF was calculated using following expression:

$$PSF = \frac{G_{array\_plane}}{G_{stc}} \tag{7}$$

where G_{array_plane} is the irradiance measured in array plane, and G_{stc} is solar irradiance under standard test condition (1000 Wm^{-2}).

Meanwhile, the mathematical model of a solar cell power output at time t is:

$$P_{pv}(t) = V_{pv}(t)I_{pv}(t) \tag{8}$$

where P_{pv} is output power of solar cell (W), V_{pv} is solar cell operating voltage (V), and I_{pv} is output current of solar cell (A) [23].

In some researches, several authors foresee energy output performance instead of power prediction in their design. As in [24], the authors use energy produced by PV array, E_{pv} in term of derating factors, η, area, A (m^2) and solar energy, E_{Sun} as follows,

$$E_{PV}(t) = E_{Sun}(t)\eta A \tag{9}$$

Moreover, some authors prefer to use energy yield as indicator for system performance. Energy yield in a period of interest can be calculated using equation [25]:

$$E_{PV} = P_{mp\_stc}\eta PSH \tag{10}$$

where *PSH* is peak sun hour at specified tilt angle over a period (h). Here PSH at specific tilt angle means the number of hours that a PV panel that is slanted at a specific tilt angle can receive $1000\ \mathrm{W/m^2}$. This mean that the maximum total energy yield that can be provided by a PV panel at a specific tilt angle.

In [26], Mellit A. et al. proposed a model for photovoltaic system using adaptive artificial neural network, that combines Levenberg-Marquardt algorithm (LM) as learning algorithm and infinite impulse response (IIR) filter to accelerate convergence of the network [26]. SAPV experimental setup at Tahifet, Algeria validated the new proposed model. Solar radiation, ambient temperature and humidity are the simulation's input, and the outputs are voltage and current for photovoltaic generator. From the result, the authors prove that the proposed model is able to simulate different weather condition, and it is possible to generate current used by the load. Later, by using the same experimental setup for PV array area as before in [26], a similar case study with new modelling technique was introduced in [27], using Adaptive Neuro-Fuzzy Interference Scheme (ANFIS). ANFIS is a method that applies learning techniques in neural network and fuzzy inferences system, where the method can exploit both data and knowledge. From global radiation, ambient temperature and clearness index as an input, voltage and current for photovoltaic generator was generated. The result was proven to have reasonable accuracy compared to measured data, and more accurate compared to ANN model.

As mentioned before, the real energy yield is always lower than theoretical energy production, due to power derating factor, η such as losses caused by dirt, shading, mismatch factor, mounting condition, manufacture tolerance, cable loss, aging and inverter efficiency. Most models correlate de-rating factor with ambient temperature. Most studies calculate temperature losses factor using the following equation,

$$f_{temp} = 1 + \left[\gamma_{pmp}(T_{cell} - T_{STC}) \right] \tag{11}$$

where γ_{pmp} is power coefficient ($\%/\mathrm{C^{-1}}$), T_{cell} is temperature measured from the back of the module, and T_{STC} is temperature in standard test condition [4].

2.2. Modelling of Battery System

Energy storage is needed to supply load when a SAPV generates energy that is not sufficient to supply a load demand, and to store surplus power when there is excess energy generated by the system. For photovoltaic systems, there are three storage medium can be used, which is battery, fuel cell and supercapacitor. The storage medium is very important as an energy source, during low radiation in night or during autonomy days [28]. The most widely used storage technology in standalone PV systems is a lead acid battery, since it has high system reliability and long time services. However, lead acid battery may be damaged by poor charging control, which may cause overcharging or under-discharging [29].

One of the most earliest battery models in a PV system was presented in [30]. The researchers modelled the battery system based on its behaviour during charging and discharging. The behaviour is described as state of charge (SOC) as follows,

$$SOC(t) = SOC(t-1)\sigma + I_b(t)\Delta t\eta_b \tag{12}$$

where σ is self-discharge rate, I_b is battery current (charge during positive and discharge during negative), and η is battery efficiency.

In recent years, most studies for system planning and sizing used different expression of SOC. In low solar radiation intensity, load demand can be met if the battery has not reached the maximum allowable depth of discharge. Simple mathematical model for battery's state of charge at time t is represented by following equation [31,32],

$$SOC(t) = SOC(t-1) + \left[\frac{N_{pv}\, P_{pv}(t) - P_{load}(t)}{V_b * C_b} \right] \tag{13}$$

where N_{pv} is number of PV module, P_{load} is electric power demand (W), V_b is battery voltage (V), and C_b is capacity of battery bank (Ah). This model ignores charging, discharging and self-discharging efficiencies. In [33–35], a more detailed model was used. In each time step, charging and discharging efficiencies were applied in calculation.

In addition to that, energy management for standalone PV-Battery-Diesel was done in [36], exclusively for a configuration where battery and diesel generators are centralized, and PV and loads are distributed. In the proposed strategy, there are three different mode operations, which are normal operation, PV power limitation, and diesel generator. During normal operation, diesel generator is not connected, and the PV array supplies the load demand and battery is either charged or discharged subject to demanded energy. While in the second mode where battery is fully charged, a new strategy to limit distributed PV generation is presented by a controller scheme to avoid being overcharged or over-currents. When power limitation is necessary, a battery inverter will increase the frequency, and then PV inverters detect frequency increment, it will reduce their generated power. The system will operates in third mode, in case when PV generated power is not enough to supply demand and battery SOC is lower than its' lowest limit. Hence, PV and diesel generator will supply demands and charge battery, until fully charged. In this research different expression for battery state of charge was used. The SOC was determined by,

$$SOC(t) = SOC(t-1) + \left[\frac{\int I_b \eta_b dt}{C_b} \right] \tag{14}$$

Meanwhile, in [37], for standalone PV system, the authors estimated battery's energy flow as in equation below,

$$E_B(t) = \begin{cases} E_B(t-1)\eta_{inv}\eta_{wire}\eta_{discharging} - E_L(t)\ E_D < 0\ (discharge) \\ E_B(t-1)\eta_{charging} + E_{pv}(t)\ E_D > 0\ (charge) \\ E_B(t-1)\ E_D = 0 \end{cases} \tag{15}$$

where

$$E_D(t) = E_{pv}(t) - E_L(t) \tag{16}$$

η is derating factor, E_B is energy stored in battery, E_{PV} is energy generated from PV generation system, E_L is energy demand, and E_D is energy difference between E_{PV} and E_L. Energy stored in battery is depend on inverter efficiency, cable losses, battery's charging and discharging efficiency, load demand and energy generated from PV system.

Besides that, there are several other methods that can be used to model a battery system. A simulation model for grid tied residential PV battery system was done in [38], and evaluated using energy assessment criteria. The authors modelled their battery system using AC coupled battery system layout, multi-crystalline PV module technologies and lithium-based battery system. Energy assessment criteria, in term of self-consumption rate, s and self-sufficiency, d of battery were evaluated, using the following equations,

$$s = \frac{E_{DU} + E_{BC}}{E_{PV}} \tag{17}$$

$$d = \frac{E_{DU} + E_{BD}}{E_L} \tag{18}$$

where E_{DU} is energy directly used by load, E_{BC} is energy to charge battery, E_{BD} is energy discharged by battery, E_{PV} is energy generated by PV panel, and E_L is load demand. Energy assessment is important to monitor battery usage, in order to prevent energy saturation. However, in this battery model the authors neglect several factors that affect a battery's efficiency, like self-discharge power, temperature effect and battery age. Besides, there is no explanation for the method to model PV generated power, valuation of battery's state of charge, and energy provided from/to grid.

Another study for optimization of battery capacity in GCPV was done, by considering battery's operation scheduling and ageing effect on economic analysis [39]. Discrete dynamic equation for battery in day's time step, (i, t), can be given as,

$$\frac{E_B(i,t) - E_B(i, t - \Delta t)}{\Delta t} = P_{Bdc}(i,t) \tag{19}$$

where P_{Bdc} is power stored in battery, E_B is energy stored in battery, and t is time. If $P_{Bdc} > 0$, it is charging since energy stored in battery is increase, and vice versa. Power feed to AC bus from battery supply, P_{Bac} is given by,

$$P_{Bac}(i,t) = \begin{cases} \eta_{Bi} P_{Bdc}(i,t), \ if \ P_{Bdc}(i,t) < 0 \\ \frac{P_{Bdc}(i,t)}{\eta_{Bi}} \ otherwise \end{cases} \tag{20}$$

where η_{Bi} is battery bidirectional converter's efficiency. Besides that, the equation for battery's ageing model is described as,

$$\Delta C(i,t) = \begin{cases} \Delta C(i, t - \Delta t) - Z \frac{P_{Bac}(i,t)}{\eta_B} \Delta t, \ if \ P_{Bac}(i,t) < 0 \\ \Delta C(i, t - \Delta t), \ if \ P_{Bac}(i,t) > 0 \end{cases} \tag{21}$$

where C is usable battery capacity, η_B is conversion capacity of battery, and Z is ageing coefficient.

A study to compare two type of storage types for grid tied PV generation application, which are lead acid storage and hydrogen storage model are simulated in [40]. Equation (21) below represent voltage of lead acid battery, V_b, where battery current, I_b is positive during charging, negative during discharge, R is internal resistance, and V_o is equilibrium voltage.

$$V_b = V_o + RI_b \tag{22}$$

Meanwhile, in [41], the authors proven that CIEMAT (Copetti) model is able to present dynamic and complex lead acid battery operation. In this paper, the Copetti battery model was simulated and compared with two experimental PV system models. The model was presented by discharge, charge and overcharge process. However, the result was slightly over approximated charging.

A model of PV pumping system, including PV array, battery and electric motor were modelled in [42,43]. However, instead of using state of charge, the authors implemented state of voltage (SOV) to model the battery system. It is because, since battery's state of charge is dependent on voltage. In SOV calculation, battery voltage, V_b was estimated using the expression below,

$$V_b = V_o + K_e ln\left[1 - \frac{Q}{C * I_b}\right] + R_b I_b \tag{23}$$

where V_o is a constant that represents battery voltage at initial condition, $C*I_b$ is battery capacity as a function of current, coefficient K_e is model parameter, Q is exchanged electric charge, R_b is battery's internal resistor, and I_b is battery's current, where $I_b > 0$ during charge and $I_b < 0$ during discharge. Expression of Q is calculated using basic charge equation as below:

$$\frac{dQ}{dt} = |I_b| \tag{24}$$

A modified battery ageing model, based on Shepherd's initial model was developed in [44], by predicting the temporal variation, and controlling deterioration of battery parameters and performance [44]. In this paper, the authors predicted charge and discharge operation using correlation of voltage, current and battery state of charge. During discharge, battery's current is lower than zero, and vice versa during charging. Both Equations (25) and (26) represent battery output voltage at each time step, U (t) during discharge and charging process. Coefficient g is the coefficient with characterise $\Delta U = f$ (Q), R is internal resistance, I is current, t is time, T is temperature, M is slope of $U = f$ (t, I, Q) characteristic, and c and d represent charge and discharge.

$$U(t) = U_d - g_d\frac{It}{C} + R_d I\left[1 + \frac{M_d It}{C(1 + C_d) - I_t}\right]$$ (25)

$$U(t) = U_c - g_c\left(1 - \frac{It}{C}\right) + R_c I\left[1 + \frac{M_d It}{C * C_c - I_t}\right]$$ (26)

3. Available Photovoltaic Software

In fact, it is crucial to predict system's performance in a given location and expected operation condition. Moreover, system size must be determined in early stage of planning and designing. System performance modelling tools and computer simulations can be useful to designers or system integrators to predict energy output, as well as analyzing possible configuration [45,46]. Nowadays, variable of software tools exist to analyze, simulate, and design PV system. Most of the systems involve solar radiation estimation, while taking account of characteristic and location of PV system. The simulation software may expedite overall design process, compared to sizing intuitively using manual calculation.

There are five categories of software application, which are simulation tools, economic evaluation tools, analysis and planning tools, site analysis tools and solar radiation maps, as shown in Figure 2. In simulation, economic evaluation, and analysis and planning tools, the software divided into two categories, which is exclusively on PV system only, and simulation on hybrid technology options.

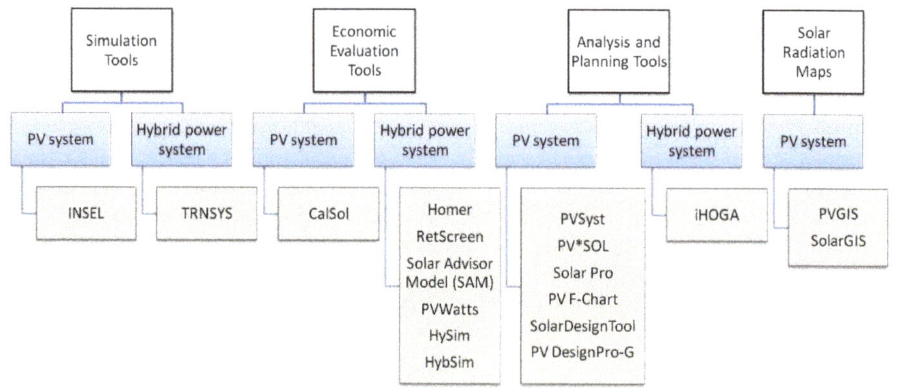

Figure 2. Classification of simulation software based on its application.

3.1. PV System Simulation Tools

Simulation tools are software programmed to simulate or predict power system performances as designed by users, using the best meteorological algorithm and/or collected meteorological database. The software were developed to simulate, analyze, monitor and visualize power system performances, but unable to optimize the system.

Integrated Simulation Environment Language (INSEL) is simulation software developed by the University of Oldenburg, Germany. INSEL is able to create system model and configurations for planning and monitoring electrical and thermal system. This software is suitable to simulate time series solar irradiance, PV power plant, solar cooling and heating system. The user will construct the proposed power system configuration by connecting blocks provided in the library. The advantage of this software is that it provides PV components database, such as PV modules, inverters, and thermal collectors. Besides, the software is also able to detect faults in the system. It is said that the software able to solve any computer simulation problem. The software has meteorological database from 2000 locations worldwide, and it can generate hourly irradiance, temperature, humidity and wind speed. The output consist graphical and numerical output. Price for full version software is 1700 euro and 85 euro for student. Trial version for this software is free and available in the website [13,47,48].

Another simulation tool available is Transient System Simulation (TRNSYS), which was developed by the University of Wisconsin, Madison, USA. Compared to INSEL, TRNSYS able to simulate cogeneration power system other than PV system and thermal system, including wind turbine, fuel cells, and batteries. The objectives of the software are to simulate low energy and HVAC system sizing, system analysis, multizone airflow analysis, electrical simulation, solar energy and thermal system design, and control scheme. This software is flexible, where the users able to modify mathematical model in its library. Besides, meteorological data and component designed by users are required as the input in simulation. A demo version for the software is available in its website, and the price for educational use is $2100 [13,14,49].

3.2. Economic Evaluation Tools

Software that classified as economic evaluation tools is able to provide economic analysis for the proposed system designed by the users. To determine whether the system is feasible or to maximize net benefit of consumption in the electricity services, user need to key in all of the cost and financial parameters as input, and then run the analysis. The analysis is needed to minimize total project costs to meet load demand or project constraints.

One suitable software for simulation and data analysis of PV system is CalSol. The founder of this program is from the Institute National de I'Energie Solaire (INES), France. This software is able to run economic analysis on grid connected, standalone and DC-grid system. Unfortunately, only French meteorological databases are available in this software. Besides, as PVWatts, this software has no PV component database and other programs unable to interconnect with this software. This software is easy to handle, suitable for pre-sizing and it is free, since the software is available via internet [3,50].

The Hybrid Optimization Model for Electrical Renewable (HOMER) is an energy modelling software, developed by National Renewable Energy Laboratory (NREL) USA. The software is suitable to design and analyze hybrid power system, including conventional generators, cogeneration, wind turbines, solar photovoltaic, hydropower, batteries, fuel cells, hydropower, biomass and others. HOMER can provide system optimization and technology options according to cost and energy resources availability. Moreover, this software able to simulate a system for 8760 h in a year and the results are presented in varies of tables and graphs. The meteorological data for the proposed site can be imported from HOMER energy website or provide specifically by the users. However, since Net Present Cost (NPC) analysis was used in economic analysis, comparison between other power system configurations in term of levelized cost of energy (LCOE) cannot be done. The developer provides a six months free version and then a renewal for 100 USD is required annually. [13,14,48,51]

In addition, RETScreen is also suitable as feasibility study tools for hybrid power generation technologies. Natural Resources Canada develops the software, and it is a Microsoft Excel-based spreadsheet model. The core of the tools consists of standardized and integrated clean energy project analysis that used in worldwide to evaluate energy production, life cycle cost, and green house gas emission reduction for various types of energy efficient and renewable energy technologies. The software also covers on grid and off grid analysis. Beside this, it has a global climate database for 6000 ground stations, and it also provides link to NASA climate database. Energy modelling analysis, cost analysis, emission analysis, financial analysis, sensitivity and risk analysis are available in this software. Unlike HOMER, the software can be used to compare levelized cost of energy (LCOE) between other power system configurations. RETScreen is a free of charge software, [13,14,48,52].

System Advisor Model (SAM) is free software, developed by National Renewable Energy Laboratory, Washington. The software is able to analyze all solar technologies, as well as provide intensive financing and cost analysis. The results from the analysis will be presented in term of levelized cost of energy (LCOE), system energy output, peak and annual system efficiency, and hourly system production, in tables and graphs. SAM can automatically download online database, including energy resources for solar, wind, bio fuel, geothermal, US incentives and US utility rates [13,53,54].

PVWatts is automated simple calculator software, it gives quick answer for the expected energy production, and cost saving in grid connected system. PVWatts is a simplified version of PVForm, and it can be used trough SAM or via the Internet. However, PVWatts only allows users to select a location within United States or pre-determined list of locations only. It is also only calculates crystalline-silicon PV modules [55,56].

Hybrid simulation (HybSim) is a hybrid energy simulator developed by Sandia National Laboratory, and suitable to simulate renewable energy such as PV, diesel generators, and battery storage at off grid system. It is able to perform financial analysis such as LCOE, life cycle cost, fuel and O&M costs, and cost comparison between different configurations. HySim is able to interconnect with other software. For example, HySim used weather and insolation data from TMY2. Unfortunately, it has not been used up until 1996 [14,48].

Meanwhile, hybrid simulation (HybSim) is a hybrid energy simulator developed and copyrighted by Sandia National Laboratory. It is suitable to simulate and analyze life cycle cost and benefit for adding renewable energy at off grid system. It is able to simulate PV, diesel generators, and battery storage. However, at this moment, only PV generation system can be generated, and wind turbine may be added in the future. For weather and insolation, HybSim use measured data for 15-min time intervals. It is able to compare cost and performance between diesel system only and hybrid system with combination of diesel, PV and batteries [14,48].

3.3. Planning and Analysis Tools

Planning and analysis tools are suitable to help users in planning, designing, sizing, optimizing sources, and defining proposed system to the highest standard. Some software provides database of PV components available from market to help users design their system in detail.

Due to diligence review in United States, PVsyst is the most commonly used for project development [45]. The founder of PVsyst is a graduate from the University of Geneva, Switzerland. PVsyst is software that able to size, design, simulate and analyze grid connected, standalone and DC-grid connected PV system. Beside Meteo Database existed in the software, this software can also import meteorological data available from web. The software also has a PV-components database that is available in the market. This system also provides variations of parameter settings possibility in the design with high precision result. However, this software does not allow interconnection with other programs. Besides, the sizing is also restricted to collector configuration. There are two types of PVsyst products, which are Pro30 (maximum 30 kW installation) and Premium. For Pro30 (unlimited installation), the price for first licence is CHF 1000, and CHF 1300 for premium product. While, for second license, the price is CHF 700 and CHF 1000, and for third licence, the price is CHF 500 and CHF 700 for both Pro30 and premium product [3,48,57,58].

PV*SOL Expert is a 3D design software for simulates and data analysis of roof-parallel and roof-integrated PV system performances. This software is suitable to plan PV system, from small off grid to large grid connected system. The developer is Solar Design Company from Powys, UK. This software also analyses shading effect on energy performances [13]. Besides, it also gives optimization configuration of PV modules among inverters. PV*SOL objectives include quick design, financial analysis and gives suitable proposal. The outputs included yield report, system efficiency, losses and economic reports. Unfortunately, interconnection with another program is not possible [59].

Laplace System Co., Ltd. (Kyoto, Japan), develops SolarPro Japan. Same as PV*SOL Expert, SolarPro is also 3D design software, and able to calculate shading losses. However, this software can only be used to simulate grid connected photovoltaic system only. Its database subsists meteorological from more than 8000 sites and PV components. SolarPro provides output in term of I–V curve, and power generation based on latitudes, longitudes, and weather conditions. Economic analysis in term of life cycle cost is also available in the software. The software price for educational use is $1900 [3,13,60].

PV F-Chart is an analysis and design program for PV system, where the company's principals are faculty members at the University of Wisconsin. The software's charge for single user is $400, and $600 for educational use. The program able to provides monthly average performance estimates hourly for the day. PV F-Chart can simulate system configurations like utility interface system, battery storage system and stand alone, restricted for PV system. Economic analysis is also included in the features, such as buy and sell cost difference, life cycle, initial investment and cash flow analysis. Its database consists of weather data from 300 sites, and can be added by users. The users can set load demand's hourly value for each month, and has statistical load variation. The output was accessible using graphical and numerical, in English and SI unit [48,61].

SolarDesignTool is an online PV design web, that available via the Internet. The software's price is $7 for lite series, $25 per month for professional series, and the users are able to have 30 days of free trial. It is suitable to simulate PV grid-tied system. This software allows users to design and configure optimal solar system and panel layout. In optimal system design, the software gives several recommendations for possible configurations. Meanwhile, in the system generator, the users also can specifically design their own system configuration, and simulate it. SolarDesignTool has climate database for US and Canada sites. However, the users still allowed to create a reusable custom location. In addition, the software is user friendly, where it can use 3D building model based on aerial and satellite imagery to create preliminary design. Other features available in the software are string configuration for string inverters and distributed MPPT inverter system, branch configuration for micro inverter, automatic optimal panel layout generation and embedded drawing tool to sketch or modify installation area [62].

PV DesignPro-G is able to design and analysis grid connected with no battery storage, from small to large-scale PV system. This software is one of three PV DesignPro included in Solar Design Studio 6.0 CD-Rom, where another two are PV DesignPro-S for stand alone, and PV DesignPro-P for water pumping. The price for the CD-Rom is $249. The software able to simulate a predesigned system with preselects climate condition, for one year, on hourly basis. It consist climate database for 239 locations in UA, Alaska, Hawaii, Puerto Rico, and Guam. Besides, the CD-Rom also provides hourly climate generator for 2132 international climates. Other than that, PV panel model was available in database for system simulation. The software also provides Annual Energy Cost analysis and Life Cycle Cost Analysis. The output was presented in tables and various chart [63].

Meanwhile, improved Hybrid Optimization by Genetic Algorithm (iHOGA) is a hybrid system optimization software that suitable to find optimum size of a proposed system. It is a C++-based software, and developed by the University of Zaragoza, Spain. It is suitable to analyze photovoltaic, wind turbine, hydroelectric turbine, fuel cells, H_2 tanks, electrolysers, storage system, and fossil fuels, for both stand-alone and grid connected system configurations. Besides, the software is able to simulate four different loads system, which are AC loads, DC loads, Hydrogen loads, and water pumping loads. Its merits are the software use multi or mono objectives optimization using genetic algorithm and sensitivity analysis with low computational time. The features are optimizing the best combination of system components and best control strategies. iHOGA also able to optimize panels slope, life cycle emission, allow probability analysis, and has purchase and selling options for grid connected. iHOGA's education version can be downloaded from the Internet for free. Unfortunately, as said in [14], the educational version has demerits, where it can only simulates daily load within 10 kWh, and sensitivity analysis, probability analysis, batteries models Coppetti and Schiffer, and net metering is not included. For professional's full version, the cost is €200 for a license, and for versions with full technical support is €300 per year [64].

In addition, PV.MY software is a MATLAB-based user friendly software tool called for optimal sizing of photovoltaic (PV) systems. The software is developed by the power research group of the national university of Malaysia. The software has the capabilities of predicting the metrological variables such as solar energy, ambient temperature and wind speed using artificial neural network (ANN), optimizes the PV module/array tilt angle, optimizes the inverter size and calculate optimal capacities of PV array, battery, wind turbine and diesel generator in hybrid PV systems. The ANN-based model for metrological prediction uses four meteorological variables, namely, sun shine ratio, day number and location coordinates. As for PV system sizing, iterative methods are used for determining the optimal sizing of three types of PV systems, which are standalone PV system, hybrid PV/wind system and hybrid PV/diesel generator system. The loss of load probability (LLP) technique is used for optimization in which the energy sources capacities are the variables to be optimized considering very low LLP. As for determining the optimal PV panels tilt angle and inverter size, the Liu and Jordan model for solar energy incident on a tilt surface is used in optimizing the monthly tilt angle, while a model for inverter efficiency curve is used in the optimization of inverter size [65] Finally, The Smart Grid Research group at Lakeside Labs has developed software called RAPS. RAPS is able to simulate a grid connected or standalone microgrid with solar, wind or other renewable energy sources. This software calculates the power generated by each source in the microgrid and then it conducts a power flow analysis. This software is helpful for optimal placement of distributed generation units in a micro grid. The software RAPS is designed for use in science and classroom with a simple to use graphical interface. It is an easy extendable framework that supports users in implementation of their own models, for grid-objects, and algorithms for grid controls [66].

3.4. Solar Radiation Maps

Solar radiation maps allow users to understand solar resources for every spot on Earth with a simple visual. There are two software programs available online for this purpose, which are PVGIS and SolarGIS. Both programs provide geographical information based on maps or satellite imaginary. Photovoltaic Geographical Information System (PVGIS) is online software that estimates solar radiation, provides solar radiation maps and calculates annual energy generated from grid connected and standalone PV system. The system was founded in Institute of Energy and Transport—European Commission and it is available on the Internet. This software provides simulation for grid connected PV system only. It is easy to handle, and as PVsyst, it able to import meteorological data. However, this software is exclusively used in Europe and Africa only and has no PV component database. This software also does not include energy de-rating factor and economic analysis. Other than that, no other software could interconnected with this program [3,67]. SolarGIS is also a free solar radiation map that is available online. SolarGIS's authors are also the main co-authors for the previous PVGIS. There are six applications for SolarGIS, which is iMaps, estimate energy output, monitor performances for existing power plants, purchase time series of meteorological data, receive real time performance monitoring for solar energy system, and data visualization on poster maps. Its database is a high resolution, and continuously updated on daily basis. The data is generated using in-house developed algorithms [68].

4. Photovoltaic Systems Optimization Criteria

To select an optimum combination to meet sizing constraint, it is necessary to evaluate power reliability and system cost analysis for the recommended system. An ideal combination for any PV system is made by the best compromise between two considered objectives, which is power reliability and system cost.

4.1. Reliability Analysis

In PV system design, especially in SAPV system, one of the most important aspects to ensure power system security is to analyze power supply availability. This is because solar energy production in one site is intermittent, and energy generated usually will not match with load demand. A reliable power system is a generation system that has sufficient power to feed load demand in a period. There are many methods to determine reliability of power system. The most popular methods to express system reliability are loss of load probability (LOLP) and loss of power supply probability (LPSP). In both methods, if the probability is 0, the load then will always be fulfilled, while if the probability is 1 then the load will never be fulfilled.

LOLP is a probability for the case when a load demand exceeds the generated power by the system. Here, the reliable PV system is defined as the system which is able to generate sufficient power (E_{PV}) to fulfil the demanded load (E_L) within time span. There are several researchers who apply this method in their proposed system's reliability analysis such as [26,69–82]. LOLP can be described as,

$$\text{LOLP} = \frac{\sum_{i=1}^{8760} Energy\ Deficit_i}{\sum_{i=1}^{8760} Energy\ Demand_j} \tag{27}$$

where,

$$Energy\ Deficit_i = \sum_{i=1}^{8760}(E_L(i) - E_{PV}(i)) \tag{28}$$

On the other hand, LPSP is defined as probability of the case when the system generates insufficient power to satisfy the load demand [83]. LPSP has been used in [84–87] to measure PV system reliability as illustrated below.

$$\text{LPSP} = \frac{\sum_{i=1}^{8760} LPS(t)}{\sum_{i=1}^{8760} E_L(t)} \tag{29}$$

where,

$$LPS(t) = E_L(t) - [(E_{PV}(t)\ \eta_{bat}) + E_B(t) - E_{B\ min}] \cdot [\eta_{inv}\ \eta_{wire}] \tag{30}$$

There are two approaches for LPSP application in standalone PV design, based on chronological simulation and probabilistic technique. Chronological simulation can present dynamic changing in system performances, for example, energy accumulation effect on battery. Nevertheless, this technique requires time series data in a certain period, and it needs more computational effort, compared to probabilistic technique [83,86,88,89]. On the other hand, probabilistic technique eliminates need for time series data and can assess long-term system performances. However, this technique not used in recent studies anymore, as it has a flaw, where researchers cannot observe dynamic changing in system performances.

Some of the researchers will analyze hourly, daily, or monthly energy generated and battery state of charge, by using PV modelling and battery storage modelling as in the previous section. Then, by using the results of energy generated and battery state of charge, reliability analysis within analysis's period, either LOLP or LPSP can be done.

4.2. System Cost Analysis

There are several economic analyses used by past researchers in order to find the optimum configuration of proposed system. This criterion was applicable at both GCPV and SAPV system. Economic feasibility analyses that frequently used in past years are Net Present Cost (NPC), Levelized Cost of Energy (LCOE), and Life Cycle Cost (LCC). Economic analysis is worth doing to determine whether the project has or has not acceptable investment. Sometimes, economic analysis was used after reliability analysis, in order to propose a system with high reliability and lowest cost [86].

NPC is the total present value of a time series and it was the development of discounter cash flow techniques. The net present worth is found by discounting all cash inflows and outflows, including cost of installation, replacement and maintenance, at an interest rate or internal rate of return (IRR) [90]. NPC can be described as,

$$\text{NPC}(i) = \sum_{n=0}^{N} \frac{A_n}{(1+i)^n} \tag{31}$$

where A_n is net flow cash at end of period n, i is IRR, and n is project lifetime.

LCOE (\$/kWh) defined as the average cost per kWh of useful electrical energy produced by the system when a lifetime, investment cost, replacement, operation and maintenance, and capital cost is considered [91]. LCOE method was frequently applied in past year's economic feasibility research, since the approach is very useful in comparing different generating technologies with different operating characteristic [6,86]. LCOE is calculated by dividing the produced electricity annualized cost on the total useful electrical energy generated. The mathematical model used to calculate LCOE is as follows,

$$\text{LCOE} = \frac{\sum_{n=1}^{N} \frac{I_n + O_n}{(1+r)^n}}{\sum_{n=1}^{N} \frac{P_n}{(1+r)^n}} \tag{32}$$

where N is economical lifetime of the system, I_n is the investment cost in year n, O_n is the maintenance and operational cost (O&M) in year n, P_n is the electricity production in year n, and r is the discount rate [2].

The third system cost analysis is life cycle cost (LCC). LCC is an estimation for sum of installation cost, operating and maintenance of an item for a period of time, and expressed in today's value [92]. Equation (33) is used to calculate LCC of a PV system,

$$\text{LCC} = C_{PV} + C_{bat} + C_{charger} + C_{inv} + C_{installation} + C_{batrep} + C_{PWO\&M} \tag{33}$$

where C_{PV} is PV array cost, C_{bat} is initial cost of batteries, $C_{charger}$ is cost of charger, C_{inv} is inverter cost, $C_{installation}$ is installation cost, C_{batrep} is battery replacement costs in present value, and $C_{PWO\&M}$ is operation and maintenance cost in present worth. $C_{PWO\&M}$'s equation is:

$$C_{PWO\&M} = \left(C_{\frac{O\&M}{Y}} \right) \left(\frac{1+i}{1+d} \right) \frac{\left[1 - \left(\frac{1+i}{1+d} \right) \right]}{\left[1 - \left(\frac{1+i}{1+d} \right)^N \right]} \tag{34}$$

Sometimes, some researchers also calculate annual basis expression of life cycle cost (ALCC), with this equation:

$$\text{ALCC} = \text{LCC} \frac{\left[1 - \left(\frac{1+i}{1+d} \right) \right]}{\left[1 - \left(\frac{1+i}{1+d} \right)^N \right]} \tag{35}$$

5. Standalone PV System Optimization Technique

To recommend an optimum configuration for SAPV system, the designer has to evaluate system design based on optimization variables. As mentioned in [6], as number of optimization variables increase, number of simulation and iteration will exponentially increase, as well as time and effort. Hence, to obtain the best system design as well as simplified sizing process, pioneer researchers introduced several techniques for system sizing calculation. In the SAPV system, there are three major methods which frequently used in former studies namely intuitive method, numerical method, and analytical method [70]. Table 1 shows a summary of merits and demerits of three main optimization techniques for better identification.

Table 1. Brief comparison on three main optimization techniques.

	Merits	Demerits
Intuitive	• Simple • Did not have to consider random nature of solar radiation	• Only suitable for rough estimation • Tend to oversize system • Cannot measure system reliability
Numerical	• Frequently used • Adequate analysis can be done • Accurate reliability analysis	• More complex • Need more time for calculation •
Analytical	• Present relationship between capacities and reliabilities • Sizing task become much simpler	• The relation cannot be applied at different sites • Had error function compared to numerical calculation

5.1. Intuitive Methods

The intuitive method is simple, easy to be implemented, and can be used to give rough suggestion for preliminary design. The sizing rules are base on designer's experience, using lowest performance either in a time period data or by directly using average value (daily, monthly, or annual) of solar irradiance. Hence, this method allocates the system to generate more power than required by a safety margins. The quantitative relationship between subsystem, such as generated power subsystem, battery's state of charge subsystem, or reliability subsystem, is not considered. Besides, this technique also does not consider random nature of solar radiation and meteorological condition. This method is not very popular because it is limited for rough estimation and preliminary design only. Besides, this method tends to give oversized design and has low reliability [12,34,79].

There are some studies for optimal sizing of a SAPV system using the intuitive method. Based on the records, the first application of intuitive technique was done in 1997 [93], where optimization of a SAPV system and tilt angle modules was done at five sites in Iran with different longitudes, latitudes and altitudes. First, the authors start optimizing the system by determining the optimum tilt angle between $0°$ to $90°$ for each sites, to maximize PV power generation. Then, PV array was chosen using the least PV size required to fulfil demand requirements, in order to minimize the cost. After that, by using statistical approach, number of successive cloudy days was decided. Based on that, battery size was determined, with expectation to fulfil demands during low intensity of solar radiation in cloudy days. However, this study only explains the PV and battery system sizing roughly without explanation of the modelling methods. Besides, the smallest PV array size was selected, and thus, it is very possible that the power generated by the system will be enough to supply the load, or unable to charge battery until having is enough charge to supply the load during cloudy days.

Afterwards in 2003, more detailed study was done using intuitive approach based on estimated monthly average of solar irradiation data in Dhaka, Bangladesh [94]. The researchers calculate array size by choosing a month with minimum solar radiation. Three designs of sizing methods to fulfil same daily load demands were developed and compared. In array sizing for first design, the authors first compared and choose a month with the worst peak sun hour (PSH) at tilt angle of site latitude, and tilt angle of $±15°$ to site latitude in summer and winter season. On the other hand, the second design's array sizing was done by manually chose worse radiation between $40°$ and $10°$ tilt angle with azimuth of $0°$, and for third design, same method was applied for $40°$ and $10°$ tilt angle with azimuth of $±45°$. Then, the battery capacity chosen is assumed be able to supply 2 to 3 autonomy days. Lastly, the design with minimum array and battery size was selected as best design. Since the comparison has too many variables, an exclusive comparison between designs is defeasible.

Later, a more neat sizing was done in 2009 [95]. Solar irradiation data in Egypt for south facing PV array tilted by site latitude was used and considered as optimum tilt angle in that particular region. To fulfil an average daily demand for a household in remote area, PV array was sized with consideration of temperature correction factor, PV efficiency, battery efficiency and inverter efficiency. Intuitively, the authors modelled battery that competent to fulfil demand continuously for largest possible period of cloudy days. Later, they analysed system's life cycle cost to calculate unit generated cost for the proposed system. However, based on findings, the system may over sized, since energy price is very high as compared to typical electricity price in Egypt.

5.2. Numerical Methods

Numerical approach is the most frequently used in optimal sizing techniques for SAPV. In this category, the design was simulated for each time step within a period. SAPV energy balance and battery's state is calculated and investigated. This technique is offer adequate and comprehensive analysis. It is very accurate, but the calculation is complex and need more time for calculation and simulation. Besides, since there are different kinds of approaches or manners applied in these techniques, the comparison between studies under this method is inconsistent. As shown in Table 1, there are 20 studies use numerical methods to determine optimal configuration in SAPV. However, before 1997, there were five studies on numerical method done.

Optimization of tilt angle, PV size and battery storage using Greek Island's monthly average meteorological data size was presented in [69]. The optimum design was set to be the combination with lowest life cycle cost, in a predefined loss of energy probability. Number of battery bank replacements was determined using life cycle cost calculation. In [96], a simple technique for SAPV sizing was developed. This study used 23 years of hourly insolation data from 20 different sites in US to develop correlations between variability in insolation and average monthly horizontal insolation. Then, sizing nomograms that give array size as function of average horizontal insolation and storage capacity was generated as function of LOLP for long time interval. In [72], a computer aided program was developed to find optimum PV tilted angle, PV array size and battery storage. The authors predicted solar radiation model based on clear sky model. The difference between PV power generated and load demand was used to size battery capacity. Meanwhile, in [71], optimization of SAPV system was done by dividing the regions into four zone, based on sky clearness index characteristic. For a given LOLP, many combinations of battery capacity and PV array peak power were determined, and the system with lowest total cost is the best configuration. Afterwards, in 1996, optimal sizing method for SAPV was applied in Corsica [97]. Similar with [71], design with the lowest cost of energy is the optimum configuration to supply 1 kWh load.

Shrestha G. et al. proposed new sizing method based on combination of PV panel sizing, battery storage sizing, charge regulator and load requirement in 1998 [73]. A stochastic model of PV generation was representing random behaviour of solar insolation. Required load demand also has been well defined in hourly basis. In this research, the system with minimum cost and minimum loss of load probability is the most favourable configuration. However, since the researchers used experimental load modelling, the LOLP and LOLH analysis are not too reliable.

Numerical code, named PHOTOV-III was presented to determine optimum configuration of SAPV system at Greece in 2004 [98]. The system objective is to operate in zero load rejections, with optimum tilt angle and minimum installation cost. In proposed system, hourly analysis using iterative technique to find all possible combinations of PV panel number and battery maximum size that able to perform zero load rejection operation. Then, based on analytical analysis, with minimum initial cost constraint, several alternatives were selected as possible optimum configuration. However, this method is not too convenience, since the possible configuration need to be selected manually.

A different simulation model for SAPV system sizing in Delhi with interconnection arrays was approached in 2005 [76]. For a predefined load, the optimum combinations of PV array and batteries for zero LOLP were calculated. Moreover, the authors compared fixed and tilted aperture arrays with single tracking aperture arrays, and simple system cost analysis was carried out on both systems.

In 2006 [99], a new sizing method was presented by using stochastic simulation for solar radiation. The proposed stochastic model is suitable to implement if original measured data is not available.

For a given reliability index, energy balance using different pairs of PV array capacity and battery capacity was done to find the optimum combination. Reliability index was calculated using ratio of hours when load is satisfied over total simulation hours. The authors analysed fifteen different sites in Greece.

A. Fragaki proposed a new sizing approach using numerical method in [100], similar to her previous work in [101] by analytical method in 2008. This study constructed sizing curves for minimum PV generator capacity and battery storage, without losing the load. Minimum PV generation required was sized at least equal to load demand, by using the worst month daily radiation. Storage prerequisite for minimum generator size was calculated from minimum battery SOC for each year based on historical solar radiation data. However, this sizing method may cause oversizing, and the PV generated power may not fully used during high solar radiation energy period.

Other than that, Celik A. et al. presented optimum sizing method based on six years of meteorological data from five sites in Turkey [78]. Meanwhile, the authors obtained load profiles from five households in Turkey. The system performance simulated using hourly solar irradiation and ambient temperature data. The scheme was simulated iteratively, by increasing PV size gradually, while battery capacity is remained constant for five days storage capacity. Hourly energy generated flow in and out from battery and its state of charge was calculated. Then, the researchers constructed sizing curves after LOLP and LCC analysis, where they calculated LCC and cost payback time for three different LOLP values for each site in life cycle assessment.

Technical and economic analysis for SAPV system in Malaysia were done in 2008 [84], by using annual daily radiation data. There are two design variables contemplated, which are tilt angle and battery's dead of discharge (DOD). For a given load demand, maximum DOD, and preset LPSP, the best consolidation of PV size and battery size was calculated iteratively for several tilt angle values. The optimum tilt angle was chosen based on the lowest unit cost value. Then, for preset LPSP and chosen tilt angle, the same analysis was repeated using several DOD values.

The same author from [84] proposed new method for SAPV system sizing in 2009 [85]. This method implemented graphical approaches, where two graphs for a given LPSP was constructed, which are array size versus battery size, and partially differentiated of system cost function graph. The Tangent point of the graphs was the optimum size for PV array and battery. This method was much better compared to his previous work, since the most advantageous system was chosen based on graphical method, where the cost function was already minimized using partially derivation.

Arun P. et al. [33] presented chance constrained programming approach to solve uncertainty problem in solar radiation. For a specific energy source, demand, desired confidence level and system characteristic, a sizing curve for all possible combinations of photovoltaic array rating and storage capacity was plotted. Then, design space for feasible design was determined from the curve. Optimum configuration for a predefined reliability level was the combination with lowest cost of energy. As mention in [82], the limitation in this study is PV model prediction using deterministic approach was too simple, and cannot represent uncertainty in solar radiation [87]. Hence, it is better for future researchers to use time series meteorological data or another PV model prediction. Moreover, preset LOLP value used is considered high compared to other researches.

Askari I.B. et al. proffered an orderly optimal sizing method for PV array and battery system in Kerman, Iran, based on the site's hourly solar radiation data [86]. Panel's tilt angle is fixed as site latitude. For a given value of LPSP, PV module size and battery was determined iteratively. Then for a given value of LPSP, chart of PV modules versus hours of battery autonomy and its LCOE's graph was plotted, and combination with minimum LCOE was selected as optimum configuration.

Later, in 2010, the authors [87] presented an optimization using the stochastic method in equipment's characteristic data (PV modules, battery, and inverters), average temperature, solar radiation and load profile analysis. The optimization was done based on economic and reliability analysis, similar to [85]. For a given LPSP, the researchers established possible combinations for PV array size and battery storage size, and then they selected optimal combination with minimum cost by partially derivation of system cost equation. Moreover, the proposed method compared with deterministic method, and their merit and demerit was listed.

Mellit A. et al. has done several studies on SAPV sizing by employed artificial neural network (ANN). The studies was done in year 2005 [74], 2007 [77], and 2010 [81], in Algeria. In [74,77], The ANN model inputs were site's latitude and longitude, and the outputs were sizing factor of PV system and storage system. Then, in [81], the model was improved to four input; latitude, longitude, altitude and LOLP, and the output was PV array sizing factor. Storage sizing factor was calculated using mathematical equation.

5.3. Analytical Methods

In SAPV system sizing, the analytical method was used to obtain a close relation or correlation in a form of equation between capacities and reliabilities. From [34], usually there are three approaches applied, which are the probability-based approach, empirical coefficient determination, and application of novel methods, such as artificial neural networks (ANN). This technique allows designers to simplify their sizing methods into a representative equation, reduce calculation or computer process and the sizing procedure is more accurate. However, this method is not very popular because the constructed expression is not flexible. It is because the expression is very exclusive and restricted for specific sites. Moreover, it is hard to produce the relationship between capacities and reliabilities. The expression of system sizing also may have small errors compared to result from iterative sizing process.

Before 1997, there were five studies used analytical method to find optimum sizing in SAPV system. Three studies recorded in 1984, one in 1987 and the last one was in 1992. In [102], an analytical method was presented to predict fraction of energy load covered by PV generation system. The researchers attained sizing's coefficients value from simulation based on long term Italian meteorological data. However, this model leads to an oversized system. Meanwhile, a similar analytical procedure was proposed in [103]. Analysis was done based on monthly average solar insolation at Italian sites as well. However, Bartoli's model produce too low sensibility to battery size, as analyzed in [70].

Loss of load probability model was derived by two-event probability density function for difference of PV generated and load demand in [104]. Battery storage size calculation for a given array size was developed. The authors constructed analytical expressions for the probability of system storage needed and auxiliary energy required to cover demands in that event. The method then extended in [105], by taken account the effect of day-to-day insolation values. In [106], a similar method was used, but the authors extended it using three states model. In this study, analytical approaches based on stochastic theory were examined and recommended.

In 1992, a new analytical model was proposed based on meteorological data from three different locations in Spain [70]. An expression to relate array capacity, storage sized and LOLP was illustrated, with four different coefficient inputs, varies based on sites. Spain coefficient maps for two LOLP values were constructed to represent SAPV system sizing.

After a long absence, in 2005, the authors proposed an ANN-based methodology to obtain LOLP curves for SAPV system in Spain by using the Multilayer Perceptron (MLP) approach. In 2003 [107], the author presented the three layer MLP structure consist two input only, which is ratio of accumulator capacity over average daily load consumption and LOLP. Then, the structure was improved with addition of yearly clearness index, as the third input. Last layer has one node, which is ratio of generator capacity over average load consumption. The LOLP curves obtained in these researches was compared with real curves and curves developed in Instituto de Energia Solar Madrid in [70]. By using different methods and consideration of clearness index, the proposed curves was more fit to the real curves, and improvement can be seen compared to previous ones.

In 2006, a sizing procedure was proposed, where a curve was constructed using combination of several climatic cycles at low daily solar radiation [101]. The study used long term daily solar radiation from London, England. The procedure started with expressed array size equation and battery size's range equation, in term of constant load. PV array sized based on assumption there were no variation of solar radiation, and load could be supplied by array sized. Meanwhile, battery sized to be able to supply load in low solar radiation time interval. Then, both equations were combined and expressed in a sizing curve. The area above the graph is the feasible design region. Then, sizing curves for three of the most prominent climatic cycles was constructed, and then new smooth line was created based on curves' tangent. Unfortunately, this developed expression is only applicable in South East England. The system may be oversized or undersized, since the authors only considered three period of climatic cycle. It seems that the missing years are neglected because relatively high solar radiation and/or short climatic cycles.

There are two approaches for SAPV system sizing at 20 different locations in Spain proposed by the same authors in 2008 [79,80]. For both researches, the authors developed sizing curves, to predict LOLP, standard deviation of LOLP, failures, and standard deviation of failures, with a given tilt angle, predetermined PV generator capacity, and battery storage capacity in specific location. In [80], a new analytical analysis on daily energy balance analysis was proposed using variable monthly demand and tilt angle (varied based on season). Based on daily incident energy on tilted PV panel and daily storage state, LOLP and number of expected failures were calculated. Then LOLP's and failures' standard deviation were calculated. Meanwhile, compared to [79], the steps used was almost the same, but this technique consider monthly average clearness index in system analysis. Both studies analyzed relation between PV generation capacity, storage capacity and LOLP analysis. Nevertheless, the second paper has further analysis in dependence between PV generation capacity, LOLP, and storage capacity with daily radiation on tilted PV panel. However, the researchers did not conduct optimization in system sizing in both publications.

After a long absence, in 2012, the authors purposed an improved technique to present an inclusive sizing for standalone system for five locations in Malaysia [82]. By using iterative approach, two graphs were plotted to show relation of PV array sizes over LOLP, and relation between PV sizes and optimum battery sizes, based on average annual daily meteorological data. By MATLAB fitting toolbox, two formulas for optimum ratio of PV capacity and battery capacity over specific load value was derived. This research attained average coefficients for different five sites. From the finding, relation between ratios of PV generated capacity and load demand is exponential with LOLP, while relationship between PV capacity and battery capacity ratio over specific load demand is linear. However, as mentioned in [12], the limitation for the method used are the sizing was done based on daily solar meteorological energy and a constant value of daily load demand. Table 2 shows a summary of optimal sizing methods presented for SAPV.

Table 2. Summary of SAPV optimal sizing methods.

Year	Authors	Technique	Reliability Analysis	System Cost Analysis	Reference
1984	Barra, L., et al.	Analaytical			[102]
1984	Bartoli, B., et al.	Analaytical			[103]
1984	Bucciarelli Jr, et al.	Analaytical			[104]
1986	Bucciarelli Jr, et al.	Analaytical			[105]
1987	Gordon, J.	Analaytical			[106]
1988	Soras, C., et al.	Numerical	LOLP	LCC	[69]
1989	Chapman, R.N.	Numerical			[96]
1992	Egido, M., et al.	Analaytical	LOLP		[70]
1995	Elsheikh Ibrahim, et al.	Numerical	LOLP		[72]
1995	Hadj Arab, A., et al.	Numerical	LOLP		[97]
1996	Notton, G., et al.	Numerical		LCOE	[97]
1997	Samimi, J., et al.	Intuitive			[92]
1998	Shrestha, G., et al.	Numerical	LOLP		[73]
2003	Bhuiyan, M., et al.	Intuitive			[94]
2004	Kaldellis, J.	Numerical			[98]
2005	Hontoria, L., et al.	Analaytical	LOLP		[75]
2005	Mellit, A., et al.	Numerical	LOLP		[74]
2005	Kaushika, N., et al.	Numerical	LOLP		[76]
2006	Markvart, T., et al.	Analaytical			[101]
2006	Balouktsis, A., et al.	Numerical			[99]
2007	Mellit, A., et al.	Numerical	LOLP		[77]
2007	Mellit, A., et al.	Numerical	LOLP		[26]
2008	Fragaki, A., et al.	Numerical	Others		[100]
2008	Celik, A., et al.	Numerical	LOLP	LCC	[78]
2008	Weixiang, Shen.	Numerical	LPSP	LCC	[84]
2008	Posadillo, R., et al.	Analaytical	LOLP		[79]
2008	Posadillo, R., et al.	Analaytical	LOLP		[80]
2009	Shen, W.	Numerical	LPSP	Others	[85]
2009	Arun, P., et al.	Numerical	Others		[33]
2009	Askari, I.B., et al.	Numerical	LPSP	LCOE	[86]
2009	Nafeh, A.	Intuitive		LCC	[95]
2010	Mellit, A.	Numerical	LOLP		[81]
2010	Cabral, C.V.T., et al.	Numerical	LPSP	Others	[87]
2012	Khatib, T, et al.	Analaytical	LOLP		[82]

6. Grid Connected PV System Optimization Technique

The system modelling for GCPV system can be the same as SAPV. Latest trend for GCPV optimum sizing is most research was done to find the optimum distribution of PV modules among inverter, and sizing within predefined space area (roof or land). However, GCPV size optimization methods cannot be categorized as SAPV. It is because most researches were done using artificial intelligence method, since in GCPV system sizing, there are many variables need to be optimized.

6.1. Numerical Methods

As mention in Section 5.2, numerical is a detail simulation is done within a period for each time step, where the calculation may become too complicated if there are too many variables to investigate. In 2005, design optimization of GCPV system on top of Federal Office Building in Carbondale [108] was presented. Optimum array size and array size was determined by maximizing array output energy, and minimizing electricity sold to grid. To evaluate ratio of energy generated used by building, the effectiveness ratio was considered. Effectiveness factor was defined as ratio of energy used from generated energy to supply load over total PV array output. Array size was determined using graph of electricity sold. The limitation of the proposed method is the authors choose the optimum PV and inverter size manually, without fine explanation for optimization criteria and inverter to PV sizing ratio.

In [2], numerical method was presented in 2013 to optimize PV system based on roof characteristics, either it is on flat or slightly tilted roof. The validation of developed simulation was done based on PVGIS, PVsyst and site measured data from several Swedish PV installations. Electricity generation profile was refined for all possible combinations constructed with different PV types, panel tilt angle, ground covering ratio, and azimuth. After evaluation of economical and physical aspects, optimum distribution under different conditions was found, and several alternatives were developed. This research also investigates three case studies, with different kind of roof characteristics. However, this study did not involve inverter-to-PV array sizing. Besides, the technique is very complicated, and need systematic calculations.

6.2. Computer Aided Method

Optimization sizing of GCPV system using computer aided design was implemented in 2006 [109]. The authors developed sizing strategies for PV installation by considering several economic issues as initial cost, payback periods and compensation of reactive energy. In modelling PV energy production, losses due to shading, conductors, tilted angle and orientation, and inverter efficiency were considered. Monthly energy generated and grid delivered power, fee, bill and income was analyzed. Many data variables were considered, which leads to a reliable results.

6.3. Genetic Algorithm (GA)

GA is an artificial intelligence technique (AI), which was inspired from evolution and inheritance trait in living organism. The algorithm imitates population's evolution, based on survival-of-the-fittest strategy. GA has three operations, which is selection, crossover and mutation. GA method has ability to derive global optimum solution with relative computational simplicity, including complicated problems with non-linear function or constraint. However, GA may suffer excessive complexity if the problem is too large [1].

In 2009 [22], the authors developed an optimum sizing method using genetic algorithm (GA) among list of commercial system devices (PV modules and inverters) in Microsoft Visual C++ software. The sizing process was started by selection of PV module and inverter model, available land area, climatic parameters, cost and economy parameters. The best number of PV modules, inverters, tilt angle, the best arrangement of PV modules among inverters, and optimum arrangement of PV modules within site area was elected based on maximum net economic profit as GA objective function. In this study, total net profit was maximised using NPC analysis, by considering total capital cost of PV modules, inverters, cost of land area, mounting structure cost, cost of installation, and maintenance cost. However, the researchers did not include inverter to PV array sizing ratio as one of optimization criteria. In addition, several power losses were not taken into account, such as temperature, dirt and cables. Moreover, the PV modules arrangement calculation is not suitable if the land area is not face to south.

In 2011 [25], a new sizing method was proposed to determine optimal PV modules and inverters by using GA. In this study, authors developed iterative sizing procedure as a benchmark to choose the best optimization methods between GA and evolutionary studies (ES). The proposed GCPV system capacity capable to fulfil energy requirement as specified by customer in kWh using preselect PV module model and inverter model. Peak Sun Hour at tilted panel angle, temperature reduction factor, manufacture's tolerance reduction factor, cable reduction factor, and inverter reduction factor was predefined was considered in sizing calculation. From possible PV modules arrangement among inverters, the highest value of inverter to array sizing factor and minimum excess factor was chosen as optimal design solution. From the methods comparison, GA method has lower percentage error compared to ES method.

In 2011, authors in [110] proposed optimization methods for large PV plant design using Greece's time series of one minute average solar radiation data. The researchers optimize the proposed system by maximizing energy production. From simulation, optimum PV module distribution, optimum PV rows, distance between rows and tilt angle was determined. Then, LCOE was analyzed for the optimal solution. The proposed method can only be used to design PV generation plant at fixed tilted panel mounted on South faced ground. PV distribution system among inverters and PV arrangement layout at predefined space area was done similar with methods in [22,111,112], except the inverter used in design are Central Inverter, Multi-string Inverter and Mini-Central Inverter.

In 2012, the same authors from [110] proposed a new optimization method for large PV plant in [113]. By using GA simulation, optimum PV module distribution, optimum PV rows, distance between rows and tilt angle was determined. However, the new technique was improved by including LCOE analysis within optimization process. The simulation result was compared to cases with non-optimized system, optimized for minimum cost only and optimized for maximum cost only.

An inclusive GA method presented in 2014, in [3] where the authors use all technical, economical and environmental criteria in their analysis. The authors developed their solar irradiance model using hourly average temperature and clearness index data Then PV model was developed to maximize energy output using field area restriction and design safety restriction, for predefined PV module model and inverter model characteristic. The PV layout design within space provided was optimized to reduce shading losses. Then, based on the configuration obtained from PV model, evaluation on economic, technical and environmental criteria was done. The authors used IMPACT2002++ for environment evaluation. All of the listed steps were conducted using five different types of PV panel technologies. The best configuration selection was based on weighted evaluation with 15 goals, which is maximizing energy output, minimizing payback time, minimizing energy payback time, and minimizing 12 environmental impacts.

6.4. Particle Swarm Optimization (PSO)

PSO is one of the metaheuristic methods using robust stochastic optimization technique based on the movement and intelligence of swarms. The technique applies concept of social interaction to problem solving. As mentioned in [111], PSO is easily programmed. Besides, the knowledge of good solution is retained and all particles able to share information between them. Meanwhile, in [112], multi-objectives was implemented, to allow optimization of two or more conflict objectives.

In 2010, Kornelakis, A., et al. developed two more optimization methods, based on particle swarm optimization (PSO) and multi-objectives particle swarm optimization technique [111,112], with several improvements compared to his previous proposed method in [22]. In [111], the authors compared his PSO-based findings with GA-based result in term of iterations. They have proved that by using PSO approach, the simulation time is shorter and lower number of iteration was obtained. Besides, GA was proved failed to locate several optimum solutions during optimisation process. In his later findings [112], environmental benefit was added in his optimization process as second objective function, while the decision variables is still the same. Particle swamp optimization was used to solve multi-objectives problem for purposed system. This technique was able to maximise both economic and environmental benefits in the system.

6.5. Evolutionary Programming (EP)

Similar to GA technique, EP also involves random process of selection, mutation and crossover in its operation. Individual fitness was defined in objectives function. EP evaluates behavioural connection between parents and offspring, with respect to similarities, differences and their performances [114].

In 2012 [4], Sulaiman, S.I., et al. proposed an optimization sizing with Evolutionary Programming Sizing Algorithm (EPSA). Unlike his previous method in [25], the optimization was able to test all available combination of PV and inverters in system database. The researchers implemented the same technique to determine optimal PV modules distributions among inverter. EPSA sizing was done using different EP modes with nonlinear step size scaling factor (NPSS), and the sizing results were compared. The proposed method can choose to optimize the system based on energy yield or net present value. Table 3 shows a summary of GCPV optimal sizing methods.

Table 3. Summary of GCPV optimal sizing methods.

Year	Authors	Technique	System Cost Analysis	Reference
2005	Gong, X., et al.	Numerical	Others	[108]
2006	Fernández-Infantes, A., et al.	Evolutionary Programming	NPC	[109]
2009	Kornelakis, A., et al.	GA	NPC	[22]
2010	Kornelakis, A.	PSO	NPC	[111]
2010	Kornelakis, A., et al.	Multi-Objective	NPC	[112]
2011	Sulaiman, S.I., et al.	GA		[25]
2011	Kerekes, Tamas, et al.	GA	LCOE	[110]
2012	Kerekes, T., et al.	GA	LCOE	[112]
2012	Sulaiman, S.I., et al.	Evolutionary Programming	NPC	[4]
2013	Näsvall, D.	Numerical	LCOE	[2]
2014	Perez-Gallardo, J.R., et al.	GA	Others	[3]

7. Sizing Constraint

In PV system design, sometimes there are limitations or constraints that need to be explored. In past studies, available space, budget and energy demand are considered before the optimization of PV system design.

7.1. Space Constraint

Sometimes, customers give specific area to mount PV modules. In past years, several researchers included area as a constraint in their sizing. There are two types of area sizing for PV system, which are roof space's area constraint and land space area constraint. As referred in [108], optimization process for grid connected PV system on the rooftop of federal office building. In the study, array size and tilt angle are optimized, and the objectives for the model are to maximize array output energy as well as minimize electricity sold to grid.

Meanwhile, in [2,4], detailed sizing on top of roof was done based on PV module orientation. As shown in Figure 3, the roof space was used by choosing between two arrangements that can maximize panel's quantity on top of the roof, where L is roof length, W is roof width, l is panel length, and w is panel width.

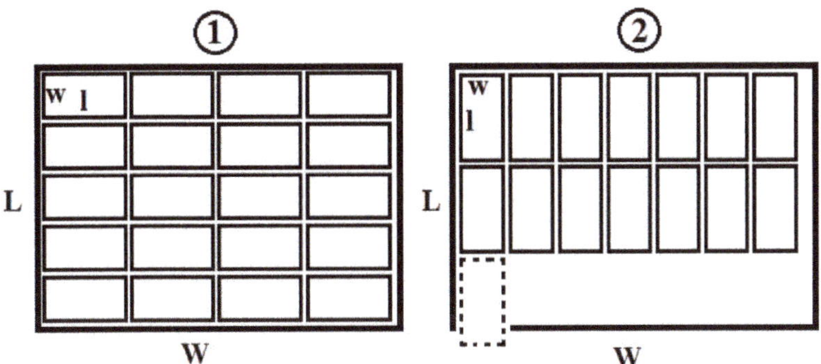

Figure 3. PV system layout options based on roof space constraint in system sizing [2].

Moreover, authors in [2] also proposed to consider PV system layout as shown in Figure 4, to allow PV panel mounted at optimum tilt angle on flat roof. Calculation was done to minimize shading losses while area was exploited at the same time, where L is roof length, W is panel width, and β is module tilt angle.

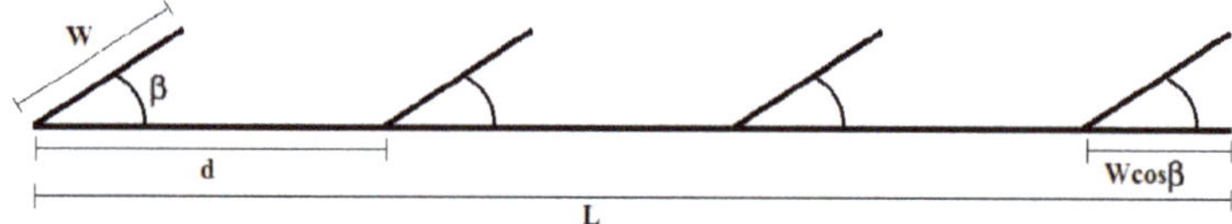

Figure 4. PV tilted system layout options based on roof space constraint in system sizing [2].

Several past studies [3,22,109–112] designed more complicated sizing in land space area, where PV modules tilt angle, maximum land space area, and shading effect was considered in order to find optimal arrangement of PV modules in the available installation area. The system layout is as shown in Figure 5 below, where D is space between collector rows, L_c is collector length, H is collector height, E_{max} is limit of collector's height from ground, W is land width, β is module tilt angle, K is rows number, L_m is module length, and H_m is module width. However, this technique has limitation, where the researchers can only use the calculation and layout for land in south direction only.

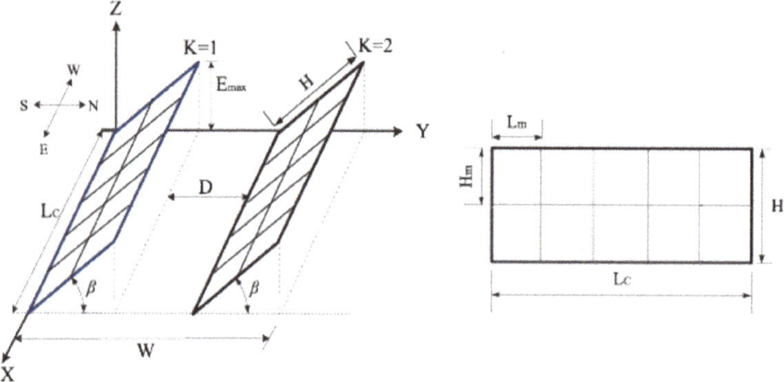

Figure 5. PV tilted system layout options based on land space constraint in system sizing [3].

7.2. Budget Constraint

In [109], a grid connected PV system was designed by considering economical element. The authors design a computer application to automatically calculate all relevant parameters in installation, such as physical, electrical, economical and ecological. The researchers included the limitation in installation budget during sizing process.

7.3. Energy Constraint

In most studies usually use one load profile as a reference or constraint in their sizing method. Energy constraint usually applied in standalone system, such as in [82,84–86]. However, grid connected system also can use electric load profile of the building for reference, as used in [25,108]. In both studies, researchers sized PV array based on energy required from customer. They selected optimal PV system based on minimization of difference between energy required and energy produced. Besides that, in [79,80], an analytical method was done in sizing for standalone PV system with variable monthly demand. This demand was varied depend on season, monthly varies demand was use in system sizing and analysis.

However, other energy constraints have been considered recently such as a bidirectional power flow in PV system that contains normal electric vehicles (EV) or vehicle to grid EV [115,116]. Similarly having mechanical energy sources combined with the PV system may also affect the constraints such as in [117].

8. Conclusions

This paper provides a review on SAPV and GCPV systems sizing procedures, which are system components modelling, optimization software available, optimization criteria, optimization method, and sizing constraint. PV modelling and battery modelling is important in system sizing optimization, in order to predict system performances. Besides, by using a suitable optimization software package, the system process is express and simpler. Other than that, the most commonly used criteria in finding optimum solution for system sizing were investigated. For SAPV methods, only three optimal sizing methods were used for the past decades. Compared to GCPV system, even though the researches is not as many as SAPV, different optimization method was used in finding optimal sizing, including artificial intelligence (AI) methods, metaheuristic methods, and multi-objective (MO) design. System constraint is an optional process in PV system sizing, based on customer's requirements. Energy constraint was commonly used in SAPV sizing. In the other hand, past researches frequently employ space constraint in GCPV. According to the review in this paper, optimal sizing of proposed system is important as early step in PV system design. An accurate sizing may prevent oversize or undersize, which leads to development of high reliability system with low cost. Besides, AI and metaheuristic methods are suggested to be included in both SAPV and GCPV system to improve sizing process and optimization results in the future.

Funding: The authors would like to acknowledge funding from An-Najah National University grant No. ANNU-1819-Sc010.

Conflicts of Interest: The authors declare no conflict of interest.

References

1. Mellit, A.; Kalogirou, S.A. Artificial intelligence techniques for photovoltaic applications: A review. *Prog. Energy Combust. Sci.* **2008**, *34*, 574–632. [CrossRef]
2. Näsvall, D. Development of a Model for Physical and Economical Optimization of Distributed PV Systems. Master's Thesis, Uppsala University, Uppsala, Sweden, 2013.
3. Perez-Gallardo, J.; Azzaro-Pantel, C.; Astier, S.; Domenech, S.; Aguilar-Lasserre, A. Ecodesign of photovoltaic grid-connected systems. *Renew. Energy* **2014**, *64*, 82–97. [CrossRef]
4. Sulaiman, S.I.; Rahman, T.K.A.; Musirin, I.; Shaari, S.; Sopian, K. An intelligent method for sizing optimization in grid-connected photovoltaic system. *Sol. Energy* **2012**, *86*, 2067–2082. [CrossRef]
5. Zainuddin, H.; Shaari, S.; Omar, A.M.; Sulaiman, S.I. Power prediction for grid-connected photovoltaic system in Malaysia. In Proceedings of the 2011 3rd International Symposium & Exhibition in Sustainable Energy & Environment (ISESEE), Melaka, Malaysia, 1–3 June 2011; pp. 110–113.
6. Zhou, W.; Lou, C.; Li, Z.; Lu, L.; Yang, H. Current status of research on optimum sizing of stand-alone hybrid solar–wind power generation systems. *Appl. Energy* **2010**, *87*, 380–389. [CrossRef]
7. Eltawil, M.A.; Zhao, Z. Grid-connected photovoltaic power systems: Technical and potential problems—A review. *Renew. Sustain. Energy Rev.* **2010**, *14*, 112–129. [CrossRef]
8. Parida, B.; Iniyan, S.; Goic, R. A review of solar photovoltaic technologies. *Renew. Sustain. Energy Rev.* **2011**, *15*, 1625–1636. [CrossRef]
9. Baños, R.; Manzano-Agugliaro, F.; Montoya, F.; Gil, C.; Alcayde, A.; Gómez, J. Optimization methods applied to renewable and sustainable energy: A review. *Renew. Sustain. Energy Rev.* **2011**, *15*, 1753–1766. [CrossRef]
10. Khare, A.; Rangnekar, S. A review of particle swarm optimization and its applications in solar photovoltaic system. *Appl. Soft Comput.* **2013**, *13*, 2997–3006. [CrossRef]
11. Mellit, A.; Kalogirou, S.; Hontoria, L.; Shaari, S. Artificial intelligence techniques for sizing photovoltaic systems: A review. *Renew. Sustain. Energy Rev.* **2009**, *13*, 406–419. [CrossRef]
12. Khatib, T.; Mohamed, A.; Sopian, K. A review of photovoltaic systems size optimization techniques. *Renew. Sustain. Energy Rev.* **2013**, *22*, 454–465. [CrossRef]
13. Lalwani, M.; Kothari, D.; Singh, M. Investigation of solar photovoltaic simulation softwares. *Int. J. Appl. Eng. Res.* **2010**, *1*, 585–601.

14. Sinha, S.; Chandel, S. Review of software tools for hybrid renewable energy systems. *Renew. Sustain. Energy Rev.* **2014**, *32*, 192–205. [CrossRef]
15. Khatib, T.; Mohamed, A.; Sopian, K. A review of solar energy modeling techniques. *Renew. Sustain. Energy Rev.* **2012**, *16*, 2864–2869. [CrossRef]
16. Demain, C.; Journée, M.; Bertrand, C. Evaluation of different models to estimate the global solar radiation on inclined surfaces. *Renew. Energy* **2013**, *50*, 710–721. [CrossRef]
17. Duffie, J.A.; Beckman, W.A. *Solar Engineering of Thermal Processes*; John Wiley & Sons: Hoboken, NJ, USA, 2013.
18. Noorian, A.M.; Moradi, I.; Kamali, G.A. Evaluation of 12 models to estimate hourly diffuse irradiation on inclined surfaces. *Renew. Energy* **2008**, *33*, 1406–1412. [CrossRef]
19. Abdulkadir, M.; Samosir, A.; Yatim, A. Modeling and Simulation based Approach of Photovoltaic system in Simulink model. *ARPN J. Eng. Appl. Sci.* **2012**, *7.5*, 616–623.
20. Rus-Casas, C.; Aguilar, J.; Rodrigo, P.; Almonacid, F.; Pérez-Higueras, P. Classification of methods for annual energy harvesting calculations of photovoltaic generators. *Energy Convers. Manag.* **2014**, *78*, 527–536. [CrossRef]
21. Sera, D.; Teodorescu, R.; Rodriguez, P. PV panel model based on datasheet values. In Proceedings of the 2007 IEEE International Symposium on Industrial Electronics, Vigo, Spain, 4–7 June 2007; pp. 2392–2396.
22. Kornelakis, A.; Koutroulis, E. Methodology for the design optimisation and the economic analysis of grid-connected photovoltaic systems. *IET Renew. Power Gener.* **2009**, *3*, 476–492. [CrossRef]
23. Nema, P.; Nema, R.; Rangnekar, S. A current and future state of art development of hybrid energy system using wind and PV-solar: A review. *Renew. Sustain. Energy Rev.* **2009**, *13*, 2096–2103. [CrossRef]
24. Khatib, T.; Mohamed, A.; Sopian, K.; Mahmoud, M. Optimal sizing of the energy sources in hybrid PV/diesel systems: A case study for Malaysia. *Int. J. Green Energy* **2013**, *10*, 41–52. [CrossRef]
25. Sulaiman, S.I.; Rahman, T.K.A.; Musirin, I.; Shaari, S. Sizing grid-connected photovoltaic system using genetic algorithm. In Proceedings of the 2011 IEEE Symposium on Industrial Electronics and Applications, Langkawi, Malaysia, 25–28 September 2011; pp. 505–509.
26. Mellit, A.; Benghanem, M.; Kalogirou, S.A. Modeling and simulation of a stand-alone photovoltaic system using an adaptive artificial neural network: Proposition for a new sizing procedure. *Renew. Energy* **2007**, *32*, 285–313. [CrossRef]
27. Mellit, A.; Kalogirou, S.A. ANFIS-based modelling for photovoltaic power supply system: A case study. *Renew. Energy* **2011**, *36*, 250–258. [CrossRef]
28. Sonnenenergie, D.G.F. *Planning and Installing Photovoltaic Systems: A Guide for Installers, Architects and Engineers*; Earthscan: Oxford, UK, 2008.
29. Huang, B.; Hsu, P.; Wu, M.; Ho, P. System dynamic model and charging control of lead-acid battery for stand-alone solar PV system. *Sol. Energy* **2010**, *84*, 822–830. [CrossRef]
30. Zahedi, A. Development of an electrical model for a PV/battery system for performance prediction. *Renew. Energy* **1998**, *15*, 531–534. [CrossRef]
31. Li, J.; Wei, W.; Xiang, J. A Simple Sizing Algorithm for Stand-Alone PV/Wind/Battery Hybrid Microgrids. *Energies* **2012**, *5*, 5307–5323. [CrossRef]
32. Piller, S.; Perrin, M.; Jossen, A. Methods for state-of-charge determination and their applications. *J. Power Sources* **2001**, *96*, 113–120. [CrossRef]
33. Arun, P.; Banerjee, R.; Bandyopadhyay, S. Optimum sizing of photovoltaic battery systems incorporating uncertainty through design space approach. *Sol. Energy* **2009**, *83*, 1013–1025. [CrossRef]
34. Arun, P.; Banerjee, R.; Bandyopadhyay, S. Sizing curve for design of isolated power systems. *Energy Sustain. Dev.* **2007**, *11*, 21–28. [CrossRef]
35. Arun, P.; Banerjee, R.; Bandyopadhyay, S. Optimum sizing of battery-integrated diesel generator for remote electrification through design-space approach. *Energy* **2008**, *33*, 1155–1168. [CrossRef]
36. Urtasun, A.; Sanchis, P.; Barricarte, D.; Marroyo, L. Energy management strategy for a battery-diesel stand-alone system with distributed PV generation based on grid frequency modulation. *Renew. Energy* **2014**, *66*, 325–336. [CrossRef]
37. Khatib, T.; Elmenreich, W. Novel simplified hourly energy flow models for photovoltaic power systems. *Energy Convers. Manag.* **2014**, *79*, 441–448. [CrossRef]

38. Weniger, J.; Tjaden, T.; Quaschning, V. Sizing of Residential PV Battery Systems. *Energy Procedia* **2014**, *46*, 78–87. [CrossRef]

39. Gitizadeh, M.; Fakharzadegan, H. Battery capacity determination with respect to optimized energy dispatch schedule in grid-connected photovoltaic (PV) systems. *Energy* **2014**, *65*, 665–674. [CrossRef]

40. Parra, D.; Walker, G.S.; Gillott, M. Modeling of PV generation, battery and hydrogen storage to investigate the benefits of energy storage for single dwelling. *Sustain. Cities Soc.* **2014**, *10*, 1–10. [CrossRef]

41. Achaibou, N.; Haddadi, M.; Malek, A. Modeling of lead acid batteries in PV systems. *Energy Procedia* **2012**, *18*, 538–544. [CrossRef]

42. Badescu, V. Dynamic model of a complex system including PV cells, electric battery, electrical motor and water pump. *Energy* **2003**, *28*, 1165–1181. [CrossRef]

43. Badescu, V. Time dependent model of a complex PV water pumping system. *Renew. Energy* **2003**, *28*, 543–560. [CrossRef]

44. Cherif, A.; Jraidi, M.; Dhouib, A. A battery ageing model used in stand alone PV systems. *J. Power Sources* **2002**, *112*, 49–53. [CrossRef]

45. Stein, J.S.; Hansen, C.W.; Cameron, C.P. *Evaluation of PV Performance Models and Their Impact on Project Risk*; Sandia National Laboratories: Albuquerque, NM, USA, 2010.

46. Menicucci, D.F.; Fernandez, J. *Users Manual for PVFORM: A Photovoltaic System Simulation Program for Stand-Alone and Grid-Interactive Applications*; Sandia National Labs: Albuquerque, NM, USA, 1989.

47. INSEL—The Graphical Programming Language for the Simulation of Renewable Energy Systems. Available online: http://www.insel.eu/ (accessed on 23 February 2018).

48. Klise, G.T.; Stein, J.S. *Models Used to Assess the Performance of Photovoltaic Systems*; Report SAND2009-8258; Sandia National Laboratories: Albuquerque, NM, USA, 2009.

49. Klein, S.A.E.A. TRNSYS 17: A Transient System Simulation Program. Available online: http://sel.me.wisc.edu/trnsys (accessed on 23 February 2010).

50. Shandilya, K.K.; Kumar, A. Available energy assessment software. *Environ. Prog. Sustain. Energy* **2013**, *32*, 170–175. [CrossRef]

51. HOMER Software. Available online: http://www.homerenergy.com/ (accessed on 23 February 2018).

52. Natural Resources Canada. Available online: http://www.retscreen.net/ang/g_photo.php (accessed on 23 February 2018).

53. Naional Renewable Energy Laboratory. Available online: https://sam.nrel.gov/ (accessed on 23 February 2018).

54. Gilman, P.; Dobos, A. SAM 2011.12. 2: General Description. *Contract* **2012**, *303*, 275–3000.

55. Marion, B.; Anderberg, M. PVWATTS-An Online Performance Calculator for Grid-Connected PV Systems. In Proceedings of the Solar Conference, Madison, WI, USA, 16–21 June 2000; pp. 119–124.

56. Marion, B.; Anderberg, M.; Gray-Hann, P. *Recent Revisions to PVWATTS: United States*; Department of Energy: Washington, DC, USA, 2005.

57. Mermoud, A. *PVSYST Version 3.2. User's Manual*; University of Geneva, University Center for the Study of Energy Problems: Geneva, Switzerland, 2002; Available online: http://www.pvsyst.com/ (accessed on 1 December 2002).

58. PVsyst: Software for Photovoltaic System. Available online: http://www.pvsyst.com/en/ (accessed on 23 February 2018).

59. Details of PVSOL Experts. Available online: http://www.valentin.de/ (accessed on 23 February 2018).

60. SolarPro. Available online: http://www.lapsys.co.jp/english/products/pro.html (accessed on 23 February 2018).

61. F-Chart Software. Available online: http://www.fchart.com/fchart/ (accessed on 23 February 2014).

62. SolarDesignTool. Available online: http://get.solardesigntool.com/feature/pv-solar-system-design/ (accessed on 23 February 2014).

63. Maui Solar Energy Software Corporation Products. Available online: http://www.mauisolarsoftware.com/ (accessed on 23 February 2014).

64. Dufo-López, R. *iHOGA*; Electrical Engineering Department, University of Zaragoza: Zaragoza, Spain, 2012.

65. Khatib, T.; Mohamed, A.; Sopian, K. A Software Tool for Optimal Sizing of PV Systems in Malaysia. *J. Model. Simul. Eng.* **2012**, *2012*, 969248. [CrossRef]

66. Pochacker, M.; Khatib, T.; Elmenreich, W. The Microgrid Simulation Tool RAPS: Description and Case Study. In Proceedings of the 2014 IEEE Innovative Smart Grid Technologies Conference—Asia (ISGT ASIA), Kuala Lumpur, Malaysia, 20–23 May 2014.

67. Huld, T.; Suri, M. *PVGIS, PV Estimation Utility*. 2010. Available online: http://re.jrc.ec.europa.eu/pvgis/apps4/pvest.php (accessed on 23 February 2018).

68. About SolarGIS. Available online: http://solargis.info/doc/ (accessed on 29 September 2018).

69. Soras, C.; Makios, V. A novel method for determining the optimum size of stand-alone photovoltaic systems. *Sol. Cells* **1988**, *25*, 127–142. [CrossRef]

70. Egido, M.; Lorenzo, E. The sizing of stand alone PV-system: A review and a proposed new method. *Sol. Energy Mater. Sol. Cells* **1992**, *26*, 51–69. [CrossRef]

71. Arab, A.H.; Driss, B.A.; Amimeur, R.; Lorenzo, E. Photovoltaic systems sizing for Algeria. *Sol. Energy* **1995**, *54*, 99–104. [CrossRef]

72. Ibrahim, O.E.E. Sizing stand-alone photovoltaic systems for various locations in Sudan. *Appl. Energy* **1995**, *52*, 133–140. [CrossRef]

73. Shrestha, G.; Goel, L. A study on optimal sizing of stand-alone photovoltaic stations. *IEEE Trans. Energy Convers.* **1998**, *13*, 373–378. [CrossRef]

74. Mellit, A.; Benghanem, M.; Arab, A.H.; Guessoum, A. An adaptive artificial neural network model for sizing stand-alone photovoltaic systems: Application for isolated sites in Algeria. *Renew. Energy* **2005**, *30*, 1501–1524. [CrossRef]

75. Hontoria, L.; Aguilera, J.; Zufiria, P. A new approach for sizing stand alone photovoltaic systems based in neural networks. *Sol. Energy* **2005**, *78*, 313–319. [CrossRef]

76. Kaushika, N.; Gautam, N.K.; Kaushik, K. Simulation model for sizing of stand-alone solar PV system with interconnected array. *Sol. Energy Mater. Sol. Cells* **2005**, *85*, 499–519. [CrossRef]

77. Mellit, A.; Benghanem, M. Sizing of stand-alone photovoltaic systems using neural network adaptive model. *Desalination* **2007**, *209*, 64–72. [CrossRef]

78. Celik, A.; Muneer, T.; Clarke, P. Optimal sizing and life cycle assessment of residential photovoltaic energy systems with battery storage. *Prog. Photovolt. Res. Appl.* **2008**, *16*, 69–85. [CrossRef]

79. Posadillo, R.; Luque, R.L. Approaches for developing a sizing method for stand-alone PV systems with variable demand. *Renew. Energy* **2008**, *33*, 1037–1048. [CrossRef]

80. Posadillo, R.; Luque, R.L. A sizing method for stand-alone PV installations with variable demand. *Renew. Energy* **2008**, *33*, 1049–1055. [CrossRef]

81. Mellit, A. ANN-based GA for generating the sizing curve of stand-alone photovoltaic systems. *Adv. Eng. Softw.* **2010**, *41*, 687–693. [CrossRef]

82. Khatib, T.; Mohamed, A.; Sopian, K.; Mahmoud, M. A new approach for optimal sizing of standalone photovoltaic systems. *Int. J. Photoenergy* **2012**, *2012*, 391213. [CrossRef]

83. Yang, H.; Zhou, W.; Lu, L.; Fang, Z. Optimal sizing method for stand-alone hybrid solar–wind system with LPSP technology by using genetic algorithm. *Sol. Energy* **2008**, *82*, 354–367. [CrossRef]

84. Weixiang, S. Design of standalone photovoltaic system at minimum cost in Malaysia. In Proceedings of the 2008 3rd IEEE Conference on Industrial Electronics and Applications, Singapore, 3–5 June 2008; pp. 702–707.

85. Shen, W. Optimally sizing of solar array and battery in a standalone photovoltaic system in Malaysia. *Renew. Energy* **2009**, *34*, 348–352. [CrossRef]

86. Askari, I.B.; Ameri, M. Optimal sizing of photovoltaic—Battery power systems in a remote region in Kerman, Iran. *Proc. Inst. Mech. Eng. Part A J. Power Energy* **2009**, *223*, 563–570. [CrossRef]

87. Cabral, C.V.T.; Diniz, A.S.A.C.; Martins, J.H.; Toledo, O.M.; Neto, L.D.V.B.M. A stochastic method for stand-alone photovoltaic system sizing. *Sol. Energy* **2010**, *84*, 1628–1636. [CrossRef]

88. Bilal, B.O.; Sambou, V.; Ndiaye, P.; Kebe, C.; Ndongo, M. Multi-objective design of PV-wind-batteries hybrid systems by minimizing the annualized cost system and the loss of power supply probability (LPSP). In Proceedings of the 2013 IEEE International Conference on Industrial Technology (ICIT), Cape Town, South Africa, 25–28 February 2013; pp. 861–868.

89. Yang, H.; Wei, Z.; Chengzhi, L. Optimal design and techno-economic analysis of a hybrid solar–wind power generation system. *Appl. Energy* **2009**, *86*, 163–169. [CrossRef]

90. Park, C.S.; Kumar, P.; Kumar, N. *Fundamentals of Engineering Economics*; Pearson/Prentice Hall: Upper Saddle River, NJ, USA, 2004.

91. Kamel, S.; Dahl, C. The economics of hybrid power systems for sustainable desert agriculture in Egypt. *Energy* **2005**, *30*, 1271–1281. [CrossRef]

92. Applasamy, V. Cost evaluation of a stand-alone residential photovoltaic power system in Malaysia. In Proceedings of the 2011 IEEE Symposium on Business, Engineering and Industrial Applications (ISBEIA), Langkawi, Malaysia, 25–28 September 2011; pp. 214–218.

93. Samimi, J.; Soleimani, E.A.; Zabihi, M. Optimal sizing of photovoltaic systems in varied climates. *Sol. Energy* **1997**, *60*, 97–107. [CrossRef]

94. Bhuiyan, M.; Asgar, M.A. Sizing of a stand-alone photovoltaic power system at Dhaka. *Renew. Energy* **2003**, *28*, 929–938. [CrossRef]

95. Nafeh, A. Design and Economic Analysis of a stand-alone PV system to electrify a remote area household in Egypt. *Open Renew. Energy J.* **2009**, *2*, 33–37. [CrossRef]

96. Chapman, R.N. Development of sizing nomograms for stand-alone photovoltaic/storage systems. *Sol. Energy* **1989**, *43*, 71–76. [CrossRef]

97. Notton, G.; Muselli, M.; Poggi, P.; Louche, A. Autonomous photovoltaic systems: Influences of some parameters on the sizing: Simulation timestep, input and output power profile. *Renew. Energy* **1996**, *7*, 353–369. [CrossRef]

98. Kaldellis, J. Optimum technoeconomic energy autonomous photovoltaic solution for remote consumers throughout Greece. *Energy Convers. Manag.* **2004**, *45*, 2745–2760. [CrossRef]

99. Balouktsis, A.; Karapantsios, T.; Antoniadis, A.; Paschaloudis, D.; Bezergiannidou, A.; Bilalis, N. Sizing stand-alone photovoltaic systems. *Int. J. Photoenergy* **2006**, *2006*, 73650. [CrossRef]

100. Fragaki, A.; Markvart, T. Stand-alone PV system design: Results using a new sizing approach. *Renew. Energy* **2008**, *33*, 162–167. [CrossRef]

101. Markvart, T.; Fragaki, A.; Ross, J. PV system sizing using observed time series of solar radiation. *Sol. Energy* **2006**, *80*, 46–50. [CrossRef]

102. Barra, L.; Catalanotti, S.; Fontana, F.; Lavorante, F. An analytical method to determine the optimal size of a photovoltaic plant. *Sol. Energy* **1984**, *33*, 509–514. [CrossRef]

103. Bartoli, B.; Cuomo, V.; Fontana, F.; Serio, C.; Silvestrini, V. The design of photovoltaic plants: An optimization procedure. *Appl. Energy* **1984**, *18*, 37–47. [CrossRef]

104. Bucciarelli, L.L., Jr. Estimating loss-of-power probabilities of stand-alone photovoltaic solar energy systems. *Sol. Energy* **1984**, *32*, 205–209. [CrossRef]

105. Bucciarelli, L.L., Jr. The effect of day-to-day correlation in solar radiation on the probability of loss-of-power in a stand-alone photovoltaic energy system. *Sol. Energy* **1986**, *36*, 11–14. [CrossRef]

106. Gordon, J. Optimal sizing of stand-alone photovoltaic solar power systems. *Sol. Cells* **1987**, *20*, 295–313. [CrossRef]

107. Hontoria, L.; Aguilera, J.; Zufiria, P. A tool for obtaining the LOLP curves for sizing off-grid photovoltaic systems based in neural networks. In Proceedings of the 3rd World Conference onPhotovoltaic Energy Conversion, Osaka, Japan, 11–18 May 2003; pp. 2423–2426.

108. Gong, X.; Kulkarni, M. Design optimization of a large scale rooftop photovoltaic system. *Sol. Energy* **2005**, *78*, 362–374. [CrossRef]

109. Fernández-Infantes, A.; Contreras, J.; Bernal-Agustín, J.L. Design of grid connected PV systems considering electrical, economical and environmental aspects: A practical case. *Renew. Energy* **2006**, *31*, 2042–2062. [CrossRef]

110. Kerekes, T.; Koutroulis, E.; Eyigun, S.; Teodorescu, R.; Katsanevakis, M.; Sera, D. A practical optimization method for designing large PV plants. In Proceedings of the 2011 IEEE International Symposium on Industrial Electronics, Gdansk, Poland, 27–30 June 2011; pp. 2051–2056.

111. Kornelakis, A.; Marinakis, Y. Contribution for optimal sizing of grid-connected PV-systems using PSO. *Renew. Energy* **2010**, *35*, 1333–1341. [CrossRef]

112. Kornelakis, A. Multiobjective Particle Swarm Optimization for the optimal design of photovoltaic grid-connected systems. *Sol. Energy* **2010**, *84*, 2022–2033. [CrossRef]

113. Kerekes, T.; Koutroulis, E.; Sera, D.; Teodorescu, R.; Katsanevakis, M. An Optimization Method for Designing Large PV Plants. *IEEE J. Photovolt.* **2012**, *3*, 814–822. [CrossRef]

Appl. Sci. **2018**, *8*, 1761

114. Bäck, T.; Rudolph, G.; Schwefel, H.-P. Evolutionary programming and evolution strategies: Similarities and differences. In Proceedings of the Second Annual Conference on Evolutionary Programming, La Jolla, CA, USA, 25–26 February 1993.
115. Morshed, M.J.; Hmida, J.B.; Fekih, A. A Probabilistic Multi-Objective Approach for Power Flow Optimization in hybrid Wind-PV-PEV Systems. *Appl. Energy* **2018**, *211*, 1136–1149. [CrossRef]
116. Islam, M.S.; Mithulananthan, N. PV based EV charging at universities using supplied historical PV output ramp. *Renew. Energy* **2018**, *118*, 306–327. [CrossRef]
117. Azizipanah-Abarghooee, R.; Niknam, T.; Bina, M.A.; Zare, M. Coordination of combined heat and power-thermal-wind photovoltaic units in economic load dispatch using chance constrained and jointly distributed random variables methods. *Energy* **2015**, *79*, 50–67. [CrossRef]

Article

Wind Loads on a Solar Panel at High Tilt Angles

Chin-Cheng Chou [1], Ping-Han Chung [2],[*] and Ray-Yeng Yang [2]

[1]   Aerospace Science and Technology Research Center, National Cheng Kung University, Tainan City 701, Taiwan; choucc@mail.ncku.edu.tw
[2]   Department of Hydraulic and Ocean Engineering, National Cheng Kung University, Tainan City 701, Taiwan; ryyang@mail.ncku.edu.tw
[*]   Correspondence: e84046082@mail.ncku.edu.tw; Tel.: +886-6975175133

Received: 11 March 2019; Accepted: 15 April 2019; Published: 17 April 2019

Featured Application: This work aims to provide a good estimation of wind loads on a solar panel to ensure proper operation under the extreme wind strength and wave climates. The data will be also useful for the design of a mooring system.

Abstract: A solar photovoltaic system consists of tilted panels and is prone to extreme wind loads during hurricanes or typhoons. To ensure the proper functioning of the system, it is important to determine its aerodynamic characteristics. Offshore photovoltaic (PV) systems have been developed in recent years. Wind loads are associated with wind, wave climates, and tidal regimes. In this study, the orientation of a single panel is adjusted to different angles of tilt (10°–80°) and angles of incidence for wind (0°–180°) that are pertinent to offshore PV panels. The critical wind loads on a tilted panel are observed at lower angles of incidence for the wind, when the angle of tilt for the panel is greater than 30°.

Keywords: offshore PV; tilt angle; wind incidence angle; wind load

---

1. Introduction

Renewable energy is an integral part of the worldwide measures to address climate change and reduce environmental pollution. Power generation from photovoltaic (PV) systems is one of the most promising substitutes to the use of fossil fuel. The total capacity in operation was 303 GW in 2016 and 402 GW in 2017, which corresponds to energy supplies of 375 TWh in 2016 and 494 TWh in 2017 (20.4% of the global renewable energy supply) [1]. PV panels are usually mounted on the rooftops of residential or commercial buildings. The systems are generally smaller than ground-mounted PV systems, for which land occupancy is a potential problem. PV systems that float on reservoirs or lakes use a pontoon structure for buoyancy, and there is a reduction in water evaporation and the temperature of solar cells. Trapani and Santafe [2] reviewed projects with floating PV systems from 2007 to 2013, and more installations are expected in the near future. Oceans cover approximately 71% of the Earth's surface, and offshore environments can take full advantage of solar energy. Offshore PV systems have been proposed recently [3,4]. However, these installations are subject to greater wind loads in severe sea wave environments.

A PV system consists of tilted panels. Any PV system design requires an accurate estimation of wind-induced loads in order to ensure proper function. Many studies determine the aerodynamic characteristics of tilted panels or their supporting structure. Naeiji et al. [5] showed that the most critical parameter is the angle at which the panel is tilted, α. Wind-induced loads are primarily due to pressure equalization for small angles of tilt and turbulence for large angles [6,7]. For a stand-alone panel that faces the direction of flow, Chung et al. [8] showed that an increase in α (15°–25°) results in a decrease in the unit sectional uplift coefficient in a uniform flow. There are strong suction forces near the front edge on the upper surface and slight wind-induced pressure on the lower surface. The mean spanwise pressure distributions have an inverted U-shape, which corresponds to the corner vortices. For the extreme cases of open terrain exposure, Stathopoulos et al. [9] showed that an increase in α (20°–45°) leads to greater suction, and peak suction occurs at $\alpha = 45°$. An increase in the intensity of the freestream turbulence results in the upstream movement of a separation bubble and side-edge vortices. More intense pressure fluctuations and bending moments have been observed [10]. Aly [11] also showed that discrepancies in the available wind tunnel data may be due to the characteristics of the inflow turbulence. Chou et al. [12] studied the effect of wind direction, β. When $\beta = 15°$–60°, there is greater suction on the upper surface near the windward corner. Chu and Tsao [13] showed that the maximum wind load occurs at $\beta = 45°$. There are maximum overturning moments for $\beta = 45°$ and 135° [14]. A sheltering effect was reported by Radu et al. [15]. The wind loads on the tilted panels are significantly reduced by the presence of neighboring upwind panels, and the degree of reduction decreases quickly. Warsido et al. [16] also had similar results.

To harness solar energy, the performance of a PV system depends significantly on the angle at which the PV panels are tilted, their orientation, and shadowing [17–19]. The optimal value for α is achieved when the sunlight is perpendicular to the surface of the PV panels. Designs for a PV system often use wind loading standards, such as the American Society Civil Engineers (ASCE 7) [7,20], in order to calculate wind loads. However, wind loads can be larger than the ASCE 7 standard [21]. For PV panels floating directly on the surface of water, Trapani et al. [3] showed that the system is subject to salt water corrosion and to the dynamics of tides, wind, and waves. In extreme winds and waves, the wind loads on the PV panels in a harsh sea environment are not the same as the load on the PV panels that are on land. This study conducts a wind load analysis using wind tunnel experiment and numerical simulation for a stand-alone panel at high α. The effect of β is also determined. The data is useful for the detailed structural design of offshore PV panels.

2. Experimental Setup

Experiments were conducted in a wind tunnel at the Architecture and Building Research Institute. The tunnel is a closed-loop type with a contraction ratio of 4.71. The constant-area test section is 2.6 m (height) × 4 m (width) × 36.5 m (length). For a stand-alone panel (60%-scale commercial module), the test configuration is shown in Figure 1. The length (L) and width (W) are 120 cm and 60 cm, respectively. At $x/L = 1.0$, the panel is 3 cm above the tunnel floor. The blockage ratio is up to 6.3%. Note that a blockage correction is required for the mean surface pressure for a tilted panel if the blockage ratio is more than 10% [22].

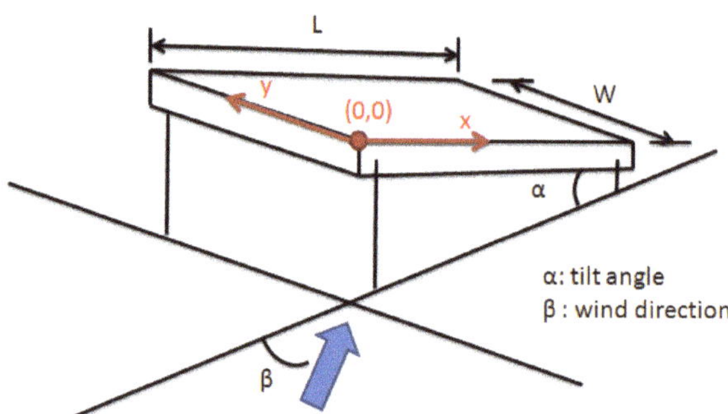

Figure 1. Test configuration for a stand-alone model.

This study determined the effect of α and β. Meteorological data (wave, wind, tide, and current) were collected from three near-shore buoys in Taiwan (Qigu, 23°05″42″N, 120°00′27″E; Hsinchu, 24°45″19″N, 120°50′12″E; and Longdong, 25°05′48″N, 121°55′19″E) [23]. In Qigu, the wind rose for the period of 2013–2017, and is shown in Figure 2. The most common values of β are 210°–225° and 315°–360°. Note that the wind direction is 30°–45° in Hsinchu, and 0° in Longdong. The inclination of the PV panels depends on the waves. The historical data for typhoons shows that the maximum wave height and wind speed in Longdong are 17.12 m (mean period = 12.5 s during typhoon Soudelor, 2015) and 26.7 m/s (mean period = 11.8 s during typhoon Soulik, 2013), respectively. The variation in α for the PV panels with respect to wind is ±45°. Therefore, the setup for α is 10°–80° (in increments of 10°). The value for β ranges from 0° (facing the direction of the wind) to 180° (in increments of 15°). Note that the angle between the PV panels and their base is not changed in sea wave environments, which is not exactly the same as the case for the experimental setup.

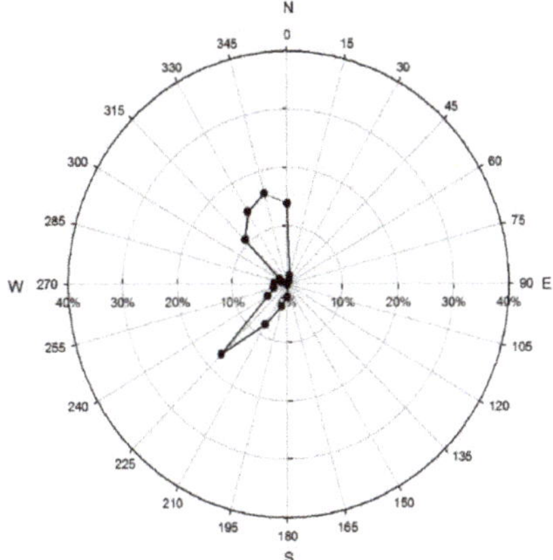

Figure 2. Probability distribution for the incidence of wind for the Qigu buoy (2013–2017).

The experiments were conducted in a uniform flow. The freestream velocity was set at 14.5 ± 0.1 m/s, measured using a Pitot-static tube, and the turbulence intensity was 0.3%. Chung et al. [24] showed that there is greater expansion on the upper surface, and more positive pressure on the lower surface of a tilted panel when the intensity of the turbulence increases. The front edge of the test model was located 2.8 m from the inlet of the working section. A Reynolds number of 1.17×10^6 is based on the length of the tilted panel. Chung et al. [25] reported that a tilted panel is not affected by the Reynolds number.

A total of 330 pressure taps were drilled on the test model and were connected to flexible polyvinyl chloride tubing that is 1.1-mm in diameter and 60-cm in length, so the phase distortion has little effect on the measured peak pressure [26]. As there are strong pressure gradients (flow separation and reattachment) near the front edge of the tilted panel, 92% pressure taps were machined on the first two-thirds of the upper and lower plate surfaces. SCANVALVE multichannel modules (Model ZOC 33/64Px 64-port; Model RAD3200 pressure transducer) were used for the surface pressure measurements. The full-scale range of the sensors was ±2490 Pa, with an accuracy of ±0.15% full scale. The sampling rate was 250 Hz and each record contained 32,768 data points. A Pitot-static tube, which was at the same height as the front edge of the tilted panel and 2.8 m from the inlet of the working section, was used to measure the static pressure, p_∞, and the dynamic pressure, q, of the incoming flow. The mean surface pressure coefficient was given as C_p $(=(p-p_\infty)/q)$. The uplift coefficient, C_L $(=\frac{1}{A}\int_A \Delta C_p \cos(\alpha)dA)$, was calculated by integrating the differential mean surface pressure distributions $(\Delta C_p = C_{p,up} - C_{p,low})$ between the upper and lower surfaces of the tilted panel.

3. Results and Discussion

3.1. Longitudinal Pressure Distributions

The C_p distributions on the centerline ($y/W = 0.5$) for $\beta = 0°$ are shown in Figure 3. The origin of the coordinates ($x/L = 0$ and $y/W = 0$) is located at the left corner of the tilted panel. The solid and hollow symbols represent the value of C_p on the upper and lower surface, respectively. Suction, which corresponds to flow separation, is observed on the upper surface for all of the test cases. For $\alpha = 30°$, the value of C_p decreases significantly near the front edge, reaches a peak value of -1.66, and approaches a more moderate level in the second half of the panel. At larger angles of tilt ($\alpha \geq 50°$), there are flattened C_p distributions, for which the value of C_p (-0.72 to -0.82) for a specific value of α varies by less than 4%. On the lower surface, there is a slight expansion near the front and rear edges for $\alpha = 10°$. Positive pressure is observed and the magnitude increases as α increases. This shows that the localized load is the most significant near the front edge for $\alpha = 30°$. There is also a slight variation in the magnitude of C_p (0.91 to 0.93) for $\alpha = 50°$–$80°$, so the uplift lift force is approximately constant for larger angles of tilt ($\alpha \geq 50°$), when $\beta = 0°$.

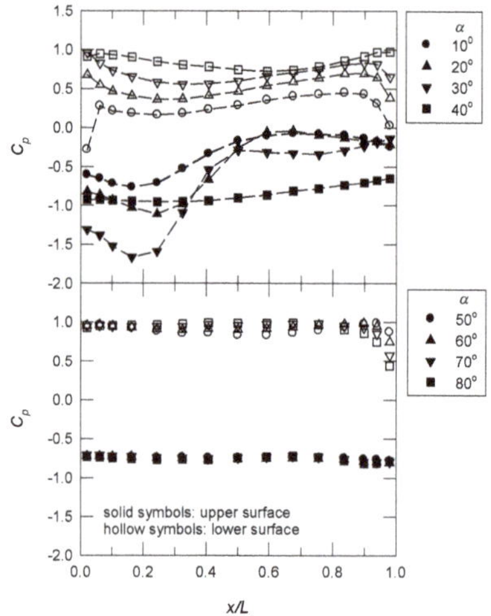

Figure 3. Mean longitudinal pressure distributions at $y/W = 0.5$; $\beta = 0°$.

The C_p distributions for $\beta = 30°$, $45°$, and $135°$ are shown in Figures 4–6. These correspond to the most common direction for the wind for the meteorological data. For $\beta = 30°$ and $45°$, there is greater suction on the upper surface for $\alpha = 10°$ and $20°$. A flow expansion in the second half of the panel is also observed, particularly for $\beta = 45°$. This agrees with the results of Chou et al. [12]. Suction near the front edge is mitigated for $\alpha = 30°$, and there is a flatter C_p distribution. The value of C_p is approximately the same for $\alpha = 40°$–$80°$ (-0.701 to -0.711), and the effect of α is minimal. Its amplitude increases slightly more than that for $\beta = 0°$ ($C_p = -0.743$ to -0.756). For $\beta = 135°$, a small degree of suction is observed near the front face, and the value of C_p increases downstream. The value of α has an obvious effect, in that the amplitude of C_p increases as α increases. There is a slight flow expansion near the rear edge for $\alpha = 10°$ and $20°$. On the lower surface for $\beta = 30°$ and $45°$, there is a fairly uniform C_p distribution for specific values of α. The value of C_p is positive, except for the case of $\alpha = 10°$ near the rear edge. This shows that the uplift force increases as α increases. Suction on the lower surface for $\beta = 135°$ produces greater downward force, particularly for $\alpha \geq 30°$.

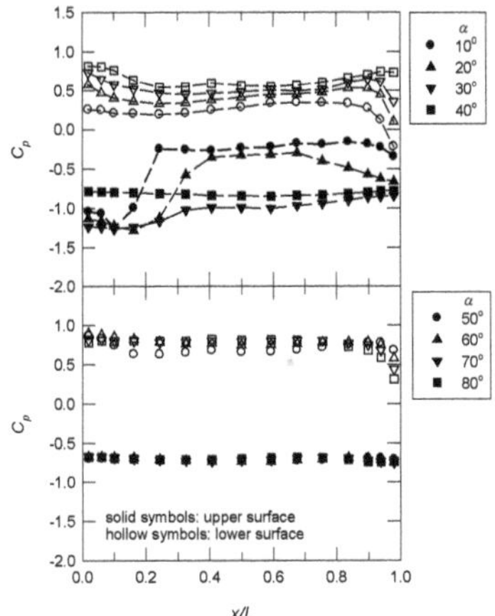

Figure 4. Mean longitudinal pressure distributions at $y/W = 0.5$; $\beta = 30°$.

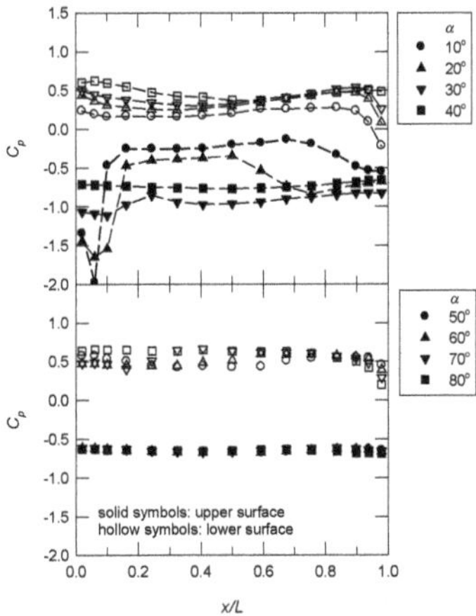

Figure 5. Mean longitudinal pressure distributions at $y/W = 0.5$; $\beta = 45°$.

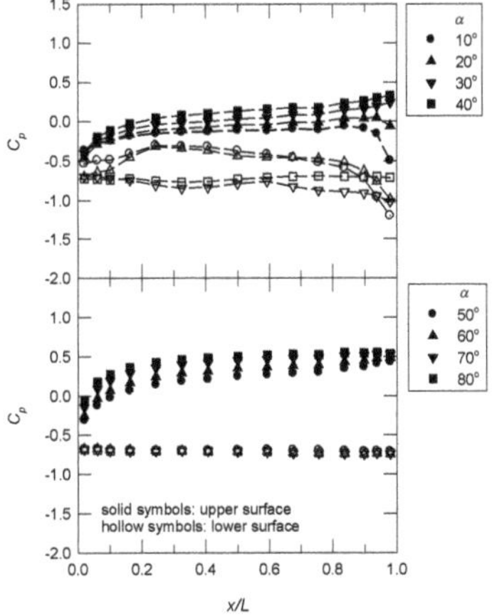

Figure 6. Mean longitudinal pressure distributions at $y/W = 0.5$; $\beta = 135°$.

3.2. Spanwise Pressure Distributions

At $x/L = 0.5$, the spanwise pressure, C_{sp}, and distributions for $\beta = 0°$ are shown in Figure 7. There is suction (negative C_{sp}) on the upper surface for all of the test cases. Inverted U-shaped distributions are observed for small α, particularly for $\alpha = 20°$ and $30°$. This produces strong corner vortices, which is in agreement with the results of Chung et al. [10]. For $\alpha \geq 40°$, the C_{sp} distributions show a small degree of variation. On the lower surface, the footprints of the corner vortices are also visible. The peak value of C_{sp} (0.29 to 0.99) at $y/W = 0.5$ increases as α increases. The increase in C_{sp} as α increases is more significant for $\alpha \leq 50°$.

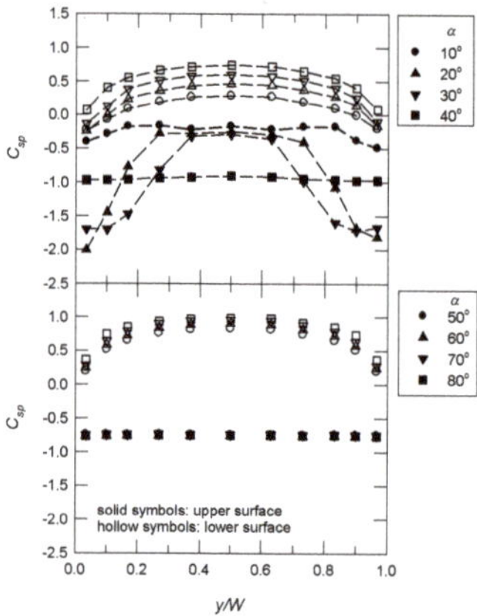

Figure 7. Mean spanwise pressure distributions at $x/L = 0.5$; $\beta = 0°$.

Figures 8 and 9 show that $\beta = 30°$ and $45°$. There are flattened C_{sp} distributions in the left half of the panel for $\alpha = 10°$ and $20°$. The expansion and compression near the right edge corresponds to the formation of a separation bubble or corner vortices. The variation in C_{sp} for $\alpha \geq 40°$ is less than 3%. An increase in the value of C_{sp} (−0.738 to −0.693) is associated with greater α (less suction). An increase in the value of C_{sp} is observed from the left to the right edges, so there is a greater uplift force near the right edge. Figure 10 shows that for $\beta = 135°$, the wind blows over the lower surface of the tilted panel. The C_{sp} distributions on the lower surface show similar patterns to those on the upper surface for $\beta = 45°$. This demonstrates that there is an increase in the downward force from the left to right edges.

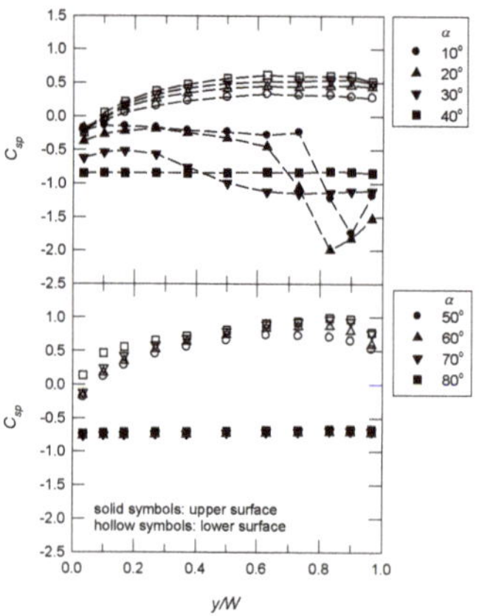

Figure 8. Mean spanwise pressure distributions at $x/L = 0.5$; $\beta = 30°$.

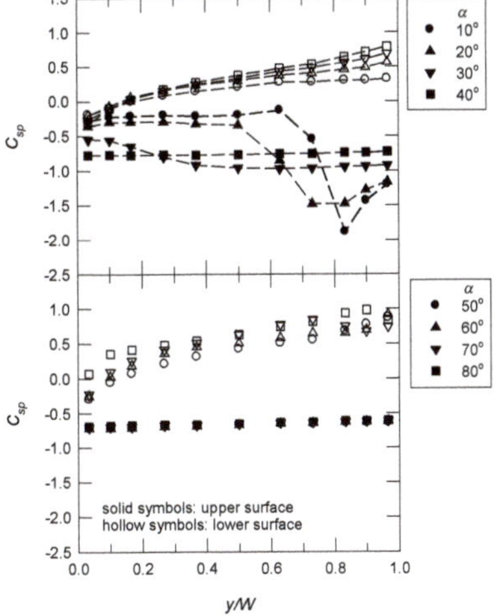

Figure 9. Mean spanwise pressure distributions at $x/L = 0.5$; $\beta = 45°$.

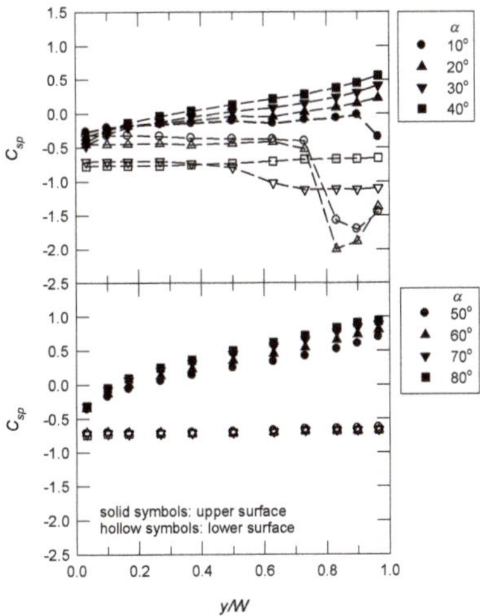

Figure 10. Mean spanwise pressure distributions at $x/L = 0.5$; $\beta = 135°$.

3.3. *The Uplift Coefficient*

The value of C_L is calculated by integrating ΔC_p. Examples of C_p distributions on the upper and lower surface ($\alpha = 30°$ and $\beta = 0°$) are shown in Figure 11. There is a symmetrical spanwise pressure distribution with respect to $y/W = 0.5$. The flow expansion and corner vortices on the first half of the panel result in a relatively large negative value of C_p for the upper surface. A positive value of C_p is observed on the lower surface. Near the front and rear edges, a more positive value for C_p corresponds to the blocking effect of the tilted panel.

The variation in C_L with respect to α and β is shown in Figure 12. The value of C_L (uplift force) is negative for $\beta < 90°$, and is relatively small for $\beta = 90°$. The lowest value for C_L is for $\alpha = 30°$ to $40°$. This is similar to the observation by Stathopoulos et al., who noted that peak suction occurs at $\alpha = 45°$ [9]. The positive value for C_L for $\beta \geq 105°$ represents a downward force. A kink is also observed at $\alpha = 50°$.

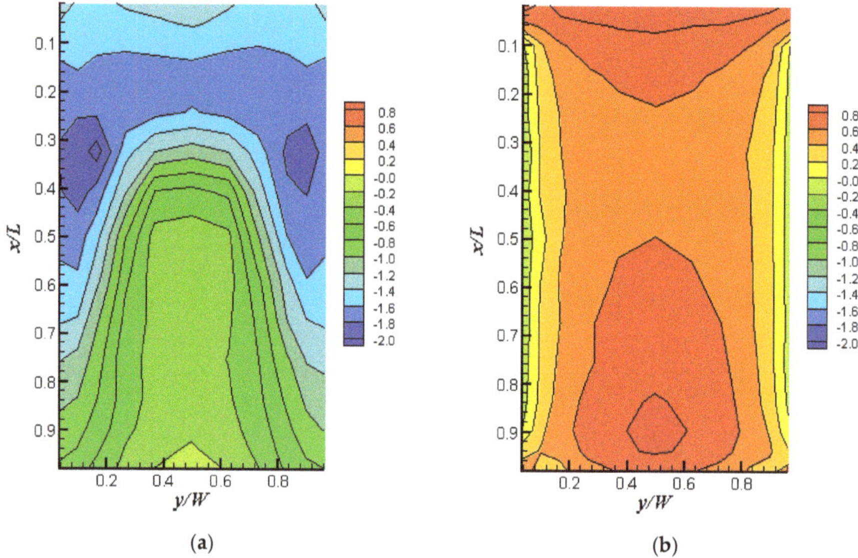

Figure 11. Pressure coefficient contours for $\alpha = 30°$ and $\beta = 0°$ for (**a**) the upper surface and (**b**) the lower surface.

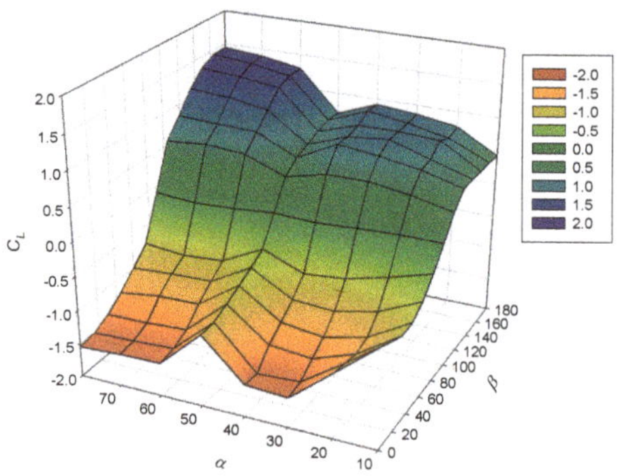

Figure 12. Uplift coefficient.

For a specific value of β, the effect of α on C_L is shown in Figure 13. For $\beta \leq 75°$, C_L decreases linearly as α increases ($\leq 30°$), following an increase for $\alpha = 50°$. The value of C_L at high α (60° to 80°) is approximately the same as that at $\alpha = 30°$ and 40°. Wind loads on a tilted panel at lower β require caution at the design stage. An opposite effect is observed for $\beta \geq 90°$. The variation of C_L with α ($\geq 60°$) is minimal. The value of C_L also varies linearly with α ($\leq 30°$) for a specific value of β. Figure 14 shows that the value of C_L increases as the value of α increases. The value of $dC_L/d\alpha$ increases from a negative to a positive value when there is an increase in β. Therefore, the uplift force is more significant at lower values of β, so $dC_L/d\alpha = -0.0542 + 7.376 \times 10^{-4}\beta - 1.67 \times 10^{-6}\beta^2$. Figure 15 shows the effect of β on C_L for a specific value of α. For $\alpha = 10°$ and 20°, the value of C_L decreases initially and the lowest value of C_L is for $\beta = 30°$, following an increase in C_L with β. There is a smaller variation in C_L for $\beta \geq 120°$. For $\alpha \geq 30°$, the value of C_L increases as the value of β increases.

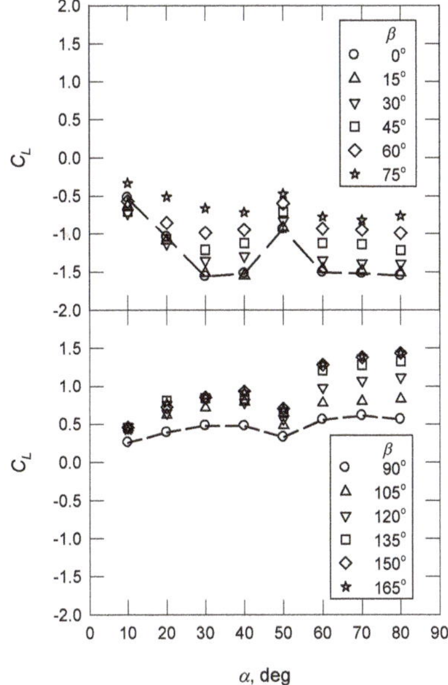

Figure 13. Uplift coefficient for a specific value of β.

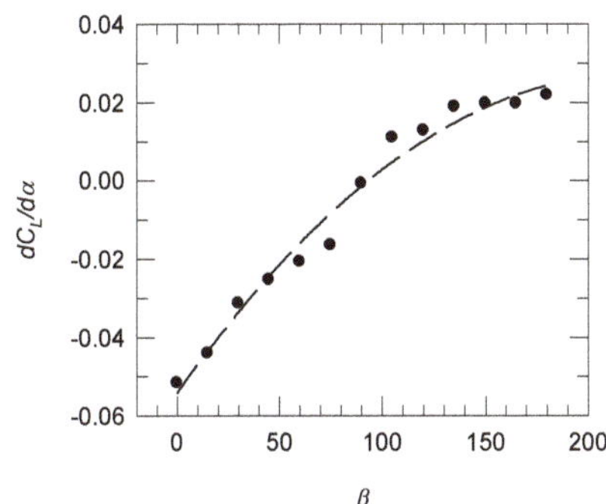

Figure 14. The effect of β on $dC_L/d\alpha$.

For the numerical study, 3D incompressible RANS simulations of wind flow over a stand-alone PV panel in full scale were performed using a steady finite volume solver of second-order accuracy (ANSYS Fluent 13) for the value of α of 10° to 40° (in increments of 10°) and the value of β of 0° to 180° (in increments of 45°). The semi-implicit method for the pressure-linked equation (SIMPLE) is used. The SST κ-ω turbulence model [27] models flows with separation reasonably well. However, it is necessary to mesh down (wall spacing; $y^+ \sim 1$). Therefore, this parametric study uses the realizable κ-ε turbulence model ($y^+ \sim 30$) [28], in which a modified transport equation for the dissipation rate is derived from an exact equation for the transport of the mean-square vorticity equation. The computational domain is an upstream fetch of 5 L and a downstream length of 10 L. The height and width are 5 L and 6 L, respectively.

The grid is created using the grid generating software, Pointwise. Once a solution is obtained and the value of y^+ for the first grid point from the wall is verified, the grid sensitivity (number of grids, G = 20–40.0 million) is performed using the value of C_p on the upper and lower surfaces for $\alpha = 10°$

and $\beta = 0°$. The difference in C_p for $G = 35$ and 40 million is less than 1.5%. Therefore, the total number of unstructured grids that is used is approximately 35 million.

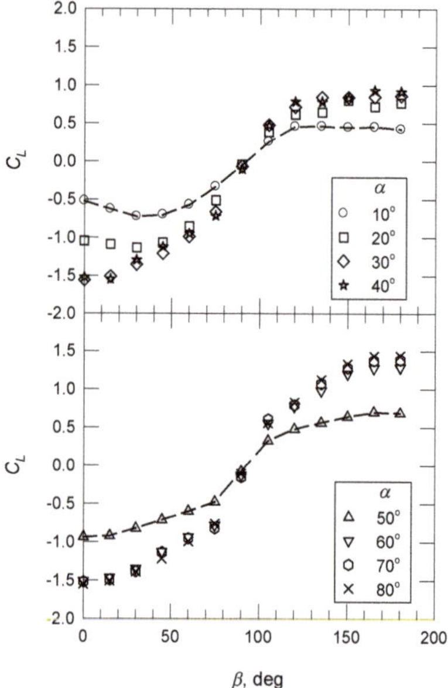

Figure 15. Uplift coefficient for a specific value of α.

At the domain inlet, there is a uniform freestream flow of 15 m/s. The results are shown in Figure 16. For $\alpha = 10°$–$40°$, the variations in the value of C_L with β is similar to that for the experimental data. The maximum difference between the numerical and experimental results is up to 18%, which occurs for the peak value of C_L or stronger corner vortices for a specific value of α (i.e., $\beta = 45°$ for $\alpha = 10°$ and $20°$). Further study of the effect of scaling on the wind flow field is required.

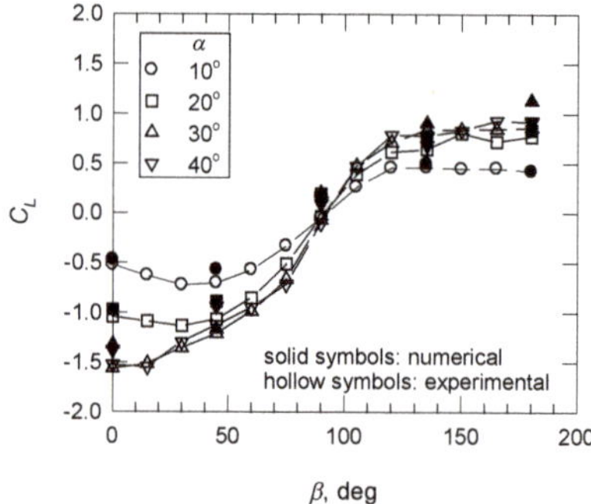

Figure 16. Uplift coefficient: numerical and experimental results.

4. Conclusions

Meteorological data were collected from three near-shore buoys in Taiwan. For a sea wave environment, this study determines the effect of α and β on wind loads on a tilted panel. At lower angles of tilt ($\leq 40°$), the experimental results agree with those of previous studies. Greater suction on the upper surface produces flow expansion and corner vortices. The uplift force increases linearly with

α. When there is an increase in β, expansion and compression are observed near the right edge, which produce a greater uplift force. The increase in the uplift coefficient with α changes from a negative to a positive value ($dC_L/d\alpha = -0.0542 + 7.376 \times 10^{-4}\beta - 1.673 \times 10^{-6}\beta^2$ for $\alpha \leq 30°$). The variation of C_L with β using a numerical simulation is similar to that for the experimental data. At high angles of tilt, there is a kink in the curve for C_L at $\alpha = 50°$. A small variation in C_L is observed at high angles of tilt ($\geq 60°$), for which the magnitude is approximately the same as that at $\alpha = 30°$ and $40°$. Caution is necessary when there are wind loads on a tilted panel at lower values of β, when the value of α is greater than $30°$.

Author Contributions: Conceptualization/methodology/data curation, C.-C.C. and P.-H.C.; writing (original draft preparation), P.-H.C.; writing (review and editing), R.-Y.Y.

Funding: This research was funded by Industrial Technology Research Institute.

Acknowledgments: The numerical simulation was conducted by K.C. Su, Aerospace Science and Technology Research Center, National Cheng Kung University.

Conflicts of Interest: The authors declare no conflict of interest.

Nomenclature

C_L	uplift coefficient
C_p	pressure coefficient in the longitudinal direction, $(p-p_\infty)/q$
$C_{p,low}$	pressure coefficient on the lower surface
C_{sp}	pressure coefficient in the spanwise direction
$C_{p,up}$	pressure coefficient on the upper surface
L	length of tilted panel
p_∞	freestream static pressure
q	dynamic pressure
W	width of tilted panel
x	coordinate in the longitudinal direction
y	coordinate in the spanwise direction
α	angle of tilt
β	wind incidence angle
ΔC_p	differential pressure, $C_{p,up} - C_{p,low}$

References

1. Mauthner, F.; Weiss, W.; Spörk-Dür, M. *Solar Heat Worldwide*; AEE-Institute for Sustainable Technologies: Gleisdorf, Austria, 2018.
2. Trapani, K.; Santafe, M.R. A review of floating photovoltaics installations: 2007–2013. *Prog. Photovolt.* **2015**, *23*, 524–532. [CrossRef]
3. Trapani, K.; Millar, D.L.; Smith, H.C. Novel offshore application of photovoltaics in comparison to conventional marine renewable energy technologies. *Renew. Energy* **2013**, *50*, 879–888. [CrossRef]
4. Sahu, A.; Yadav, N.; Sudhakar, K. Floating photovoltaic power plant: A review. *Renew. Sustain. Rev.* **2016**, *66*, 815–824.
5. Naeiji, A.; Raji, F.; Zisis, I. Wind loads on residential scale rooftop photovoltaic panels. *J. Wind. Eng. Ind. Aerodyn.* **2017**, *168*, 228–246. [CrossRef]
6. Cao, J.; Yoshida, A.; Saha, P.K.; Tamura, Y. Wind loading characteristics of solar arrays mounted on flat roofs. *J. Wind. Eng. Ind. Aerodyn.* **2013**, *123*, 214–225. [CrossRef]
7. Kopp, G.A.; Banks, D. Use of the Wind Tunnel Test Method for Obtaining Design Wind Loads on Roof-Mounted Solar Arrays. *J. Struct. Eng.* **2013**, *139*, 284–287. [CrossRef]
8. Chung, K.-M.; Chang, K.-C.; Chou, C.-C. Wind loads on residential and large-scale solar collector models. *J. Wind. Eng. Ind. Aerodyn.* **2011**, *99*, 59–64. [CrossRef]
9. Stathopoulos, T.; Zisis, I.; Xypnitou, E. Local and overall wind pressure and force coefficients for solar panels. *J. Wind. Eng. Ind. Aerodyn.* **2014**, *125*, 195–206. [CrossRef]

10. Chung, K.-M.; Chou, C.-C.; Chang, K.-C.; Chen, Y.-J. Effect of a vertical guide plate on the wind loading of an inclined flat plate. *Wind. Struct. Int. J.* **2013**, *17*, 537–552. [CrossRef]

11. Aly, A.M. On the evaluation of wind loads on solar panels: The scale issue. *Sol. Energy* **2016**, *135*, 423–434. [CrossRef]

12. Chou, C.-C.; Chung, K.-M.; Chang, K.-C. Wind Loads of Solar Water Heaters: Wind Incidence Effect. *J. Aerodyn.* **2014**, *2014*, 1–10. [CrossRef]

13. Chu, C.-R.; Tsao, S.-J. Aerodynamic loading of solar trackers on flat-roofed buildings. *J. Wind. Eng. Ind. Aerodyn.* **2018**, *175*, 202–212. [CrossRef]

14. Jubayer, C.M.; Hangan, H. Numerical simulation of wind effects on a stand-alone ground mounted photovoltaic (PV) system. *J. Wind. Eng. Ind. Aerodyn.* **2014**, *134*, 56–64. [CrossRef]

15. Radu, A.; Axinte, E. Wind forces on structures supporting solar collectors. *J. Wind. Eng. Ind. Aerodyn.* **1989**, *32*, 93–100. [CrossRef]

16. Warsido, W.P.; Bitsuamlak, G.T.; Barata, J.; Chowdhury, A.G. Influence of spacing parameters on the wind loading of solar array. *J. Fluids Struct.* **2014**, *48*, 295–315. [CrossRef]

17. Guo, M.; Zang, H.; Gao, S.; Chen, T.; Xiao, J.; Cheng, L.; Wei, Z.; Sun, G. Optimal tilt angle and orientation on photovoltaic modules using HS algorithm in different climates in China. *Appl. Sci.* **2017**, *7*, 1028. [CrossRef]

18. Lau, K.; Tan, C.; Yatim, A. Effects of ambient temperatures, tilt angles, and orientations on hybrid photovoltaic/diesel systems under equatorial climates. *Renew. Sustain. Rev.* **2018**, *81*, 2625–2636. [CrossRef]

19. Castellano, N.N.; Parra, J.A.G.; Valls-Guirado, J.; Manzano-Agugliaro, F. Optimal displacement of photovoltaic array's rows using a novel shading model. *Appl. Energy* **2015**, *144*, 1–9. [CrossRef]

20. ASCE 7-10. *Minimum Design Loads for Buildings and Other Structures*; American Society of Civil Engineers: Reston, VA, USA, 2010.

21. Bender, W.; Waytuck, D.; Wang, S.; Reed, D. In situ measurement of wind pressure loadings on pedestal style rooftop photovoltaic panels. *Eng. Struct.* **2018**, *163*, 281–293. [CrossRef]

22. Chung, K.-M.; Chen, Y.-J. Effect of High Blockage Ratios on Surface Pressures of an Inclined Flat Plate. *J. Eng. Arch.* **2016**, *4*, 1–16. [CrossRef]

23. Central Weather Bureau (CWB), Ministry of Transportation and Communication. Available online: https://www.cwb.gov.tw/V7/climate/marine_stat/wave.htm (accessed on 22 May 2018).

24. Chung, K.M.; Chou, C.C.; Chang, K.C.; Chen, Y.J. Wind loads on a residential solar water heater. *J. Chin. Inst. Eng.* **2013**, *36*, 870–877. [CrossRef]

25. Chung, K.; Chang, K.-C.; Liu, Y. Reduction of wind uplift of a solar collector model. *J. Wind Eng. Ind. Aerodyn.* **2008**, *96*, 1294–1306. [CrossRef]

26. Irwin, H.; Cooper, K.; Girard, R. Correction of distortion effects caused by tubing systems in measurements of fluctuating pressures. *J. Wind Eng. Ind. Aerodyn.* **1979**, *5*, 93–107. [CrossRef]

27. Menter, F.R. Two-equation eddy-viscosity turbulence models for engineering applications. *AIAA J.* **1994**, *32*, 1598–1605. [CrossRef]

28. Shih, T.H.; Liou, W.W.; Shabbir, A.; Yang, Z.; Zhu, J. A new κ-ε eddy-viscosity model for high Reynolds number turbulent flows–Model development and validation. *Comput. Fluids* **1995**, *24*, 227–238. [CrossRef]

Article

Numerical Analysis for Thermal Performance of a Photovoltaic Thermal Solar Collector with SiO$_2$-Water Nanofluid

Ali J. Chamkha [1,2] and Fatih Selimefendigil [3]*

[1] Mechanical Engineering Department, Prince Sultan Endowment for Energy and Environment, Prince Mohammad Bin Fahd University, Al-Khobar 31952, Saudi Arabia; achamkha@pmu.edu.sa
[2] RAK Research and Innovation Center, American University of Ras Al Khaimah, Ras Al Khaimah P.O. Box 10021, UAE
[3] Department of Mechanical Engineering, Celal Bayar University, Manisa 45140, Turkey
* Correspondence: fthsel@yahoo.com; Tel.: +90-223-6201-2370

Received: 17 October 2018; Accepted: 5 November 2018; Published: 11 November 2018

Abstract: Numerical analysis of a photovoltaic-thermal (PV/T) unit with SiO$_2$-water nanofluid was performed. The coupled heat conduction equations within the layers and convective heat transfer equations within the channel of the module were solved by using the finite volume method. Effects of various particle shapes, solid volume fractions, water inlet temperature, solar irradiation and wind speed on the thermal and PV efficiency of the unit were analyzed. Correlation for the efficiencies were obtained by using radial basis function neural networks. Cylindrical shape particles were found to give best performance in terms of efficiency enhancements. Total efficiency enhances by about 7.39% at the highest volume fraction with cylindrical shape particles. Cylindrical shape particle gives 3.95% more enhancement as compared to spherical ones for the highest value of solid particle volume fraction. Thermal and total efficiency enhance for higher values of solid particle volume fraction, solar irradiation and lower values of convective heat transfer coefficient and inlet temperature. The performance characteristics of solar PV-thermal unit with radial basis function artificial neural network are found to be in excellent agreement with the results obtained from computational fluid dynamics modeling.

Keywords: PV-thermal collector; nanofluid; particle shape; finite volume method

---

1. Introduction

Nanofluids are composed of base fluid such as water, ethylene glycol or mineral oil and added solid nano-sized particles. They have been extensively used in different thermal engineering applications [1–14]. The nano-sized particle could be metallic or non-metallic such as Cu, Ag, CuO, Al$_2$O$_3$, TiO$_2$, SiO$_2$ with average particle size less than 100 nm. Higher thermal conductivity of the nanoparticles increase the thermal conductivity of the heat transfer fluid and enhances the thermal performance. Size, shape and type of the particles are effective for the thermal conductivity enhancement of nanofluids. Thermophysical properties are derived from theoretical or experimental studies for nanofluids containing various particle types, shapes and sizes for different temperatures. Generally, a small amount of particle addition of the base fluid results in higher heat transfer enhancements. Application of the nanofluids for the thermal engineering systems are diverse such as in refrigeration, microelectromechanical systems (MEMs), cooling of nuclear reactors, thermal management of fuel cells, cooling of hydrogen storage, heat exchangers and many others. In the refrigeration application, nano additives are added to compressor oil to increase the coefficient of performance. In some applications, solid nano particles are added to the refrigerants. In heat exchanger design, more compact and lightweight structures can be designed when heat transfer fluid has a higher thermal conductivity with the addition of nanoparticles.

Application of nanotechnology in the field of renewable energy is growing. There are many studies related to the nanofluids application in solar power. A review for the application of the nanofluid in solar energy was presented in the study by Mahian et al. [15]. Using nanofluids in solar collectors and solar water heaters and their impacts on the efficiency and environmental effects were also discussed. Mahian et al. [16] performed analytical study for the performance of a solar collector with various types of nanofluids such as Cu/water, Al_2O_3/water, TiO_2/water, and SiO_2/water nanofluids with particle size of 25 nm. System with Cu/water nanofluid has lowest entropy generation rate whereas Al_2O_3/water nanofluid has the highest heat transfer coefficient as compared to other nanofluids. In the study by Meibodi et al. [17], an experimental investigation was performed for a flat plate solar collector with SiO_2/ethylene glycol (EG-water nanofluid. Various mass flow rates and particle volume fraction up to 1% were tested. It was observed that, despite the low conductivity of SiO_2 nanoparticles, solar collector efficiency was found to be enhanced with nanofluid. Chen et al. [18] studied the effects of inclusion of Au nanoparticles for the photo-thermal conversion performance numerically and experimentally for various solar intensities and particle volume fractions. The absorption efficiency was found to increase with higher nanoparticle volume fractions. Effects of SiO_2 nanoparticles in solar collector tubes were numerically and experimentally studied by Yan et al. [19]. Heat transfer rate was found to be higher for nanofluid and, due to nanofluid agglomeration, the heat transfer rate deteriorates for longer operation times.

In the Photovoltaic/Thermal modules (PV/T), heat and electricity are produced by using photovoltaic and heat extraction units. A review study for the application of nanofluids in PV/T systems and discussions about effective parameters and effectiveness of nanofluids were presented in [20]. Al-Waeli et al. [21] performed an experimental study for the determination of effective thermophyscial properties of water containing SiC nanoparticles that was used as a cooler for PV/T system. It was observed that thermal conductivity enhancements are about 8.2% for the temperature range of 25–60 °C. The electrical efficiency with 3 wt % of SiC nanofluid results in electrical efficiency enhancements of 24.1% and it was observed that the nanofluids were stable for long use. In the study by Hassani et al. [22], a new cascade PV/T module was proposed with separate channels. Two nanofluids were used to enhance the electrical and thermal performance of the PV/T module. Jing et al. [23] investigated the effects of silica/water nanofluids on the efficiency of PV/T module. Various sizes of nanoparticles, concentrations and flow velocity were considered. Optimum operational parameters for the economical considerations were also obtained.

In the present study, efficiency of a PV/T module with SiO_2-water nanofluid was numerically investigated for nanoparticle properties (shape, volume fraction) and for different operating conditions. Despite the low conductivity of SiO_2 nanoparticles as compared to other particles, its low cost, favorable physical and chemical properties makes it attractive for usage with water. Artificial neural networks with radial basis functions are used to obtain the correlations for efficiencies of PV-thermal module.

2. Mathematical Modeling

Figure 1 shows a schematic representation of a PV-thermal module which is composed of several layers and a channel in which SiO_2-water nanofluid is flowing throughout. Thermophysical properties of the layers in the PV-thermal module is given in Table 1.

An energy balance between the solar irradiance and heat transfer to the heat transfer fluid with nanoparticles is considered. Within the layers of the PV-module, the heat conduction equation is used.

Within the layers of the PV-thermal module steady state, the heat conduction equation is valid and is given by the following equation:

$$\nabla \cdot \left(k_{layer} \nabla T \right) = 0.$$

(1)

Navier–Stokes and energy equations for the fluid flow and heat transfer in the channel are given by the following equations:

$$\nabla.(\mathbf{u}) = \mathbf{0},\tag{2}$$

$$\rho\mathbf{u}.\nabla\mathbf{u} = -\nabla p + \mu\nabla^2\mathbf{u},\tag{3}$$

$$\rho c_p\mathbf{u}.\nabla T = \nabla.(k\nabla T).\tag{4}$$

The PV cell electrical efficiency is given by the following equation:

$$\eta_{pv} = \eta_{T_{ref}}\left[1 - \beta_{ref}\left(T_{pv} - T_{ref}\right)\right].\tag{5}$$

Thermal efficiency is defined as the ratio of the energy gained by the collector divided by the total incident energy

$$\eta_{th} = \frac{\dot{m}c_p\left(T_{out} - T_{in}\right)}{G}.\tag{6}$$

- For the upper surface, the heat flux boundary condition with incident radiation and convective heat loss due to wind (heat transfer coefficient h and wind speed are related) is considered:

$q' = q - hA(T_{upper} - T_\infty)$.
- Among the layers of of the PV-module, heat flux continuity is utilized, $q_{layer,n+1} = q_{layer,n}$.
- At the inlet of the channel, temperature and velocity are uniform, $u = u_0$, $v = 0$, $T = T_c$.
- At the exit of the channel, gradients in the x-direction are set to zero, $\frac{\partial u}{\partial x} = 0$, $\frac{\partial v}{\partial x} = 0$, $\frac{\partial T}{\partial x} = 0$.

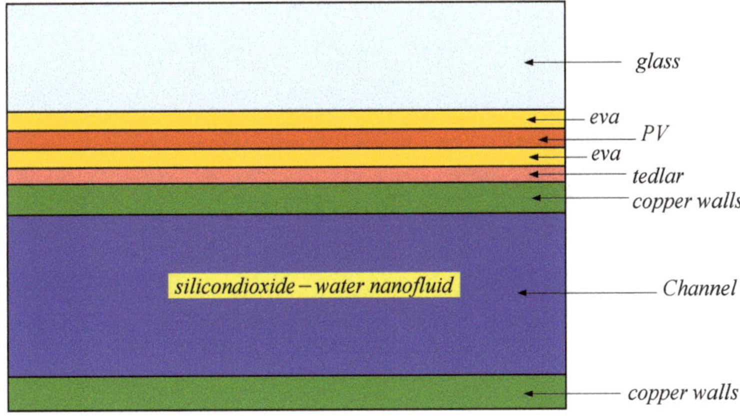

Figure 1. Schematic representation of the layers in the PV-thermal module.

Table 1. Thermophysical properties of layers of the PV-thermal module.

Name	Thickness (mm)	Density (kg/m^3)	Thermal Conductivity (W/m K)	Heat Capacity (J/kg K)
glass	3.2	2515	0.98	820
eva	0.45	960	0.31	2090
PV	0.2	2330	150	712
tedlar	0.35	1162	0.23	1465
aluminum	1	2700	160	900

2.1. Nanofluid Thermophysical Properties

SiO_2-water nanofluid was used in this study and the thermo-physical properties are given in Table 2 [24]. The effective density, specific heat, thermal expansion coefficient of the nanofluid are given by the following formulas:

$$\rho_{nf} = (1 - \phi)\rho_f + \phi\rho_p, \tag{7}$$

$$(\rho c_p)_{nf} = (1 - \phi)(\rho c_p)_f + \phi(\rho c_p)_p, \tag{8}$$

$$(\rho\beta)_{nf} = (1 - \phi)(\rho\beta)_f + \phi(\rho\beta)_p, \tag{9}$$

where the subscripts f, nf and p denote the base fluid, nanofluid and solid particle, respectively.

The effective thermal conductivity of the nanofluid includes the effect of Brownian motion. In this model, the effects of particle size, particle volume fraction and temperature dependence are taken into account and it is given by the following formula [25]:

$$k_{nf} = k_{st} + k_{Brownian}, \tag{10}$$

where k_{st} is the static thermal conductivity as given by [26]

$$k_{st} = k_f \left[\frac{(k_p + 2k_f) - 2\phi(k_f - k_p)}{(k_p + 2k_f) + \phi(k_f - k_p)} \right]. \tag{11}$$

The interaction between the nanoparticles and the effect of temperature are included in the models as

$$k_{Brownian} = 5 \times 10^4 \times 1.9526 \times (100\phi)^{-1.4594} \phi\rho_f c_{p,f} \sqrt{\frac{\kappa_b T}{\rho_p d_p}} f'(T, \phi), \tag{12}$$

where the function f' is given in [25].

The effective viscosity model of the nanofluid was given in [27]

$$\mu_{nf} = \mu_f \frac{1}{\left(1 - 34.87 \left(\frac{d_p}{d_f}\right)^{-0.3} \phi^{1.03}\right)}, \tag{13}$$

where the average particle size of the fluid is given as [27]:

$$d_f = \left(\frac{6M}{N\pi\rho_f}\right)^{1/3}, \tag{14}$$

with M and N denoting the molecular weight and Avogadro number.

Table 2. Thermophysical properties of base fluid and SiO_2 nanoparticle [28].

Property	Water	SiO_2
ρ (kg/m^3)	998.2	2200
c_p (J/kg K)	4812	703
k (W/mK)	0.61	1.2
μ (N s/m^2)	0.001003	-

2.2. Nanoparticle Shape Effect

The above given correlations in Equations (10)–(14) are used for the description of effective thermal conductivity for spherical particles. The effective thermal conductivity and viscosity of the nanofluid using non-spherical nanoparticle shape are defined using the following formulas:

$$k_{nf} = k_f \left(1 + C_k \phi\right), \quad \mu_{nf} = \mu_f \left(1 + A_1 \phi + A_2 \phi^2\right), \tag{15}$$

where the constant coefficients for different nanoparticle shapes are defined as in Table 3 [28,29].

Table 3. Constant coefficients for the effect of nanoparticle shape to the thermal conductivity and viscosity of the nanofluid [28,29].

Nanoparticle Type	C_k	A_1	A_2
cylindrical	3.95	13.5	904.4
bricks	3.37	1.9	471.4
blades	2.74	14.6	123.3

2.3. Solution Method

The finite volume method was used to solve the governing equations along with the boundary conditions. A general convection-diffusion equation for a scalar transport variable Ψ has the following form:

$$\nabla.\left(\rho \mathbf{u}\Psi\right) = \nabla.\left(\Gamma \nabla \Psi\right) + b \tag{16}$$

for velocity $\mathbf{u}$, source term b and diffusion coefficient Γ. Integration of the PDE over a control volume and using Gauss divergence theorem yields:

$$\int_A (\mathbf{n}).\left(\rho \mathbf{u}\Psi\right) dA = \int_A (\mathbf{n}).\left(\Gamma \nabla \Psi\right) dA + \int_{CV} b\, dV. \tag{17}$$

After using suitable discretization schemes for convective and diffusion terms, the resulting algebraic equation at the node point p surrounded by neighboring relevant nodes (subscript n) is written as:

$$a_p \phi_p = \sum a_n \phi_n + s. \tag{18}$$

A QUICK scheme is used to discretize the convective terms in the momentum and energy equations while a SIMPLE algorithm is used for velocity–pressure coupling. The resulting system of algebraic equations was solved using the Gauss–Siedel point-by-point iterative method and algebraic multigrid method. The normalized residual is calculated as:

$$R^\phi = \frac{\sum_{all\ cells} \left|a_p \phi_p - a_n \phi_n - s\right|}{\sum_{all\ cells} \left|a_p \phi_p\right|}. \tag{19}$$

When the residuals for all dependent variables become less than 10^{-5}, an iterative solution is stopped. Under-relaxation factors are used to enhance the converge speed of the solution and the under-relaxation parameters for u, v, and T are all set to 0.6, whereas the under-relaxation parameter for pressure correction is set to 0.32.

2.4. Grid Independence and Code Validation

The grid independent test for various numbers of elements was performed. High gradients in the boundary layers are resolved by using finer meshes near the walls.Thermal efficiency and PV efficiency for different number of elements are demonstrated in Table 4. G3 with 66,056 triangular elements are used in the subsequent computations. Validation of the present code is performed by using the numerical results of [30]. Forced convection in a cavity was considered at Reynolds number of 500. The comparison results for the local Nusselt number distribution along the walls of the cavity are shown in Figure 2.

Table 4. Grid independence test (q = 1000 W/m^2, h = 5 W/m^{2}K, $\phi = 0.05$, T_in = 30 °C).

Grid Name	Number of Elements	Thermal Efficiency (%)	PV Efficiency (%)
G1	10,816	50.32	12.52
G2	19,457	48.25	12.50
G3	66,056	47.20	12.49
G4	144,934	47.13	12.49

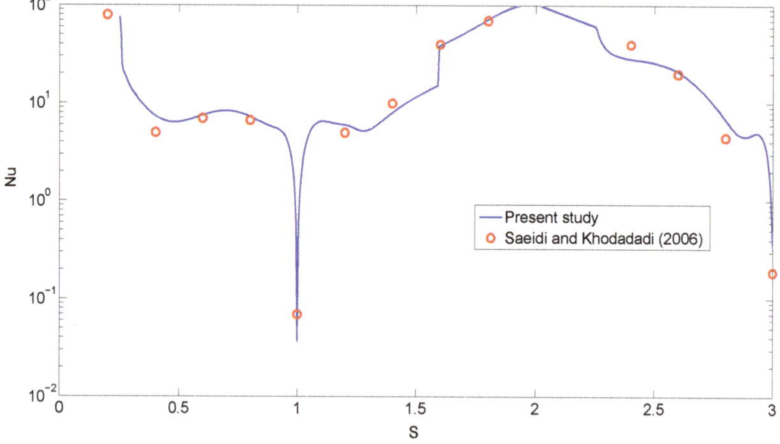

Figure 2. Code validation study.

3. Results and Discussion

Effects of nanoparticle addition to the water in the channel of a PV-thermal module on the thermal and PV-efficiency was numerically investigated. SiO$_2$ nanoparticles different shapes and solid particle volume fractions were used. Figure 3 shows the velocity and temperature distribution in the PV-module. In the channel, a laminar velocity profile is developed and its maximum value is seen in the mid of the channel which has a value of 0.025 m/s for the fixed value of (q = 1000 W /m^2, h = 5 W /m^2 K, $\phi = 0.02$ with cylindrical shape particles). For this flow velocity, Reynolds number remains less than 2100 in the channel. Thermal gradients are seen in the layers of the PV module, which is due to the different thicknesses and thermal conductivities of these layers.

Figure 4 shows the effects of nanoparticle volume fraction (ϕ) and particle type on the variation of thermal and total efficiency of the PV-thermal module. Both efficiencies enhance with higher ϕ values. Among different particle shapes, cylindrical ones perform best. Discrepancy between cylindrical shape and other shapes increases for higher particle volume fractions. Total efficiency increases by about 7.39% at the highest volume fraction ($\phi = 0.05$) with cylindrical shape particles. As compared to spherical shape particle, cylindrical ones gives 3.95% more enhancement in the total efficiency for the highest particle volume fraction.

As the inlet temperature of water-SiO$_2$ nanofluid increases, thermal and total efficiency deteriorate as shown in Figure 5. The rate of deterioration is higher for the thermal efficiency and up to 40% in the reduction of the efficiency is seen when nanofluid temperature is increased from 10 °C to 50 °C.

Figure 6 demonstrates the effects of solar irradiation and solid nanoparticles volume faction on the variation of efficiencies (h = 5 W /m^2 K, T_{in} = 30 °C with cylindrical shape particles). Thermal efficiency increases with higher values of solar irradiation while the PV-efficiency decreases and higher efficiency values are achieved for higher ϕ values. Higher surface temperature is obtained for higher values of solar irradiation and PV-efficiency decreases, which are defined in Equation (5). Thermal efficiency increases by about 9.17% and 9.82% for water and for nanofluid with highest volume fraction. There is a negligible effect of the particle addition on the PV-efficiency enhancements.

Convective loss is characterized by the convective heat transfer coefficient dependent upon the wind speed v. As the value of heat transfer coefficient enhances, thermal efficiency decreases as it is shown in Figures 7 and 8. However, PV-efficiency enhances with higher h values since the PV-layer surface temperature decreases, but the rate of enhancement is not significant.The discrepancy between thermal efficiency for heat transfer coefficient of h = 5 W /m^2 K and h = 10 W /m^2 K is 12.5% and 9.28% for water and for nanofluid with $\phi = 0.05$. Adding nanoparticle results in higher thermal efficiency enhancement for the highest value of heat transfer coefficient, which is 11.11% with the highest volume fraction of cylindrical particles.

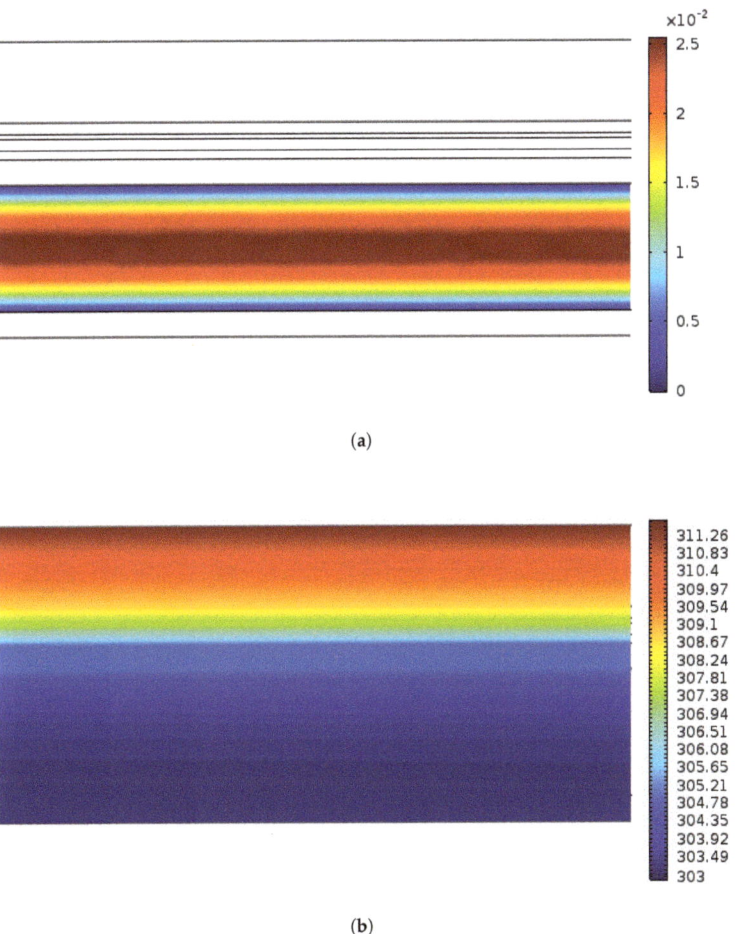

Figure 3. velocity field (**a**) and temperature (**b**) in the Photovoltaic-thermal module module, (q = 1000 W/m^2, h = 5 W/m^{2}K, $\phi = 0.02$, cylindrical shape).

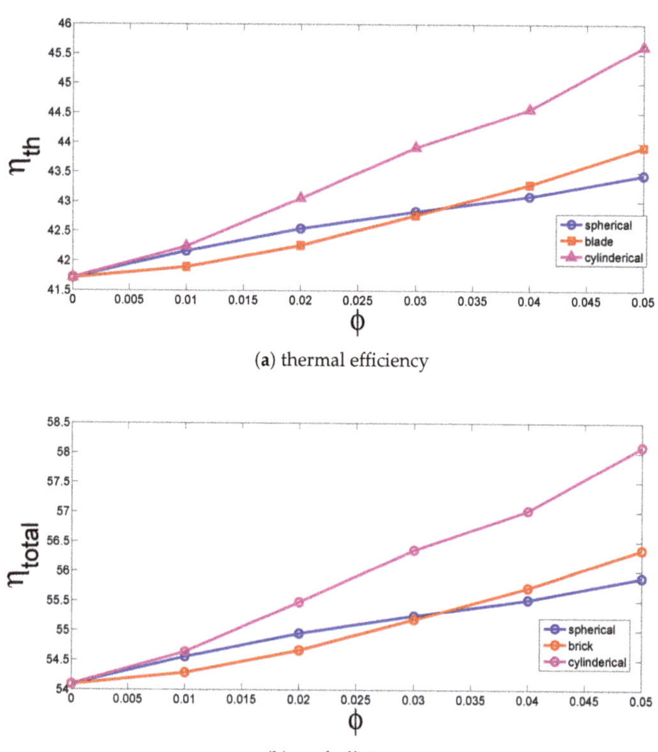

(**a**) thermal efficiency

(**b**) total efficiency

Figure 4. Effects of the particle shape and solid volume fraction on the variation of efficiencies (q = 1000 W/m^2, h = 5 W/m^{2}K, $T_i n$ = 30 °C)

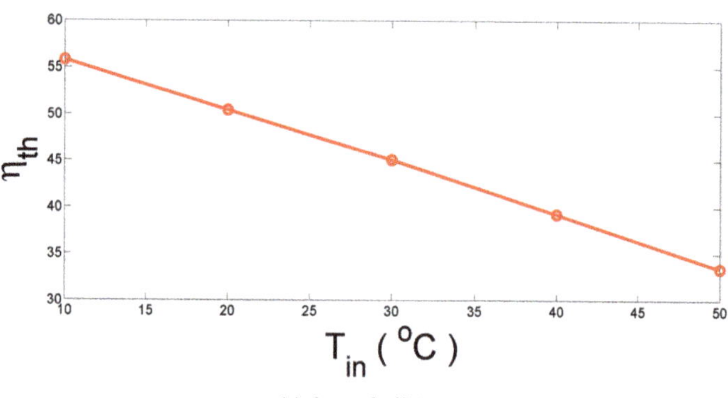

(**a**) thermal efficiency

Figure 5. *Cont.*

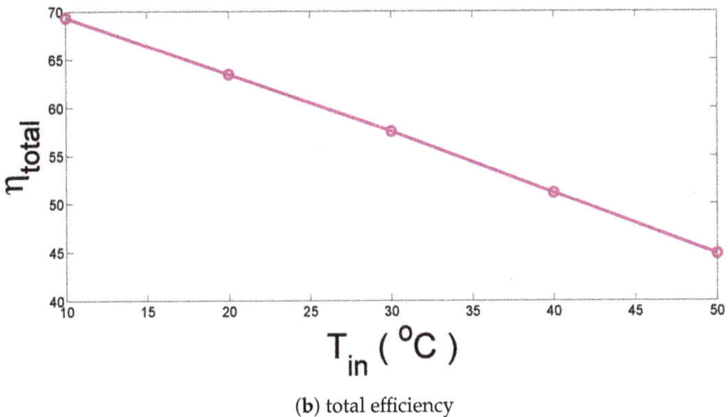

(**b**) total efficiency

Figure 5. Effects of the inlet temperature on the variation of efficiencies (q = 1000 W/m^2, h = 5 W/m^{2}K, $\phi = 0.02$, cylindrical shape).

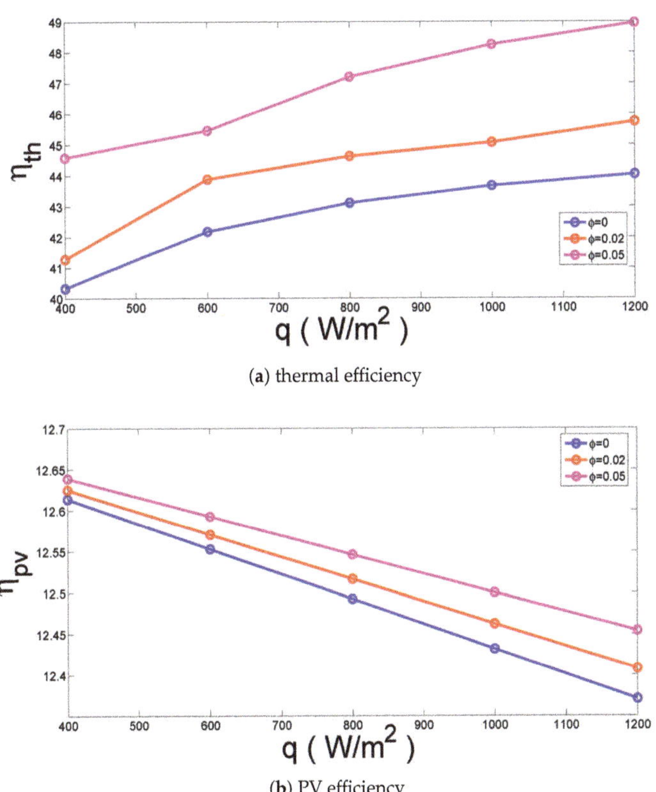

(**a**) thermal efficiency

(**b**) PV efficiency

Figure 6. *Cont.*

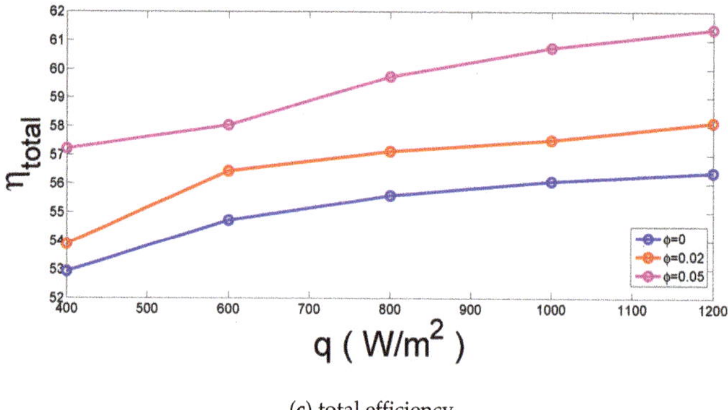

(**c**) total efficiency

Figure 6. Effects of the solar radiation on the variation of efficiencies for various solid particle volume fraction (h = 5 W/m^{2}K, T_in = 30 °C, cylindrical shape).

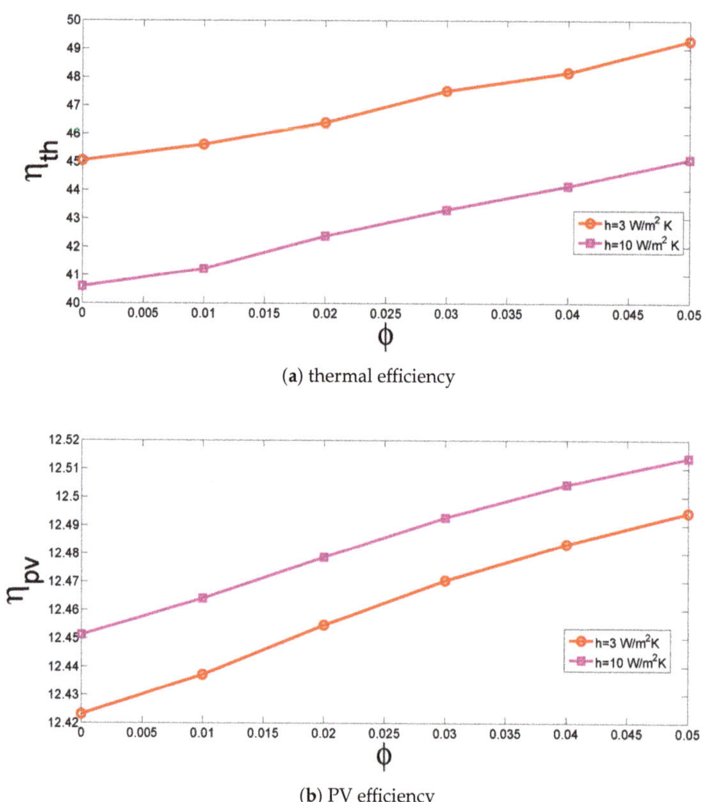

(**a**) thermal efficiency

(**b**) PV efficiency

Figure 7. Efficiency versus solid particle volume fraction for two values of external heat transfer coefficient (q = 1000 W/m^2, T_{in} = 30 °C, cylindrical shape).

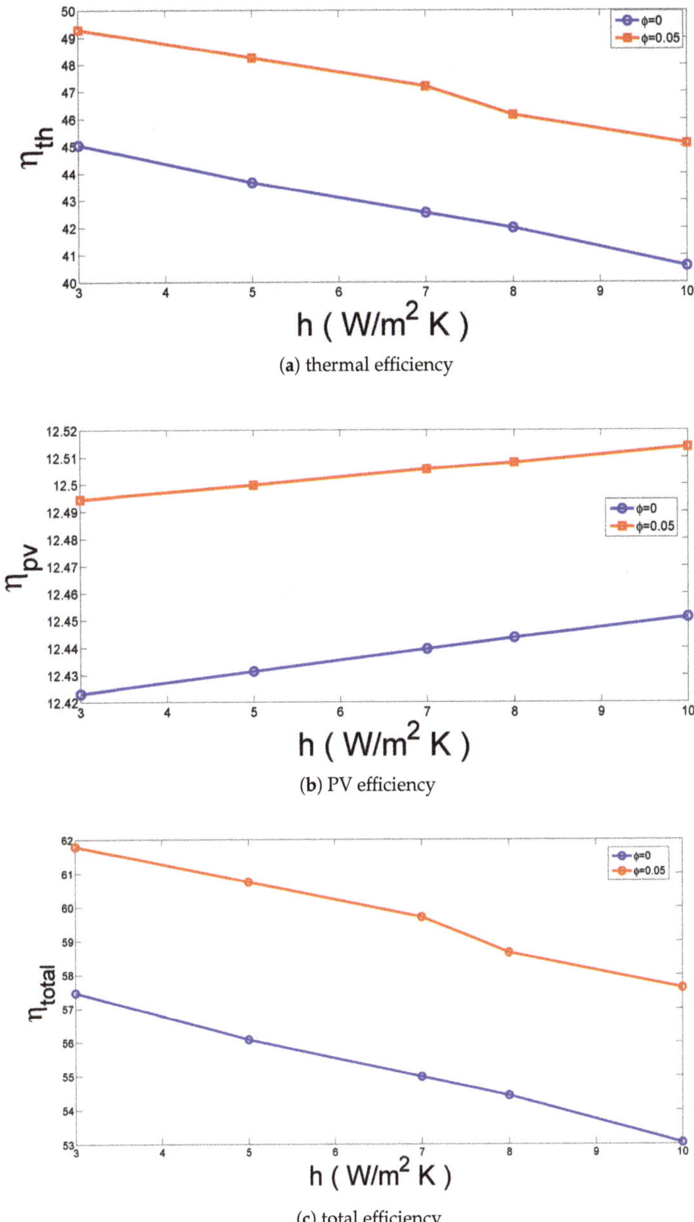

Figure 8. Efficiency versus external heat transfer coefficient for two values of nanoparticle volume concentration (q = 1000 W/m^2, T_{in} = 30 °C, cylindrical shape).

3.1. Efficiency Correlation with Artificial Neural Networks

Artificial neural networks (ANN) or other practical prediction methods can be used to obtain the correlations for efficiencies of PV-thermal module or thermal engineering systems [31–41]. Radial basis function networks consist of three-layer network structures that include input, hidden and output layers. The hidden layer nodes are radial basis functions. The outputs are calculated by a weighted average sum of the radial basis functions, which can be given as [37]:

$$y(x_i) = \sum_{k=1}^{N} w_k \Psi \left(||x_i - d_k|| \right) + b. \tag{20}$$

Radial basis function response decreases monotonically from a center point with distance. Gaussian function is a radial basis function which has central point c and smoothness parameter σ which controls the shape of the function. It is given in the following form:

$$f(x) = e^{\left(-(x-c)^2 / \sigma^2 \right)}. \tag{21}$$

A schematic representation of network topology is given in Figure 9 with three inputs: (solid particle volume fraction (ϕ), convective heat transfer coefficient (h) and solar irradiation (q)) and two outputs (thermal and PV-efficiency).

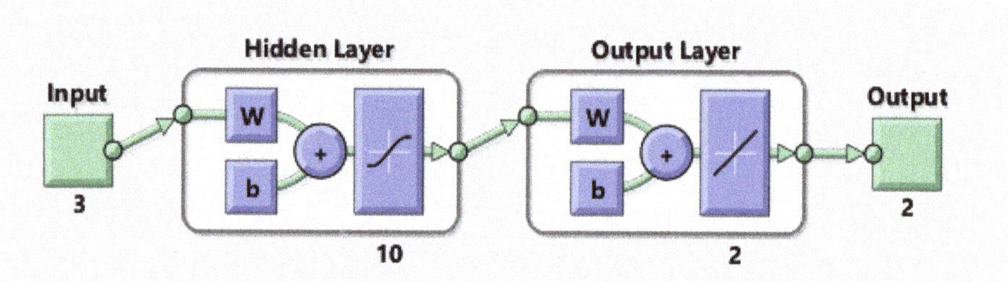

Figure 9. Schematic representation of the layers in the PV-thermal module.

In order to evaluate the performance of trained networks, different error measures can be used. Mean square error (MSE) and coefficient of determination (R^2) performance parameters can be given as:

$$\text{MSE} = \frac{1}{N} \sum_{k=1}^{N} (y^{CFD} - y_k^{ANN})^2, \tag{22}$$

$$R^2 = 1 - \frac{\sum_{k=1}^{N} (y_k^{CFD} - y_k^{ANN})^2}{\sum_{k=1}^{N} (y_k^{CFD} - \bar{y})^2}, \tag{23}$$

where y_k^{ANN}, y_k^{CFD}, N and $\bar{y}$ represent the predicted value form ANN, CFD value, sample number and the mean value of CFD values, respectively.

The MATLAB Neural Network Toolbox (Version 2010, The Mathworks, Natick, USA) was used to select the number of hidden layers, number of neurons in each layer, training algorithm [42]. Feed-forward network structure with one hidden layer and a linear output layer was selected. The number of the neurons of ANN model was taken as 10. The feed-forward network structure with hidden layers and the linear output layer was selected and Levenberg–Marquardt back-propagation was used as the training algorithm. The random data division property of MATLAB is used and 70% of the data was used for estimation while 15% was used for validation and 15% of the data was used for testing purposes. Table 5 shows the number of samples for training, validation and testing, mean squared error (MSE) values and regression R values. A higher R value denotes a higher correlation between the outputs and target values.

Table 6 represents the comparison results of CFD data between the predicted data by artificial neural networks for various values of pertinent parameters. The difference between the actual CFD data and established artificial neural network model is very small. This modeling strategy with ANN is useful for this system in order to obtain the performance predictions of a PV-thermal module in a fast and cheap way as compared to a high fidelity CFD computation, but it still requires some of the data from CFD computations for training.

Table 5. Number of samples, mean square error (MSE) and correlation coefficients in the artificial neural network (ANN) modeling.

	Number of Samples	MSE	R
Training	378	3.89×10^{-6}	0.99992
Validation	81	4.60×10^{-6}	0.99991
Testing	81	3.53×10^{-6}	0.99993

Table 6. Performance predictions of solar PV-thermal module with ANN.

ϕ	h (W/m^{2}K)	q (W/m^2)	$\eta_{thermal}$ (CFD)	$\eta_{thermal}$ (ANN)	η_{PV} (CFD)	η_{PV} (ANN)
0	3	400	41.716	41.953	12.609	12.594
0	3	800	44.497	44.436	12.485	12.444
0	3	1200	45.424	45.31	12.361	12.408
0	4	700	43.305	43.281	12.519	12.507
0	4	1100	44.497	44.576	12.397	12.417
0	5	600	42.18	41.973	12.553	12.557
0	5	1000	43.663	43.636	12.431	12.454
0	6	500	40.604	40.399	12.586	12.593
0	6	900	42.952	42.708	12.465	12.479
0	7	400	38.24	38.28	12.619	12.63
0	7	800	41.716	41.797	12.499	12.491
0	7	1200	42.875	42.967	12.38	12.417
0	8	700	40.524	40.696	12.532	12.527
0	8	1100	42.222	42.148	12.414	12.427
0.015	3	400	42.47	42.85	12.617	12.606
0.015	3	800	45.365	45.387	12.503	12.478
0.015	3	1200	46.33	46.291	12.389	12.379
0.015	4	700	44.124	44.235	12.535	12.529
0.015	4	1100	45.629	45.532	12.421	12.427
0.015	5	600	43.113	42.926	12.566	12.568
0.015	5	1000	44.786	44.604	12.454	12.46
0.015	6	500	41.697	41.235	12.596	12.601
0.015	6	900	43.757	43.674	12.485	12.477
0.015	7	400	38.609	38.995	12.626	12.634
0.015	7	800	42.47	42.634	12.516	12.513
0.015	7	1200	43.757	43.799	12.406	12.424
0.015	8	700	41.367	41.526	12.547	12.539
0.015	8	1100	43.172	42.991	12.437	12.44
0.02	3	600	44.62	44.91	12.565	12.555
0.02	3	1000	46.404	46.344	12.455	12.428
0.02	4	500	43.727	43.395	12.595	12.589
0.02	4	900	45.611	45.438	12.486	12.479
0.02	5	400	41.273	41.261	12.625	12.623
0.02	5	800	44.62	44.356	12.516	12.52
0.02	5	1200	45.363	45.515	12.408	12.421
0.02	6	700	43.345	43.185	12.546	12.56
0.02	6	1100	44.62	44.647	12.438	12.434
0.02	7	600	41.645	41.829	12.576	12.579
0.02	7	1000	43.727	43.765	12.469	12.456
0.02	8	500	40.158	39.914	12.605	12.598
0.02	8	900	42.637	42.828	12.499	12.493
0.025	3	400	44.038	43.705	12.624	12.61

Table 6. *Cont.*

ϕ	h (W/m^{2}K)	q (W/m^2)	$\eta_{thermal}$ (CFD)	$\eta_{thermal}$ (ANN)	η_{PV} (CFD)	η_{PV} (ANN)
0.025	3	800	45.981	46.237	12.516	12.501
0.025	3	1200	47.06	47.149	12.409	12.386
0.025	4	700	45.148	45.11	12.546	12.551
0.025	4	1100	46.158	46.375	12.44	12.443
0.025	5	600	44.038	43.801	12.575	12.583
0.025	5	1000	45.593	45.517	12.47	12.464
0.025	6	500	42.484	42.021	12.604	12.608
0.025	6	900	44.326	44.522	12.499	12.495
0.025	7	400	40.153	39.743	12.632	12.639
0.025	7	800	43.391	43.495	12.528	12.526
0.025	7	1200	44.902	44.648	12.425	12.435
0.025	8	700	42.188	42.4	12.557	12.542
0.025	8	1100	43.803	43.829	12.454	12.458
0.035	3	400	45.282	44.557	12.629	12.617
0.035	3	800	47.023	47.139	12.528	12.527
0.035	3	1200	48.184	48.136	12.427	12.414
0.035	4	700	45.779	46.11	12.556	12.573
0.035	4	1100	47.498	47.371	12.455	12.447
0.035	5	600	45.282	44.79	12.583	12.592
0.035	5	1000	46.675	46.538	12.484	12.479
0.035	6	500	43.192	42.898	12.61	12.614
0.035	6	900	45.669	45.523	12.512	12.514
0.035	7	400	40.057	40.62	12.637	12.642
0.035	7	800	44.411	44.556	12.539	12.532
0.035	7	1200	45.862	45.682	12.441	12.452
0.035	8	700	42.794	43.424	12.566	12.545
0.035	8	1100	44.965	44.845	12.469	12.475
0.04	3	600	46.84	46.837	12.582	12.58
0.04	3	1000	48.178	48.229	12.483	12.472
0.04	4	500	44.966	45.202	12.609	12.608
0.04	4	900	47.286	47.305	12.511	12.528
0.04	5	400	42.156	42.872	12.636	12.641
0.04	5	800	46.171	46.335	12.538	12.558
0.04	5	1200	47.509	47.53	12.441	12.434
0.04	6	700	44.737	45.254	12.565	12.563
0.04	6	1100	46.718	46.678	12.469	12.467
0.04	7	600	44.163	43.741	12.591	12.572
0.04	7	1000	45.769	45.72	12.496	12.501
0.04	8	500	41.755	41.638	12.617	12.596
0.04	8	900	44.61	44.85	12.523	12.524
0.045	3	400	46.023	45.272	12.634	12.622
0.045	3	800	48.324	48.05	12.537	12.549
0.045	3	1200	49.092	49.222	12.441	12.43
0.045	4	700	47.338	47.102	12.564	12.579
0.045	4	1100	48.534	48.38	12.468	12.454
0.045	5	600	46.023	45.742	12.59	12.595
0.045	5	1000	47.864	47.522	12.495	12.499
0.045	6	500	44.182	43.729	12.616	12.617
0.045	6	900	46.023	46.579	12.521	12.526
0.045	7	400	41.421	41.485	12.641	12.641
0.045	7	800	46.023	45.646	12.548	12.536
0.045	7	1200	46.79	46.784	12.455	12.466
0.045	8	700	44.708	44.436	12.574	12.549
0.045	8	1100	46.023	45.908	12.481	12.495
0.05	3	600	47.203	47.64	12.589	12.58
0.05	3	1000	49.301	49.157	12.495	12.499
0.05	4	500	46.154	45.935	12.614	12.613

Table 6. *Cont.*

ϕ	h (W/m^{2}K)	q (W/m^2)	$\eta_{thermal}$ (CFD)	$\eta_{thermal}$ (ANN)	η_{PV} (CFD)	η_{PV} (ANN)
0.05	4	900	47.786	48.193	12.521	12.538
0.05	5	400	44.581	43.556	12.639	12.647
0.05	5	800	47.203	47.351	12.547	12.554
0.05	5	1200	48.078	48.497	12.454	12.449
0.05	6	700	46.454	46.239	12.572	12.558
0.05	6	1100	47.68	47.661	12.48	12.488
0.05	7	600	45.455	44.587	12.597	12.573
0.05	7	1000	47.203	46.745	12.506	12.518
0.05	8	500	41.959	42.418	12.622	12.602
0.05	8	900	45.455	45.935	12.531	12.531

4. Conclusions

In this study, a numerical simulation of a PV-thermal module with SiO$_2$-water nanofluid was performed. It was observed that cylindrical shape particles give the best performance in terms of efficiency enhancement. Total PV/T module efficiency enhances by about 7.39% at the highest volume fraction with cylindrical shape particles. As compared to spherical ones, up to 4% more in the efficiency enhancement was observed with cylindrical shape particles. Thermal and total efficiency increase for higher solid particle volume fraction, higher values of solar irradiation, lower values of convective heat transfer coefficient and inlet temperature. Adding nanoparticles is advantageous for the case where convective heat transfer coefficient is high. Finally, correlation based on radial basis artificial neural networks was obtained for thermal and PV-efficiency of the PV-thermal module. The performance characteristics of solar PV-thermal module with ANN are compared with those obtained using the CFD modeling and have been to be in excellent agreement

Author Contributions: F.S. performed the numerical simulations and wrote some sections of the manuscript. A.J.C. prepared some other sections of the paper and analyzed the results. All of the authors contributed equally for reviewing and revising the manuscript.

Funding: This research received no external funding.

Conflicts of Interest: The authors declare no conflict of interest.

Abbreviations

b bias term
c center point
d particle size
G incident energy
h local heat transfer coefficient
k thermal conductivity
M molecular weight
N Avogadro number
n unit normal vector
p pressure
R residual
T temperature
u, v x–y velocity components
w weight of neural network
x, y Cartesian coordinates

Greek Characters

α thermal diffusivity
β thermal diffusivity
η efficient
θ non-dimensional temperature
μ dynamic viscosity
ρ density of the fluid
σ smoothing parameter
ϕ solid volume fraction
Ψ radial basis function

Subscripts

c cold
h hot
nf nanofluid
n neigbour
pv photo-voltaic
ref reference
th thermal

References

1. Nielda, D.; Kuznetsov, A. Forced convection in a parallel-plate channel occupied by a nanofluid or a porous medium saturated by a nanofluid. *Int. J. Heat Mass Transf.* **2014**, *70*, 430–433. [CrossRef]
2. Abu-Nada, E.; Chamkha, A.J. Mixed convection flow in a lid-driven inclined square enclosure filled with a nanofluid. *Eur. J. Mech. B/Fluids* **2010**, *29*, 472–482. [CrossRef]
3. Chamkha, A.J.; Abu-Nada, E. Mixed convection flow in single- and double-lid driven square cavities filled with water-Al_2O_3 nanofluid: Effect of viscosity models. *Eur. J. Mech. B/Fluids* **2012**, *36*, 82–96. [CrossRef]
4. Mahmoudi, A.H.; Pop, I.; Shahi, M. Effect of magnetic field on natural convection in a triangular enclosure filled with nanofluid. *Int. J. Therm. Sci.* **2012**, *59*, 126–140. [CrossRef]
5. Oztop, H.F.; Abu-Nada, E. Numerical study of natural convection in partially heated rectangular enclosures filled with nanofluids. *Int. J. Heat Fluid Flow* **2008**, *29*, 1326–1336. [CrossRef]
6. Selimefendigil, F.; Oztop, H.F. Pulsating nanofluids jet impingement cooling of a heated horizontal surface. *Int. J. Heat Mass Transf.* **2014**, *69*, 54–65. [CrossRef]
7. Sheikholeslami, M.; Gorji-Bandpy, M.; Ganji, D.; Soleimani, S.; Seyyedi, S. Natural convection of nanofluids in an enclosure between a circular and a sinusoidal cylinder in the presence of magnetic field. *Int. Commun. Heat Mass Transf.* **2012**, *39*, 1435 –1443. [CrossRef]
8. Selimefendigil, F.; Oztop, H.F. Identification of forced convection in pulsating flow at a backward facing step with a stationary cylinder subjected to nanofluid. *Int. Commun. Heat Mass Transf.* **2013**, *45*, 111–121. [CrossRef]
9. Sridhara, V.; Satapathy, L.N. Al_2O_3-based nanofluids: A review. *Nanoscale Res. Lett.* **2011**, *6*, 1–16. [CrossRef] [PubMed]
10. Selimefendigil, F.; Oztop, H.F. Influence of inclination angle of magnetic field on mixed convection of nanofluid flow over a backward facing step and entropy generation. *Adv. Powder Technol.* **2015**, *26*, 1663–1675. [CrossRef]
11. Gherasim, I.; Roy, G.; Nguyen, C.T.; Vo-Ngoc, D. Experimental investigation of nanofluids in confined laminar radial flows. *Int. J. Therm. Sci.* **2009**, *48*, 1486–1493. [CrossRef]
12. Selimefendigil, F.; Oztop, H.F.; Abu-Hamdeh, N. Mixed convection due to rotating cylinder in an internally heated and flexible walled cavity filled with SiO_2-water nanofluids: Effect of nanoparticle shape. *Int. Commun. Heat Mass Transf.* **2016**, *71*, 9–19. [CrossRef]
13. Selimefendigil, F.; Oztop, H.F. Effects of Nanoparticle Shape on Slot-Jet Impingement Cooling of a Corrugated Surface With Nanofluids. *J. Therm. Sci. Eng. Appl.* **2017**, *9*, 021016. [CrossRef]
14. Selimefendigil, F.; Oztop, H.F. Laminar Convective Nanofluid Flow Over a Backward-Facing Step With an Elastic Bottom Wall. *J. Therm. Sci. Eng. Appl.* **2018**, *10*, 041003. [CrossRef]

15. Mahian, O.; Kianifar, A.; Kalogirou, S.A.; Pop, I.; Wongwises, S. A review of the applications of nanofluids in solar energy. *Int. J. Heat Mass Transf.* **2013**, *57*, 582–594. [CrossRef]

16. Mahian, O.; Kianifar, A.; Sahin, A.Z.; Wongwises, S. Performance analysis of a minichannel-based solar collector using different nanofluids. *Energy Convers. Manag.* **2014**, *88*, 129–138. [CrossRef]

17. Meibodi, S.S.; Kianifar, A.; Niazmand, H.; Mahian, O.; Wongwises, S. Experimental investigation on the thermal efficiency and performance characteristics of a flat plate solar collector using SiO_2/EG-water nanofluids. *Int. J. Heat Mass Transf.* **2015**, *65*, 71–75. [CrossRef]

18. Chen, M.; He, Y.; Huang, J.; Zhu, J. Investigation into Au nanofluids for solar photothermal conversion. *Int. J. Heat Mass Transf.* **2017**, *108*, 1894–1900. [CrossRef]

19. Yan, S.; , Wang, F.; Shi, Z.; Tian, R. Heat transfer property of SiO_2/water nanofluid flow inside solar collector vacuum tubes. *Appl. Therm. Eng.* **2017**, *118*, 385–391. [CrossRef]

20. Yazdanifard, F.; Ameri, M.; Ebrahimnia-Bajestan, E. Performance of nanofluid-based photovoltaic/thermal systems: A review. *Renew. Sustain. Energy Rev.* **2017**, *76*, 323–352. [CrossRef]

21. Al-Waeli, A.H.; Sopian, K.; Chaichan, M.T.; Kazem, H.A.; Hasan, H.A.; Al-Shamania, A.N. An experimental investigation of SiC nanofluid as a base-fluid for a photovoltaic thermal PV/T system. *Energy Convers. Manag.* **2017**, *142*, 547–558. [CrossRef]

22. Hassani, S.; Taylor, R.A.; Mekhilef, S.; Saidur, R. A cascade nanofluid-based PV/T system with optimized optical and thermal properties. *Energy* **2016**, *112*, 963–975. [CrossRef]

23. Jing, D.; Hu, Y.; Liu, M.; Wei, J.; Guo, L. Preparation of highly dispersed nanofluid and CFD study of its utilization in a concentrating PV/T system. *Sol. Energy* **2015**, *112*, 30–40. [CrossRef]

24. Vajjha, R.; Das, D. Experimental determination of thermal conductivity of three nanofluids and development of new correlations. *Int. J. Heat Mass Transf.* **2009**, *52*, 4675–4682. [CrossRef]

25. Koo, J.; Kleinstreuer, C. Laminar nanofluid flow in microheat-sinks. *Int. J. Heat Mass Transf.* **2005**, *48*, 2652–2661. [CrossRef]

26. Maxwell, J. *A Treatise on Electricity and Magnetism*; Oxford University Press: Oxford, UK, 1873.

27. Corcione, M. Heat transfer features of buoyancy-driven nanofluids inside rectangular enclosures differentially heated at the sidewalls. *Int. J. Therm. Sci.* **2010**, *49*, 1536–1546. [CrossRef]

28. Vanaki, S.M.; Mohammed, H.A.; Abdollahi, A.; Wahid, M.A. Effect of nanoparticle shapes on the heat transfer enhancement in a wavy channel with different phase shifts. *J. Mol. Liquids* **2014**, *196*, 32–42. [CrossRef]

29. Timofeeva, E.; Routbort, J.; Singh, D. Particle shape effects on thermophysical properties of alumina nanofluids. *J. Appl. Phys.* **2009**, *106*, 014304. [CrossRef]

30. Saeidi, S.; Khodadadi, J. Forced convection in a square cavity with inlet and outlet ports. *Int. J. Heat Mass Transf.* **2006**, *49*, 1896–1906. [CrossRef]

31. Chine, W.; Mellit, A.; Lughi, V.; Malek, A.; Sulligoi, G.; Pavan, A.M. A novel fault diagnosis technique for photovoltaic systems based on artificial neural networks. *Renew. Energy* **2016**, *90*, 501–512. [CrossRef]

32. Selimefendigil, F.; Bayrak, F.; Oztop, H.F. Experimental analysis and dynamic modeling of a photovoltaic module with porous fins. *Renew. Energy* **2018**, *125*, 193–205. [CrossRef]

33. Selimefendigil, F.; Oztop, H.F. Numerical Study and POD-Based Prediction of Natural Convection in a Ferrofluids-Filled Triangular Cavity with Generalized Neural Networks. *Numer. Heat Transf. Part A Appl.* **2015**, *67*, 1136–1161. [CrossRef]

34. Mellit, A.; Saglam, S.; Kalogirou, S. Artificial neural network-based model for estimating the produced power of a photovoltaic module. *Renew. Energy* **2013**, *60*, 71–78. [CrossRef]

35. Huang, C.; Bensoussan, A.; Edesess, M.; Tsui, K.L. Improvement in artificial neural network-based estimation of grid connected photovoltaic power output. *Renew. Energy* **2016**, *97*, 838–848. [CrossRef]

36. Selimefendigil, F.; Polifke, W. Nonlinear, Proper-Orthogonal-Decomposition-Based Model of Forced Convection Heat Transfer in Pulsating Flow. *AIAA J.* **2014**, *52*, 131–145. [CrossRef]

37. Bonanno, F.; Capizzi, G.; Graditi, G.; Napoli, C.; Tina, G. A radial basis function neural network based approach for the electrical characteristics estimation of a photovoltaic module. *Appl. Energy* **2012**, *97*, 956–961. [CrossRef]

38. Selimefendigil, F.; Oztop, H.F. A Fuzzy-Pod Based Estimation of Unsteady Mixed Convection in a Partition Located Cavity with Inlet and Outlet Ports. *Int. J. Comput. Methods* **2015**, *12*, 1350107. [CrossRef]

39. Selimefendigil, F.; Oztop, H.F. Soft Computing Methods for Thermo-Acoustic Simulation. *Numer. Heat Transf. Part A Appl.* **2014**, *66*, 271–288. [CrossRef]
40. Aminossadati, S.; Kargar, A.; Ghasemi, B. Adaptive network-based fuzzy inference system analysis of mixed convection in a two-sided lid-driven cavity filled with a nanofluid. *Int. J. Therm. Sci.* **2012**, *52*, 102–111. [CrossRef]
41. Selimefendigil, F.; Oztop, H.F. POD-based reduced order model of a thermoacoustic heat engine. *Eur. J. Mech. B/Fluids* **2014**, *48*, 135–142. [CrossRef]
42. *Matlab: The Language of Technical Computing*; The Math Works Inc.: Natick MA, USA, 2000.

 *applied sciences*

Article

Experimental Efficiency Analysis of a Photovoltaic System with Different Module Technologies under Temperate Climate Conditions

Slawomir Gulkowski *, Agata Zdyb and Piotr Dragan

Institute of Renewable Energy Engineering, Faculty of Environmental Engineering, Lublin University of Technology, Nadbystrzycka 40B, 20-618 Lublin, Poland; a.zdyb@pollub.pl (A.Z.); piotr.dragan@wisznice.pl (P.D.)
* Correspondence: s.gulkowski@pollub.pl; Tel.: +48-81538-4654

Received: 30 October 2018; Accepted: 26 December 2018; Published: 3 January 2019

Featured Application: This research contributes to guiding planners and investors in the sizing, locating and selection of module types for photovoltaic (PV) installations. It can be useful for the prediction of electric energy production by different PV technologies at high latitude under temperate climate conditions.

Abstract: This study presents a comparative analysis of energy production over the year 2015 by the grid connected experimental photovoltaic (PV) system composed by different technology modules, which operates under temperate climate meteorological conditions of Eastern Poland. Two thin film technologies have been taken into account: cadmium telluride (CdTe) and copper indium gallium diselenide (CIGS). Rated power of each system is approximately equal to 3.5 kWp. In addition, the performance of a polycrystalline silicon technology system has been analyzed in order to provide comprehensive comparison of the efficiency of thin film and crystalline technologies in the same environmental conditions. The total size of the pc-Si system is equal to 17 kWp. Adequate sensors have been installed at the location of the PV system to measure solar irradiance and temperature of the modules. In real external conditions all kinds of modules exhibit lower efficiency than the values provided by manufacturers. The study reveals that CIGS technology is characterized by the highest energy production and performance ratio. The observed temperature related losses are of the lowest degree in case of CIGS modules.

Keywords: photovoltaic systems; thin film modules; performance assessment; PV efficiency

---

1. Introduction

The European Directive (EU) 2015/1513 amending Directive 2009/28/EC on the promotion of the use of energy from renewable sources sets a binding target of 20% final energy consumption in EU from renewable sources by 2020 [1,2]. One of the renewable energy systems (RES) is solar energy which can be utilized in photovoltaic (PV) systems that are non-polluting and do not generate greenhouse gases or wastes which have to be stored. This kind of system has no moving parts, so the maintenance cost is very little [3,4]. Due to its advantages, photovoltaic (PV) energy production has experienced a rapid growth all over the world in recent years, and, according to the policies of particular countries, a number of photovoltaic (PV) power plants increases [5–9].

Producers of PV modules provide characteristic parameters measured under Standard Test Conditions (STC) which are defined as follows: solar irradiance $G = 1000\ \text{W/m}^2$, module temperature 25 °C, AM 1.5, and wind speed less than 5 m/s. The testing measurements are usually made in laboratories with the use of solar simulators. Under real conditions, the results of PV performance can be different because of the influence of specific environmental parameters in the given location [10], such as global solar radiation intensity and spectrum [11], ambient temperature, relative humidity, wind speed, and dust concentration in the air [12–17]. All these factors strongly depend on meteorological conditions, characteristic for given climate [18,19]. Moreover, it is important that the influence of environmental conditions is complex and final energy production depends on the overlapping of several effects, which are difficult to analyze individually [20]. In general, the most important factors are solar irradiation and temperature, since they directly influence the energy production, and PV modules temperature, which affects modules efficiency. For this reason, outdoor analysis of daily and seasonal variations of PV module performance allow the energy production to be realistically estimated and also limit the over or under sizing in prospective plants.

In recent years much research on outdoor measurements of PV systems has been done, but the majority of them refer to warm climate conditions [21–23]. Nonetheless, spreading of PV technology all over the world implies the need for this kind of study also in locations characterized by lower insolation level such as temperate climate [24,25].

It is also worth noting that most of work on outdoor photovoltaic (PV) performance is devoted to the investigation of pc-Si technology, since this is the one which is the most popular on the PV market. However, the dropping price of thin film technology (e.g., a-Si, CdTe, CIGS) is a motivation to testing of various kinds of thin film solar modules under real external conditions. The following types of PV technologies were studied experimentally under different climatic conditions:

- a-Si, HIT, mc-Si—India [26],
- a-Si, HIT, pc-Si—India [27],
- a-Si, c-Si, CdTe—Turkey [5],
- mc-Si—Greece [2],
- mc-Si, pc-Si, a-Si, CIGS, CdTe–Italy [28],
- c-Si, a-Si, HIT—Italy [29],
- mono-Si, CIS—Turkey [30],
- a-Si, c-Si, CIGS—United Kingdom [11],
- a-Si, c-Si, CIGS, CdTe—Germany [31],
- mono-Si, poli-Si, a-Si, CIGS, CIS, CdTe—Netherlands [20].

All these investigations add new knowledge about the influence of external conditions on solar modules built with different semiconductor materials and allow for selecting the best technology type in the given location. At present, there are no published results of this kind of comparative study carried out in the Polish climate. Thus, the goal of this work is to characterize the solar conditions in Poland and evaluate the outdoor performance of three different PV technologies at the same location. This paper presents the comparison of traditional polycrystalline silicon (pc-Si) PV technology and two thin film technologies: cadmium telluride (CdTe) and copper-indium-gallium-diselenide (CIGS).

Monthly solar irradiation at the given location has been calculated and the energy yield for each type of PV technology has been compared. Daily efficiency and average yield, as well as direct current (DC) performance ratio, have been calculated and analyzed. The effects of the temperature on performance of different studied PV technologies have been also investigated, since the relevance of temperature as the parameter affecting modules efficiency depends on the type of semiconductor materials used. The attempt has been made to discuss the obtained results with research performed in other places under similar climate conditions.

2. Methodology

2.1. Experimental Photovoltaic Installation

The experimental PV installation consisted of different, fully commercial modules located in the East of Poland (Latitude 51 °51 and Longitude 23 °10) is shown in Figure 1. Technologies of the modules used in the analysis and their nominal power were as follows: cadmium telluride

(3.3 kWp), copper-indium-gallium-diselenide (3.72 kWp), and polycrystalline silicon (17 kWp). Detailed specification of each kind of module based on the PV manufacturer data sheet is presented in Table 1. The modules are oriented to the south and tilted at the optimum angle for the given latitude equal to 34°. The installation is connected to the grid using the inverters. Both thin film installations are connected with the use of inverters with one maximum power point tracker (MPPT), while for pc-Si PV string the inverter equipped with two MPPT systems is used. The distance between the rows of panels equals to 6.3 m.

Figure 1. Experimental installation consisting of different PV (photovoltaic) module technologies.

Table 1. PV modules specification data.

PV Technology	P_{max} [W]	Area [m^2]	Temp. Coefficient of P_{max}, β [%/°C]	Efficiency η_r [%]
Cadmium telluride (CdTe)	75	0.72	−0.25	10.6
Copper indium gallium diselenide (CIGS)	155	1.25	−0.31	12.6
Polycrystalline silicon (pc-Si)	250	1.55	−0.4	15.4

At the location of the experimental system, solar radiation intensity on PV module plane, as well as the temperature of the modules, were measured. A solar radiation sensor based on the monocrystalline silicon solar cell is located in the centre of the installation and tilted at the same angle as the modules. Pt1000 resistance temperature detectors are attached to the back of the module. Detailed specification of the devices used is listed in Table 2.

Table 2. Specification of the devices used for irradiance and temperature measurements.

Solar Radiation Sensor	Temperature Sensor
Monocrystalline silicon cell (5 × 3.3 cm) Temperature range: −20 °C to +70 °C Radiation range: 0 to max. 1400 W/m^2 Tolerance of the irradiance sensor: +/−5%	Range: −40 °C–70 °C

Inverters, as well as both mentioned sensors, were connected to the central data-logging computer system for synchronous data collection. All parameters, such as solar irradiance, DC generated electric power, and module temperature of each PV technology studied were measured at each 5 min during the considered year of 2015 and stored for the analysis. All necessary computations were carried out with the use of Matlab/Simulink software (MathWorks, Natick, MA, USA).

2.2. Methods of Experimental Data Analysis

Calculations for performance evaluation of PV modules were carried out according to the International Electrotechnical Commission standard IEC 61724-1 [32]. The daily energy (E_{DC}^d) produced by each technology PV array was calculated from the following equation:

$$E_{DC}^d = \sum_{sunrise}^{sunset} P_{DC}^d(t) \cdot \Delta t,$$

(1)

where: Δt—time step of measurements (5 min), $P_{DC}^d(t)$—power generated at a particular time of the day measured as a product between the current and the voltage at the inverter inlet.

Daily irradiation was calculated on the basis of irradiance G (W/m^2) according to the formula:

$$H^d \left(\frac{kWh}{m^2 \cdot day} \right) = \sum_{sunrise}^{sunset} G(t) \cdot \Delta t,$$

(2)

where $G(t)$—irradiance measured at a particular time t of the day.

The effect of the differences in energy production caused by different size of installation of each technology array was eliminated by dividing the daily DC energy production (E_{DC}^d) by the nominal power output of the installation at STC (P_M^{STC}). The daily yield is given by the ratio between daily energy produced by each technology PV array and output peak power P_M^{STC} of the modules under STC:

$$Y^d(kWh/kWp) = \frac{E_{DC}^d}{P_M^{STC}}.$$

(3)

The effect of insolation can be expressed by reference yield Y_R^d, which is defined as the ratio of daily irradiation and irradiance at STC according to the formula:

$$Y_R^d = \frac{H^d}{G^{STC}},$$

(4)

where: H^d—daily irradiation expressed in kWh/m^2, G^{STC}—irradiance under standard test conditions (in kW/m^2).

For an ideal PV system operating under STC, the calculations based on the Equations (3) and (4) should provide the same results. Nonetheless, the power losses in the PV system (modules and other particular components) lead to lowering of the Y^d value in comparison with Y_R^d. In order to evaluate real energy production relation to the ideal scenario, performance ratio (PR) of the system was calculated:

$$PR = \frac{Y^d}{Y_R^d}.$$

(5)

Another parameter used for the assessment of real performance of the modules was efficiency, defined as a quotient between the energy production in reference period and solar irradiation received by the module:

$$\eta = \frac{E_{DC}^d}{H^d \cdot Area} \cdot 100\%,$$

(6)

where *Area* is the total area of the modules.

3. Results

Monthly irradiation on the module plane at the considered location, in a warm summer continental climate according to Köppen's climate classification, is shown in Figure 2. As can be seen, the most sunny period, beneficial for energy production from photovoltaics, was June to August in 2015. There are significant differences in the irradiation level in plane of module during the year: from a minimum of 20 kWh/m^2 in January to a maximum of 191 kWh/m^2 registered in August. The monthly average daily irradiation on module plane ranged from only 0.67 kWh/m^2 in January to 6.18 kWh/m^2 in August. In general, about 80% of yearly solar irradiation in Poland is received during the sunny and warm half of the year, from April to August.

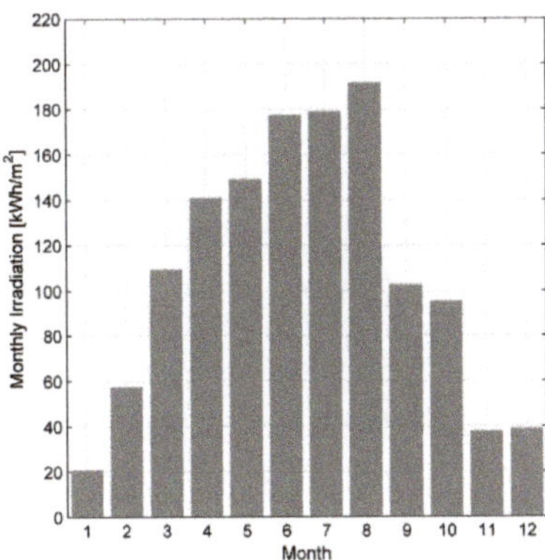

Figure 2. Monthly irradiation on module plane measured in 2015.

In order to better explain the on-site solar irradiance conditions, Figure 3 shows the accumulated irradiance distribution of incident global irradiation measured for period under study. As can be seen, irradiance conditions characterized by low values were the most frequent ones, resulting in median equal to 193 W/m^2. About 36% of the measurements have an irradiance lower than 100 W/m^2 and almost 98% of registered values are below 1000 W/m^2, however, about 17% of the results were characterized by a good or very good irradiance conditions, which varied from 700 W/m^2 to 1200 W/m^2.

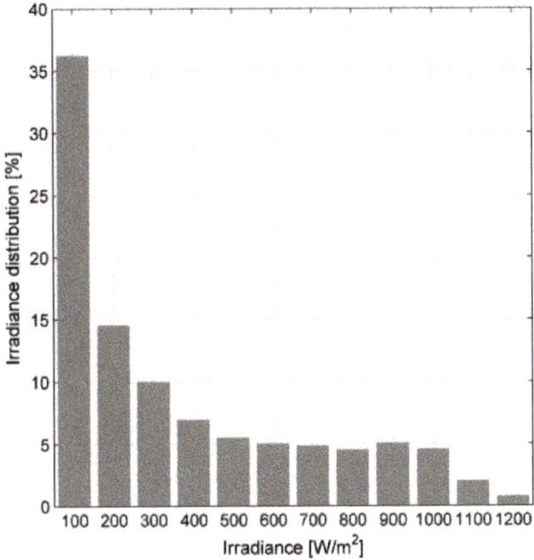

Figure 3. Accumulated percentage of collected incident global irradiation distribution according to irradiance levels measured in 2015.

Simultaneously to irradiance measurements, the temperature of the modules was monitored in the given location (Figure 4). The values of temperature of the modules depend mainly on the irradiation and ambient temperature and have a great impact on the efficiency of the modules. Registered values of the modules temperature varied from about 0 °C in cloudy winter days to 60 °C in summer sunny days. The median of the module temperature for the studied period is equal to 30 °C.

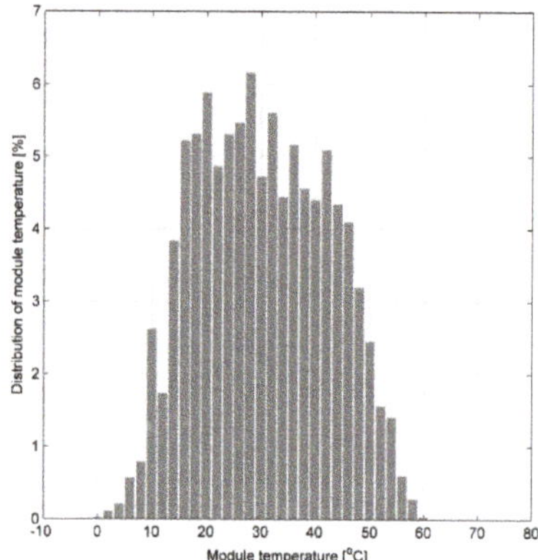

Figure 4. Percentage distribution of module temperature measured in 2015.

The obtained results (Figures 2–4) indicate that due to seasonal variations, in considered location under moderate climate, STC used by manufacturers are rarely met. Only a small portion of instantaneous irradiance measurements are around 1000 W/m^2 and module temperature values quite often reach 15–48 °C in the analyzed period.

An example of a hot sunny day, characterized by the high solar irradiance level reaching 1000 W/m^2 when the modules are heated by solar radiation up to 55 °C, shows that under this kind of real conditions, the power generated by pc-Si installation is lower compared to thin film technologies (Figure 5).

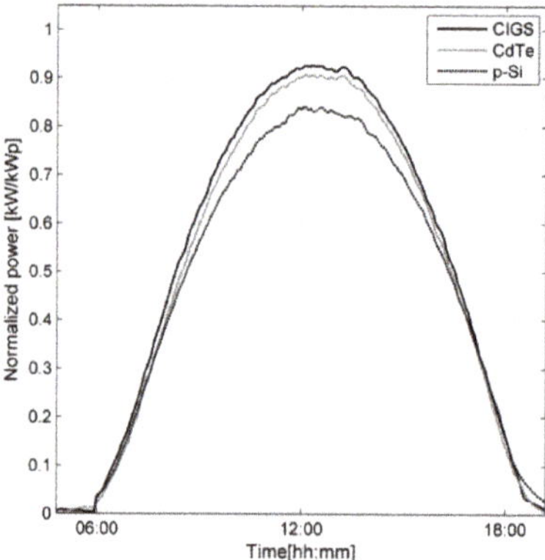

Figure 5. Normalized DC power generated by the PV systems of each studied technology. Maximum irradiance and temperature recorded for that day were 1000 W/m^2 and 55 °C, respectively.

For different analyzed PV technologies, the dependency of daily energy output as a function of daily average module temperature was shown in Figure 6. The increase of modules temperature caused by high incident solar irradiance during sunny days results in the decrease of energy production for all investigated technologies. As can be seen, thin film modules exhibit higher energy yield values than polycrystalline silicon ones. In particular, CIGS modules were characterized by the highest energy production in these specific—high temperature—conditions. The difference in daily energy production between CIGS and pc-Si modules during summer sunny days varies from 3.7% to 8.4%, while the CdTe modules produced from 1.7% to 6.1% more energy than pc-Si modules. The reason why CIGS installation exhibits better performance on sunny days can be explained as a result of double effect. Firstly, the temperature coefficient of power is lower than in case of pc-Si modules. Secondly, power gains due to spectral effects reported in the literature, being those gains for CIGS technology more remarkable at high latitudes in the north Hemisphere, as in the case of Reference [31].

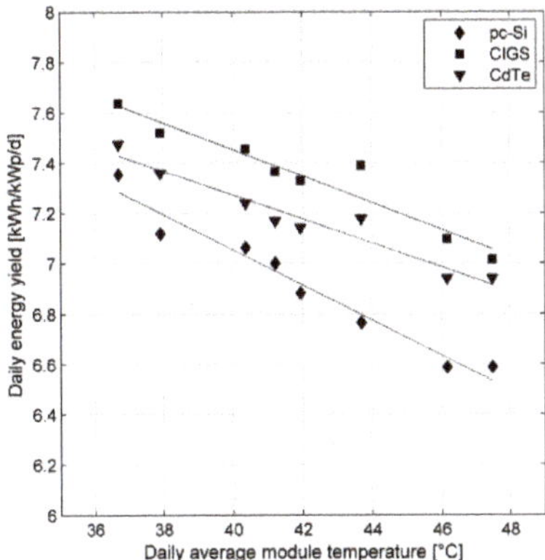

Figure 6. Daily energy yield vs. daily average temperature of the different technology modules calculated for sunny summer days of 2015 with similar solar irradiation levels.

Furthermore, CdTe modules, because of the lower power temperature coefficient and characteristic for this technology in high latitudes spectral gain in summer, have a better response than pc-Si, as shown in Figure 6 [31,33]. Nevertheless, the efficiency of CdTe modules is lower, and, finally, their yield is not as good as CIGS technology but better than pc-Si.

Experimental efficiency of each studied PV technology calculated as the average of daily efficiencies (Equation (6)) measured over the period February 2015–November 2015 can be seen in Figure 7. Efficiencies of pc-Si, CIGS and CdTe technologies were found to be 14.5%, 11.0%, and 8.7%, respectively. The results of calculations based on experimental data indicate lower efficiency values for all considered PV technologies in comparison to those obtained under STC presented in manufacturer datasheet (Table 1). The highest difference (17.9%) was noted for CdTe technology. In the case of CIGS and pc-Si technologies, these differences were found to be 12.7% and 5.8% respectively.

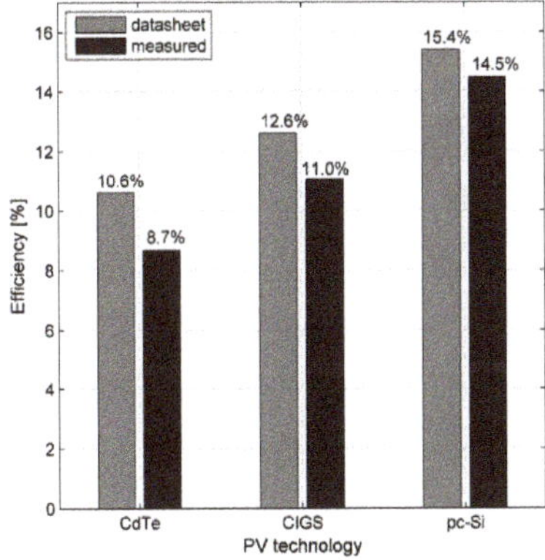

Figure 7. Comparison of the photovoltaic (PV) modules' efficiencies presented in manufacturer datasheet with the efficiencies obtained experimentally over the period February 2015–November 2015.

Analysis of daily energy efficiency calculated for chosen sunny days of 2015 with relatively high solar irradiation level (winter sunny days are characterized by lower values of solar irradiation in comparison with summer sunny days), presented in Figure 8, shows the linear decrease of daily efficiency with the increase of the daily average modules temperature. Temperature coefficients of efficiency for studied types of PV technology were derived from experimental data according to the following equation [34]:

$$\eta = \eta_r - \mu(T_c - T_r),\tag{7}$$

in which: η_r—reference module efficiency at reference temperature T_r = 25 °C, μ—temperature coefficient, T_c—measured temperature of the module. The temperature coefficient value is a slope of a line that represents the efficiency changes for each module type in Figure 8. The values of temperature coefficient, both obtained from experimental data and calculated based on manufacturer information (Table 1) as $\mu = \beta\eta_r$, are presented in Table 3.

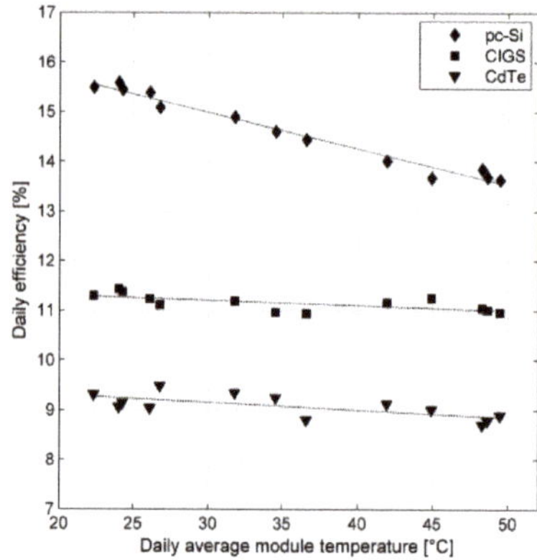

Figure 8. Effect of the module temperature on daily efficiency of each PV technology calculated for a chosen sunny day of 2015.

Table 3. Calculated and experimental values of the modules temperature coefficient.

Photovoltaic (PV) Technology	Temperature Coefficients Calculated According to Manufacturer Data [%/°C]	Temperature Coefficients Obtained Basing on Experimental Data [%/°C]
Cadmium telluride (CdTe)	0.0265	0.0152
Copper indium gallium diselenide (CIGS)	0.0391	0.0102
Polycrystalline silicon (pc-Si)	0.0616	0.0728

The comparison of the data obtained by two methods shows that the temperature coefficient resulting from experimental data exhibits a higher value for pc-Si modules than the value calculated according to the datasheet. The experimentally estimated value of pc-Si modules temperature coefficient is the highest compared to other technologies, which is directly connected to the strong influence of temperature on pc-Si energy production (Figure 6). The temperature coefficients of both thin film technologies are lower than these for pc-Si. In realistic operational conditions, their temperature coefficients achieve even lower values than the values determined from the datasheet, which is favorable and confirms the observations of high energy production by these technologies presented in Figure 6.

Figure 9 shows the dependency of DC output power on the PV modules temperature under real outdoor conditions for selected sunny days of 2015. The effect was determined by measuring DC output power and instantaneous module temperature at incident solar irradiance equal to 1000 W/m^2 (+/−0.6%). The lowest module temperature at irradiance of 1000 W/m^2 was found to be 28 °C (registered at 11.18 on 23rd of March 2015). The highest value was equal to 49 °C (registered at 12.09 on 2nd of July 2015). Experimentally obtained power was compared with the maximum power point

(MPP) value at 25 °C, taken from the manufacturer datasheet. Calculated results depict a strong dependency of output power on the temperature only for pc-Si modules. The variation of output power for pc-Si technology modules is around 18 W (7%) with the increase of temperature from 28 °C to almost 50 °C. In terms of thin film modules, the decrease of power with temperature is much smaller (0.2–2.0 %).

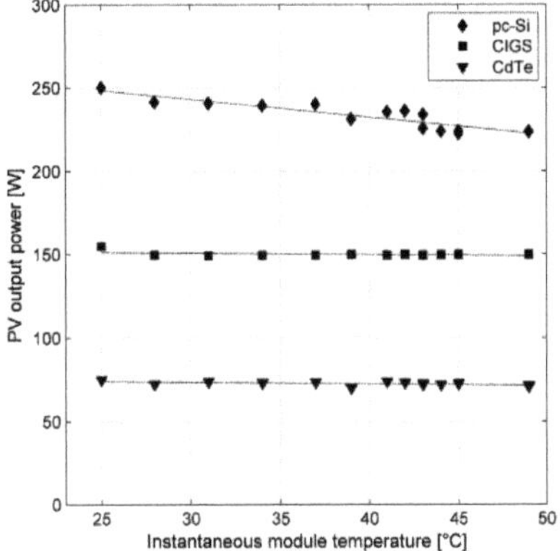

Figure 9. Photovoltaic (PV) output power of the different technology modules measured at 1000 W/m^2 vs. instantaneous temperature of the module.

The daily average energy yield of 1 kWp of each technology is presented in Figure 10. The shape of the energy yield profile is similar to the shape of irradiation profile for the studied year, as shown in Figure 2. The highest energy production was noticed in summer months for all technologies due to the best irradiance conditions. During the summer period (June–August) CIGS modules produced about 3% more energy than polycrystalline Si modules, which is in agreement with observations of temperature influence in this period. CdTe modules exhibit the lowest energy production, even in summer, in spite of the small value of temperature coefficient. Energy production by CdTe technology is lower in comparison with pc-Si of 1.1–8.9% which is probably caused by its small efficiency. An exception is June, when CdTe produced 0.2% more energy per kWp than pc-Si.

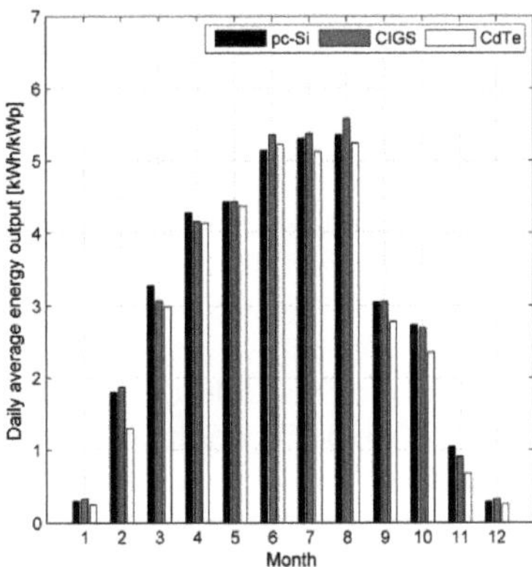

Figure 10. Comparison of daily average energy yield of each technology modules in 2015.

Monthly performance ratio (PR) calculated for each studied module technology is shown in Figure 11. PR values of CIGS array were found to be higher than that of pc-Si (from 2.5% in June to 3.5% in August). Higher energy yield and performance ratio of pc-Si modules were noticed in colder period of the year, i.e., from March to May (up to 4.8%), and from October to November (up to 4.6%). In terms of CdTe modules, lower monthly PR values were noticed in comparison to pc-Si technology for the whole year (from about 1% in May to even 14% in October), with the exception of June. During winter (December–February) PR values drop significantly due to occurrence of snow cover and decrease of inverters' efficiency at low input power levels. These two problems make it difficult to interpret the PR results obtained in winter, and thereby to assess the particular PV technologies in this period. Most of the months (March–October) are characterized by better insolation (Figure 2) and PR value above 80% (even 90% for pc-Si)

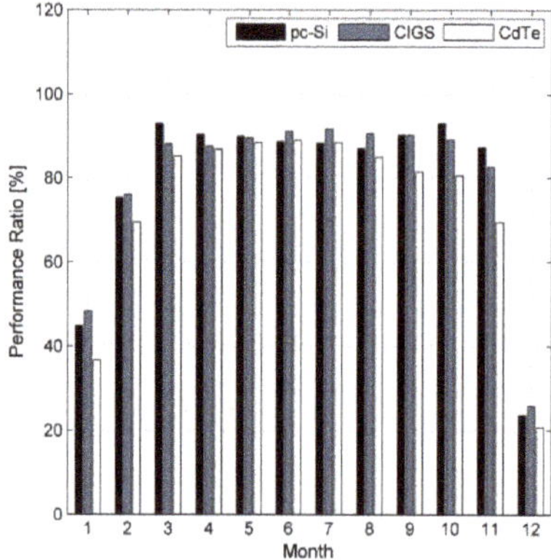

Figure 11. Comparison of calculated monthly performance ratio (PR) of each technology of the modules.

4. Discussion

The performance of solar modules depends on real external conditions, which is clearly visible in the results presented in this work, based on the measurements collected under temperate climate which is characterized by significant changes in both daily and yearly insolation and temperature. The role of solar radiation intensity on charge carriers generation and parameters of PV cells is direct and obvious. Nonetheless, it is worth remembering that most of the radiation absorbed by the cells is not converted into electricity, but it increases their temperature, thereby reducing the efficiency.

The influence of the temperature on the performance of solar modules, revealed in the presented experimental results, is different for various types of modules, since they are built with different semiconductors materials. In general, the semiconductor bandgap should correlate well with the solar radiation spectrum to achieve its maximum absorption. The optimum value of the bandgap, within the range of 1.0–1.6 eV results in good efficiency, which is observed even for single solar cells [35]. Temperature increase leads to narrowing of the semiconductor bandgap and also the intensification of recombination processes, in which current carriers disappear.

The bandgaps of the semiconductors used in the studied modules were different: 1.1 eV for polycrystalline silicon, the material with a crystalline fraction of over 95%, and 1.45–1.5 eV in case of cadmium telluride [36–38]. Copper indium gallium selenide (CIGS) is a solid solution characterized by a higher bandgap value of around 1.04–1.68 eV, which depends on the exact composition of the material [39]. In the presented investigations, the role of the bandgap value is thus visible since the use of wider bandgap semiconductors extend their operational temperature [40]. The broader bandgap of CIGS may result in better resistance of this material to temperature increase.

The comparison of the presented results with studies conducted by other authors is rather difficult, even taking into account investigations performed at similar latitude, since the technical parameters of the modules and other devices differ, as do weather conditions in summer and winter. Exemplary studies performed in the UK show better performance of CIGS in winter due to the spectrum [11]. The investigations focused on the assessment of the spectral impact on different PV technologies, performed in Germany [31], indicate spectral gains of 2.4% for CdTe, 1.1% for c-Si (that can be compared with pc-Si), and 0.6% for CIGS during the entire year. Temperature coefficients of power estimated in the Netherlands for different PV technologies are higher than technical data for CdTe and CIGS and similar for poly-Si [20]. Considering the results of the investigations in the mentioned countries, it is worth remembering that the weather conditions in both winter and summer differ from those in Poland in spite of similar latitude.

In general, all kinds of modules in the analyzed solar plant of the whole nominal power of 1.4 MWp produced 1530.23 MWh in the entire 2015 year. This achieved result is remarkable, considering the role of the climatic zone in which the PV plant is situated, however, further long-term measurements are necessary. The degradation process of modules has to be observed and taken into account in long term comprehensive assessment. All kinds of modules are sensitive to some external factors, such as humidity that leads to the failure of electric parts, snow, dust and other pollutions that cause hot spot appearance, and even damage of the modules [41,42]. However, it is also worth emphasizing that external conditions influence particular types of modules in different ways, since the construction of the traditional and thin film modules is not the same. Internal processes in various semiconductor materials also contribute to the degradation.

5. Conclusions

The performance of three different PV technologies was characterized under moderate climate of Eastern Europe, in which about 80% of irradiation in plane of module is received in the warm half of the year. In this kind of climate, both standard test conditions (STC) and nominal operating cell temperature (NOCT) are not met, since only a small portion of irradiance achieves values within the range 800–1000 W/m^2. Module temperature is distributed around 30 °C.

Under this kind of external conditions, daily and monthly energy efficiency calculations revealed its lower values in comparison to efficiency measured under STC for each type of modules. The difference in normalized energy production among three studied PV technologies as well as the decrease of energy output with the increase of module temperature were shown. The highest impact of the temperature on generated power was registered in the case of pc-Si installation compared to thin film. CIGS and CdTe modules exhibit high resistance to temperature rise, which is indicated by temperature coefficients whose experimentally obtained values are even lower than the values based on the manufacturer data. On hot summer days, CIGS installation produced more energy than the polycrystalline one. Daily average yield, as well as the performance ratio of each technology installation, also indicated better performance of CIGS technology during summer months.

The presented results thus indicate that under temperate climate operating conditions, CIGS thin film technology is a valuable alternative to popular pc-Si.

Author Contributions: Conceptualization, S.G. and A.Z.; Data curation, P.D.; Formal analysis, A.Z.; Investigation, P.D.; Methodology, S.G.; Software and computing, S.G.; Supervision, A.Z.

Funding: This work was supported by Polish Ministry of Science and Higher Education. The solar plant was partially financed by EU project RPLU.06.02.00-06-086/12-00.

Acknowledgments: Authors would like to thank José Vicente Muñoz Diez and Gustavo Nofuentes Garrido from University of Jaén, Spain for valuable discussion.

Conflicts of Interest: The authors declare no conflict of interest.

References

1. Directive (EU) 2015/1513 of the European Parliament and of the Council of 9 September 2015 Amending Directive 2009/28/EC on the Promotion of the Use of Energy from Renewable Sources. Available online: http://data.europa.eu/eli/dir/2015/1513/oj (accessed on 20 December 2018).

2. Gaglia, A.G.; Lykoudis, S.; Argiriou, A.A.; Balaras, C.A.; Dialynas, E. Energy efficiency of PV panels under real outdoor conditions—An experimental assessment in Athens, Greece. *Renew. Energy* **2017**, *101*, 236–243. [CrossRef]

3. Dale, M. A comparative Analysis of Energy Costs of Photovoltaic, Solar Thermal, and Wind Electricity Generation Techniques. *Appl. Sci.* **2013**, *3*, 325–337. [CrossRef]

4. Rehman, S.; El-Amin, I. Performance evaluation of an off-grid photovoltaic system in Saudi-Arabia. *Energy* **2012**, *46*, 451–458. [CrossRef]

5. Ozden, T.; Akinoglu, B.G.; Turan, R. Long term outdoor performances of three different on-grid PV arrays in central Anatolia—An extended analysis. *Renew. Energy* **2017**, *101*, 182–195. [CrossRef]

6. Raugei, M.; Frankl, P. Life cycle impacts and costs of photovoltaic systems: Current state of the art and future outlooks. *Energy* **2009**, *34*, 392–399. [CrossRef]

7. Dinçer, F. The analysis on photovoltaic electricity generation status, potential and policies of the leading countries in solar energy. *Renew. Sustain. Energy Rev.* **2011**, *15*, 713–720. [CrossRef]

8. Olchowik, J.M.; Dragan, P.; Gembarzewski, O.; Gulkowski, S.; Szymczuk, D.; Tomaszewski, R. The Reasons of the Delays in Introducing in Poland Law Regulations Favorable for Photovoltaics. In Proceedings of the 28th EUPVSEC, Paris, France, 30 September–4 October 2013; pp. 4676–4679.

9. Olchowik, J.M.; Cieslak, K.; Gulkowski, S.; Mucha, J.; Sordyl, M.; Zabielski, K.; Szymczuk, D.; Zdyb, A. Progress of development of PV systems in South-Eastern Poland. In Proceedings of the 35th IEEE Photovoltaic Specialists Conference, Honolulu, HI, USA, 20–25 June 2010.

10. Canete, C.; Carretero, J.; Sidrach-de-Cardona, M. Energy performance of different photovoltaic module technologies under outdoor conditions. *Energy* **2014**, *65*, 295–302. [CrossRef]

11. Gottschalg, R.; Betts, T.R.; Eeles, A.; Williams, A.R.; Zhu, J. Influences on the energy delivery of thin film photovoltaic modules. *Sol. Energy Mater. Sol. Cells* **2013**, *119*, 169–180. [CrossRef]

12. Kaldellis, J.K.; Kokala, A. Quantifying the decrease of the photovoltaic panels' energy yield due to phenomena of natural air pollution disposal. *Energy* **2010**, *35*, 4862–4869. [CrossRef]

13. Kaldellis, J.K.; Kapsali, M. Simulating the dust effect on the energy performance of photovoltaic generators based on experimental measurements. *Energy* **2011**, *36*, 5154–5161. [CrossRef]

14. Skoplaki, E.; Palyvos, J.A. On the temperature dependence of photovoltaic module electrical performance: A review of efficiency/power correlations. *Sol. Energy* **2009**, *83*, 614–624. [CrossRef]

15. Schwingshackl, C.; Petitta, M.; Wagner, J.E.; Belluardo, G.; Moser, D.; Castelli, M.; Zebisch, M.; Tetzlaff, A. Wind effect on PV module temperature: Analysis of different techniques for an accurate estimation. *Energy Procedia* **2013**, *40*, 77–86. [CrossRef]

16. Tanesab, J.; Parlevliet, D.; Whale, J.; Urmee, T. Seasonal effect of dust on the degradation of PV modules performance deployed in different climate areas. *Renew. Energy* **2017**, *111*, 105–115. [CrossRef]

17. Olchowik, J.M.; Gulkowski, S.; Cieslak, K.; Banas, J.; Jozwik, I.; Szymczuk, D.; Zabielski, K.; Mucha, J.; Zdrojewska, M.; Adamczyk, J.; et al. Influence of temperature on the efficiency of monocrystalline silicon solar cells in the South-eastern Poland conditions. *Mater. Sci.-Pol.* **2006**, *24*, 1127–1132.

18. Huld, T.; Gottschalg, R.; Beyer, H.G.; Topič, M. Mapping the performance of PV modules, effects of module type and data averaging. *Sol. Energy* **2010**, *84*, 324–338. [CrossRef]

19. Makrides, G.; Zinsser, B.; Phinikarides, A.; Schubert, M.; Georghiou, G.E. Temperature and thermal annealing effects on different photovoltaic technologies. *Renew. Energy* **2012**, *43*, 407–417. [CrossRef]

20. Louwen, A.; de Waal, A.C.; Schropp, R.E.I.; Faaij, A.P.C.; van Sark, W.G.J.H.M. Comprehensive characterization and analysis of PV module performance under real operating conditions. *Prog. Photovolt. Res. Appl.* **2017**, *25*, 218–232. [CrossRef]

21. Fuentes, M.; Nofuentes, G.; Aguilera, J.; Talavera, D.L.; Castro, M. Application and validation of algebraic methods to predict the behaviour of crystalline silicon PV modules in Mediterranean climates. *Sol. Energy* **2007**, *81*, 1396–1408. [CrossRef]

22. Sidrach-de-Cardona, M.; López, L.M. Performance analysis of a grid-connected photovoltaic system. *Energy* **1999**, *24*, 93–102. [CrossRef]

23. Muñoz, J.V.; Nofuentes, G.; Fuentes, M.; de la Casa, J.; Aguilera, J. DC energy yield prediction in large monocrystalline and polycrystalline PV plants: Time-domain integration of Osterwald's model. *Energy* **2016**, *114*, 951–960. [CrossRef]

24. Grzesiak, W.; Mackow, P.; Maj, T.; Polak, A.; Klugman-Radziemska, E.; Zawora, S.; Drabczyk, K.; Gulkowski, S.; Grzesiak, P. Innovative system for Energy collection and management integrated within a photovoltaic module. *Sol. Energy* **2016**, *132*, 442–452. [CrossRef]

25. Zdyb, A.; Krawczak, E. The influence of external conditions on the photovoltaic modules performance. In *Environmental Engineering V*, 1st ed.; Pawlowska, M., Pawlowski, L., Eds.; CRC Press Taylor & Francis Group: Boca Raton, FL, USA, 2017; pp. 261–266. ISBN 9781138031630.

26. Bora, B.; Kumar, R.; Sastry, O.S.; Prasad, B.; Mondal, S.; Tripathi, A.K. Energy rating estimation of PV module technologies for different climatic conditions. *Sol. Energy* **2018**, *174*, 901–911. [CrossRef]

27. Sharma, V.; Kumar, A.; Sastry, O.S.; Chandel, S.S. Performance assessment of different solar photovoltaic technologies under similar outdoor conditions. *Energy* **2013**, *58*, 511–518. [CrossRef]

28. Belluardo, G.; Ingenhoven, P.; Sparber, W.; Wagner, J.; Weihs, P.; Moser, D. Novel method for the improvement in the evaluation of outdoor performance loss rate in different PV technologies and comparison with two other methods. *Sol. Energy* **2015**, *117*, 139–152. [CrossRef]

29. Aste, N.; Del Pero, C.; Leonforte, F. PV technologies performance comparison in temperate climates. *Sol. Energy* **2014**, *109*, 1–10. [CrossRef]

30. Kesler, S.; Kivrak, S.; Dincer, F.; Rustemli, S.; Karaaslan, M.; Unal, E.; Erdiven, U. The analysis of PV power potential and system installation in Manavgat, Turkey—A case study in winter season. *Renew. Sustain. Energy Rev.* **2014**, *31*, 671–680. [CrossRef]

31. Dirnberger, D.; Blackburn, G.; Müler, B.; Reise, C. On the impact of solar spectral irradiance on the yield of different PV technologies. *Sol. Energy Mater. Sol. Cells* **2015**, *132*, 431–442. [CrossRef]

32. IEC 61724-1. *Photovoltaic System Performance Monitoring—Guidelines for Measurement, Data Exchange, and Analysis (Part 1)*; Technical Report 1; International Electrotechnical Commission (IEC): Geneva, Switzerland, 2017.

33. Polo, J.; Alonso-Abella, M.; Ruiz-Arias, J.A.; Balenzategui, J.L. Worldwide analysis of spectral factors for seven photovoltaic technologies. *Sol. Energy* **2017**, *142*, 194–203. [CrossRef]

34. Mattei, M.; Notton, G.; Cristofari, C.; Muselli, M.; Poggi, P. Calculation of the polycrystalline PV module temperature using a simple method of energy balance. *Renew. Energy* **2016**, *31*, 553–567. [CrossRef]

35. Goetzberger, A.; Hebling, C. Photovoltaic materials, past, present, future. *Sol. Energy Mater. Sol. Cells* **2000**, *62*, 1–19. [CrossRef]

36. Tobias, I.; del Canizo, C.; Alonso, J. Crystalline Silicon Solar Cells and Modules. In *Handbook of Photovoltaic Science and Engineering*, 2nd ed.; Luque, A., Hegedus, S., Eds.; Wiley: Hoboken, NJ, USA, 2003; pp. 255–306.

37. Khana, N.A.; Rahmanb, K.S.; Aris, K.A.; Ali, A.M.; Misran, H.; Akhtaruzzaman, M.; Tiong, S.K.; Amin, N. Effect of laser annealing on thermally evaporated CdTe thin films for photovoltaic absorber application. *Sol. Energy* **2018**, *173*, 1051–1057. [CrossRef]

38. Olusol, O.I.; Madugu, M.L.; Dharmadas, I.M. Investigating the electronic properties of multi-junction ZnS/CdS/CdTe graded bandgap solar cells. *Mater. Chem. Phys.* **2017**, *191*, 145–150. [CrossRef]

39. Shafarman, W.N.; Stolt, L. Cu(InGa)Se2 solar cells. In *Handbook of Photovoltaic Science and Engineering*, 2nd ed.; Luque, A., Hegedus, S., Eds.; Wiley: Hoboken, NJ, USA, 2011; pp. 567–616.

40. Dupré, O.; Vaillon, R.; Green, M.A. Physics of the temperature coefficients of solar cells. *Sol. Energy Mater. Sol. Cells* **2015**, *140*, 92–100. [CrossRef]

41. Lorenzo, E.; Moretón, R.; Luque, I. Dust effects on PV array performance: In-field observations with non-uniform patterns. *Prog. Photovolt. Res. Appl.* **2014**, *22*, 666–670. [CrossRef]

42. Gulkowski, S.; Żytkowska, N.; Dragan, P. Temperature distribution analysis of different technologies of PV modules using infrared thermography. *E3S Web Conf.* **2018**, *49*, 00044. [CrossRef]

Article

Solar Tower Power Plants of Molten Salt External Receivers in Algeria: Analysis of Direct Normal Irradiation on Performance

Abdelkader Rouibah [1], Djamel Benazzouz [1], Rahmani Kouider [2], Awf Al-Kassir [3], Justo García-Sanz-Calcedo [3,*] and K. Maghzili [4]

[1] Mechanics Laboratory of Systems and Solids, Faculty of Engineering Sciences (FSI), M'Hamed Bougara University, Boumerdès 35000, Algeria; rouibah_a@yahoo.fr (A.R.); dbenazzouz@yahoo.fr (D.B.)
[2] Department of Engineering, University of Bab Ezzouar, Algiers 16311, Algeria; kouiderrah1@gmail.com
[3] Industrial Engineer School, University of Extremadura, 06006 Badajoz, Spain; aawf@unex.es
[4] Faculty of technology, Ziane Achour University, Djelfa 17000, Algeria; khaledmeghzili2@gmail.com
* Correspondence: jgsanz@unex.es; Tel.: +34-924-289-600

Received: 3 July 2018; Accepted: 17 July 2018; Published: 25 July 2018

Abstract: The increase of solar energy production has become a solution to meet the demand of electricity and reduce the greenhouse effect worldwide. This paper aims to determine the performance and viability of direct normal irradiation of three solar tower power plants in Algeria, to be installed in the highlands and the Sahara (Béchar, El Oued, and Djelfa regions). The performance of the plants was obtained through a system advisor model simulator. It used real data gathered from appropriate meteorological files. A relationship between the solar multiple (*SM*), power generation, and thermal energy storage (*TES*) hours was observed. The results showed that the optimal heliostat field corresponds to 1.8 *SM* and 2 *TES* hours in Béchar, 1.2 *SM* and 2 *TES* hours for El Oued, and 1.5 *SM* and 4 *TES* hours for Djelfa. This study shows that there is an interesting relationship between the solar multiple, power generation, and storage capacity.

Keywords: solar tower power plants; direct normal irradiation; energy projects; system advisor model

---

1. Introduction

Algeria highlighted its solar potential where as of the most important heritages in the World, because more than 2 million km^2 receive an annual insulation of about 2.5 kWh/m^2 [1,2]. The renewable energy program involves the installation of renewable power of the order of 22 GW by the year 2030 for the national market (i.e., more than 37% of national electricity production with maintaining the export option as a strategic objective) [3].

The world market for solar thermodynamics (CSP) is estimated at 14 GW in 2020 and 72 GW on the horizon of 2035 in very strong growth compared to the capacity installed in 2012 which amounts to 2.8 GW [4]. This strategic choice is motivated by the potential of solar energy from the "Sahara and High Plateau". This energy constitutes the major axis of the program which is devoted to solar thermal energy [5,6].

The geographic location of Algeria has several advantages to using solar energy. Algeria is situated in the center of North Africa between the 38–35° of latitude north and 8–128° longitude east, and it has an area of 2,381,741 km^2. The Sahara represents 86% of the area of the country. The climate is transitional between maritime (north) and semi-arid to arid (middle and south).

Solar power systems have a quasi-zero proportional cost: there is no fuel, only expenses (maintenance, guarding, repairs, etc.) which depend very little on the production. However, it is necessary to take into account its investment costs [7] which are much higher compared to fossil techniques or other renewable energies [8].

In solar tower power plants, since the solar energy is insufficiently dense, it is necessary to concentrate it by means of reflecting mirrors in order to obtain operating temperatures for the production of electricity. The solar radiation can be concentrated on a linear or point receiver. The receiver absorbs the energy reflected by the mirror and transfers it to a thermodynamic heat transfer fluid [9,10].

The performance of the solar system is characterized by its concentration factor. This coefficient makes it possible to evaluate the intensity of the solar concentration [11]. Whenever the concentration factor is high, the temperature reached will be high. Online concentration systems generally have a lower concentration factor less than that of point concentrators [12].

Ho et al. [13] reviewed central receiver designs for concentrating solar power applications with high-temperature power cycles. Boudaoud et al. [14] carried out a technical and economic analysis of electricity costs and the economic feasibility of solar tower power plants in Algeria. Behar et al. [15] evaluated a wide range of clear sky solar radiation models based on theoretical input parameters for the Algerian climate in order to estimate the performance of solar energy projects for which meteorological and radiometric measurement stations are not available. Mihoub et al. [16] proposed a methodology to have an optimal design with a better configuration of the future Algerian solar tower power plants with objectives, the minimization of the electricity costs (LCOE), and the maximization of annual production of electricity.

Quaschning [17] realized a technical and economic system comparison of photovoltaic and concentrating solar thermal power systems depending on annual global irradiation. He concluded that the electricity generation cost much below 0.10 €/kWh for solar thermal systems and about 0.12 €/kWh for solar photovoltaic can be expected in 10 years in North Africa. In addition, Zhu et al. [18] concluded that introduction of a solar tower field increasing leveled cost of electricity; it contributes to the reduction of CO_2 capture cost compared to the case of standard coal-fired power plants.

Toro et al. [19] studied the thermo-economic design evaluation and optimization of the Central Receiver Concentrated Solar Plants, allowing for improvement of the thermodynamic and economic efficiency of the systems, as well as decreasing the exergy and exergy-economic cost of their products.

Eddine Boukelia et al. [20] made a review of considerations on the assessments for concentrating solar power potential of Algeria. The analysis showed the competitive viability of CSP plants. Algeria has the key prerequisites to make economical CSP power generation, including high-quality insolation and appropriate land, in addition to water availability and extensive transmission and power grid.

Boudaoud et al. performed a technic economic assessment of a solar tower power pilot plant located in Tipaza, near Algiers. Using the economical, technical, meteorological, and radiometric data, they have carried a simulation of the solar tower power plants (STPP). The results showed that for a net annual energy of about 1 MW, the leveled cost of electricity is about 0.1/kWh, which is relatively high in comparison with the leveled cost of fossil power plant (0.04/kWh) [21].

Larbi et al. [22] showed that solar chimney power plants can produce from 140 to 200 kW of electricity on a site like Adrar (Algeria) during the year, according to an estimate made on the monthly average of sunning. This production is sufficient for the needs of the isolated areas.

Viebahn et al. studied two virtual sites in Algeria and in Spain; they showed a long-term reduction of electricity generating costs to figures between 4 and 6 ct/kWh in 2050. Although the greenhouse gas emissions of current CSP systems showed a good performance (31 g CO_2-equivalents/kWh) compared with fossil-fired systems (130–900 CO_2-eq/kWh), they could further be reduced to 18 g CO_2-eq/kWh in 2050 [23].

Abbas et al. performed a techno-economic assessment of 100 MW of three types of concentrating solar thermal power plants for electricity generation located in one typical site of the Saharan environment of Algeria (Tamanrasset) [24].

The exploitation of Algeria's solar potential complements rural electrification programs. Currently, the use of renewable energies can reach regions far from the national electricity grid.

The parameters and performance of the solar tower power plants of molten salt external receivers to be installed in the North of Algeria are currently not defined. The savings potential is very high and it is planned to install several plants, without the location yet being defined. Most of the studies focus on the south of Algeria, whose solar radiation is higher than the north.

The aim of this paper is to analyze direct normal irradiation on the performance of solar tower power plants of molten salt external receivers in the North of Algeria to optimize the configuration of concentrating solar power (CSP). A comparison study between the three power stations located in the Algerian regions of Béchar, El Oued, and Djelfa, was presented. Each plant is equipped with a molten salt storage mode, the receiver is of external type and the implantation of the field of heliostats that has been defined for an annual production of 20 MW.

This research is useful for prioritizing, sizing, and locating new installations and for determining the technical parameters to be used in the construction projects of solar tower power plants of molten salt external receivers. It will also be useful to define the ideal location based on the solar radiation that maximizes the yield of the CSP plant.

2. Methodology

A numerical simulation under a system advisor model (SAM) based on direct normal irradiance (DNI) with real and satellite data was carried out to optimize the parameters characterizing these performances. The influence of these parameters on each other made it possible to choose an optimal CSP configuration.

In the research, SAM software (Version 2017.9.5, National Renewable Energy Laboratory. Golden, CO, USA, 2017) was used. The SAM is a software to model and simulate the performance of energy parameters and the economics of systems to facilitate the decision-making process in the field of renewable energies [25].

The solar radiation intensity is an important factor in the evaluation of CSP plants. Direct normal irradiance is the amount of direct normal solar radiation received per unit area. There are three techniques to assess the evolution of DNI over time for a given location [26,27].

Optimization of the design of the heliostats field is a trade-off between optical performance and cost, so this process includes both optical and economic analysis. This implantation can be performed by determining the optimum values of the radial spacing ΔR and the azimuth spacing ΔA_Z.

There are various optimization procedures to establish these two geometric position parameters. One of the most effective procedures is the radial offset arrangement [28]. The evaluation of the radial and azimuthally distance can be evaluated by empirical Equations (1) and (2) [29]. These parameters also depend on the angle (α) between the heliostat, the ground, and the tower, as shown in Figure 1 [30,31].

$$\Delta R HM = 63.0093 - 0.587313 \cdot \theta + 0.018423909 \cdot \theta^2 + \left(2.808733 - 0.1480498 \cdot \theta + 0.001489201 \cdot \theta^2\right) \cdot \cos \alpha \qquad (1)$$

$$\Delta A_Z WM = 2.46812 - 0.0401054 \cdot \theta + 0.000923594 \cdot \theta^2 + \left(0.17344593 - 0.009112590 \cdot \theta + 0.00012761 \cdot \theta^2\right) \cdot \cos \alpha \qquad (2)$$

where ΔR is the radial distance between heliostats (m), HM is the heliostat height meters, θ is the receiver elevation angle from heliostat, α is the heliostat loft angle in degree, ΔA_Z is the azimuthal distance between heliostats (m), and WM is the heliostat width meters (m).

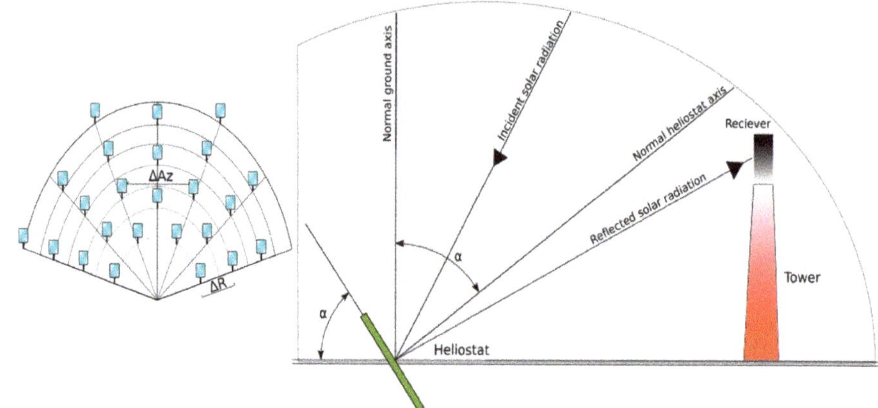

Figure 1. Implementation of the heliostats field. Representation of the optical angle α.

Table 1 shows the location parameters for different regions.

Table 1. Location parameters for different regions.

Parameter	Design Parameters	Djelfa	El Oued	Béchar
Location	DNI (W/m^2)	1050.00	750.00	700.00
	Latitude (°)	34.68	33.50	31.50
	Longitude (°)	3.25	6.78	−2.25
	Elevation (m)	1144.00	69.00	816.00

The size of the heliostat field influences the optical performance and depends on the desired power and temperature of the heat transfer fluid at the output. The total incident thermal energy is given by the following Equation (3):

$$Q_h = I_d \cdot A_h \cdot N_h \tag{3}$$

where I_d is the direct normal irradiation, A_h is the surface of the heliostat, and N_h is the number of the heliostat, and 144 m^2 was considered the surface of the heliostat field.

The efficiency of the field η_h is defined by the following Equation (4):

$$\eta_h = \frac{Q_{rec}}{Q_{inc}} = \frac{Q_{rec}}{I_d \times A_h \times N_h} \tag{4}$$

where Q_{rec} is the heat flow of the receiver and Q_{inc} is heat flow of the incident.

The efficiency is calculated considering losses due to different effects (cosine, shading, blocking, overflow, reflection, dispersion) and it is given by the following Equation (5) [32]:

$$\eta_h = \eta_{cos} \cdot \eta_{omb} \cdot \eta_{bloc} \cdot \eta_{deb} \cdot \eta_{ref} \cdot \eta_{disp} \tag{5}$$

where η_{cos} is the losses due to cosine effect, η_{omb} is the losses due to shading effect, η_{bloc} is the losses due to blocking effect, η_{deb} is the losses due to overflow; η_{ref} is the losses due to reflection and η_{disp} is the losses due to dispersion.

The model of the receiver of the present study is an external type. It consists of a large number of vertically disposed pipes through which a heat transfer fluid is pumped in the vertical direction. Inside the pipe three types of heat transfer are identified (convection, conduction, and radiation) and the exchange with the outside by radiation (solar and radiation losses), by convection (losses at the body of the receiver) and by conduction (losses through thermal bridges). Figure 2 shows the different heat exchanges of the receiver with the external environment.

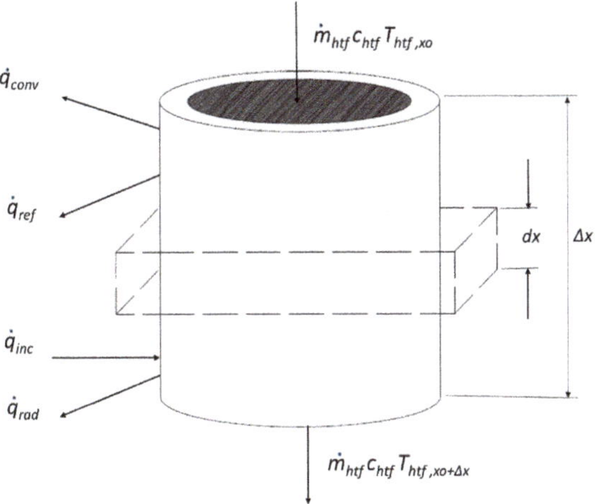

Figure 2. Energy balance of the external receiver.

The heat flux of the receiver can be expressed by Equation (6):

$$Q_{rec} = q_{htf} + q_{conv} + q_{rad} + q_{ref} \tag{6}$$

$$Q_{rec} = S_i \cdot I_d \tag{7}$$

where q_{htf} is the heat flow of molten salt; q_{conv}: loss of convection heat flow; q_{rad}: loss of radiation heat flux; q_{ref}: loss of reflection flux; S_i: total surface.

The incidence of solar radiation I_d on the receiver is evaluated by the flux map radiation model. The distribution of the radiation flux is integrated by combining the effect of the different losses occurring in a heliostats filed layout (cosine effect, shadowing effect, blocking effect, atmospheric attenuation, spillage and flux image profile), based on direct radiation from 950 W/m² [26,33]. The energy absorbed by the heat transfer fluid (q_{htf}) is given by the following Equation (8):

$$q_{htf} = m_{htf} \cdot C_{htf} \cdot \left(T_{htf(x+dx)} - T_{htf(x)} \right) = US_i \cdot \left(T_{st} - T_{htf} \right) \tag{8}$$

$$US_i = \frac{1}{R_{cond} + R_{conv}} \tag{9}$$

$$R_{cond} = \frac{\ln \frac{D_{ot}}{D_{it}}}{2 \cdot \pi \cdot L_t \cdot K_t \cdot N_t} \tag{10}$$

$$R_{conv} = \frac{2}{\pi \cdot h_{htf} \cdot L_t \cdot D_t \cdot N_t} \tag{11}$$

where q_{htf} is the heat flow of molten salt (W), m_{htf} is the molten salt flow rate (kg/s), C_{htf} is the heat capacity of the molten salt fluid (kJ/kg·K), T_{htf} is the inlet temperature of the molten salt at x position (K), US_i is heat transfer conductance coefficient (W/K), T_{st} is receiver temperature at the surface (K), R_{cond} is heat transfer resistance by conduction (K/W), R_{conv} is heat transfer resistance by convection (K/W), D_{ot} is outer diameter of the tube (m), D_{it} is the inner diameter of the tube (m), L_t is length of the tube (m), K_t is thermal conductivity of the receiver tube (W/m·K), N_t is total number of the receiver tube, and D_t diameter of the tube (m).

The losses by convection are given by Equation (12):

$$q_{conv} = S_i \cdot h_{conv} \cdot (T_{st} - T_{ic-air}) \tag{12}$$

where q_{conv} is the loss of convection heat flux (W), S_i is total surface/Surface total (m^2), h_{conv} is the convective heat losses from each receiver tube (W/m^2·K), T_{st} is the receiver temperature at the surface (K), and T_{ic-air} is the temperature of the air in the inner cavity (K).

The radiation losses q_{rad} have a negligible value because the absorber has a high absorption of short waves of solar radiation and the same for losses by reflection q_{ref} due to the less emissivity of the long thermal waves.

The performance of a good configuration of the solar tower system is based on several parameters such as power generation injected to the grid, incident solar radiation, and storage capacity. The capacity factor and the multiple solar characterizing the performance of a central solar tower system.

The ratio of the energy generated by the system in partial time E_{gp} and the energy generated in full-time E_{gf} determines the capacity factor [27] and is given by Equation (13):

$$CF = \frac{E_{gp}}{365.24 \cdot E_{gf}} \tag{13}$$

where CF is capacity factor, E_{gp} is the energy generated in part-time (W), and E_{gf} is the energy generated in full-time (W)

The ratio of energy to design point (thermal power produced by the field of heliostats q_{sf} for different DNI values), and the thermal power required by the power block under nominal conditions q_{pb} determines the solar multiple (*SM*). It is expressed by Equation (14), [34].

$$SM = \frac{q_{sf}}{q_{pb}} \tag{14}$$

where SM is solar multiple factor, q_{sf} is the energy generated by the field of heliostats (W), and q_{pb} is the energy required by the power block (W). For a system without a storage mode, $SM = 1$.

Table 2 shows component characteristics and design parameters of the solar tower system used in the research.

Annual meteorological database that known as the reference year test (TRY) or typical meteorological year (TMY) was used. It consists of measured values, which are statistically selected from the annual individual values measured over a long period. The file formats used are file extensions: TMY2, TMY3, EPW, and CSV. To optimize the performance of the solar tower system of different regions, one needs to optimize the solar fields by the variation of the solar multiple (*SM*) in function of thermal energy storage (*TES*) hours, in order to optimize the dimensions of the system and maximize the production of electricity and the capacity factor of the solar tower system.

Table 2. Characteristics of the components of the solar tower system.

Parameter	Design Parameters	Value
Field of heliostats	Surface of the heliostat (m^2)	144.00
Tower and receiver	Diameter of pipes (mm) Thickness of pipes (mm) Type of pipe material (stainless steel)	60.00 1.25
Fluid	Heat transfer fluid (HTF) type	60% $NaNO_3$, 40% KNO_3
Coolant	Input temperature (°C) Output temperature (°C)	565.00 290.00
Power block	Design turbine output (MWe) Thermodynamic cycle efficiency (%) Operating pressure of the boiler (bar) Type of cooling capacitor	820.00 37.00 100.00 Air
Energy storage	Type of storage Load storage in full hours	2 tanks 0–12 h

3. Results

The results obtained in the research, grouped according to the regions analyzed, are as follows.

3.1. Region of Béchar

In Figure 3a, it can be observed that the electrical production per square meter of heliostats increase proportionally with the *SM* except the decrease recorded in the interval (1.4–1.5) due to the increase in the surfaces of the heliostats mirrors and the decrease in the length of the tower, which are influenced by the losses due to the effects of the heliostat field as indicated in Figure 3b. Beyond *SM* = 1.6, the increase in the solar field area has no influence on the evolution of electrical production which converges and increases slightly due to the effects corresponding to the enormous expansion of the solar field and atmospheric attenuation.

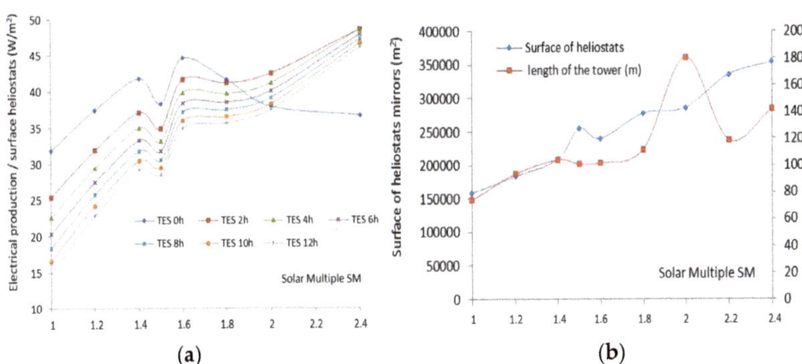

Figure 3. (a) Solar multiple effect (*SM*) on electrical production/surface heliostats under different values of *TES* (Béchar). (b) Solar multiple effect on the surface of the heliostats mirrors and length of the tower (Béchar).

For *SM* values of 2.0, 2.2, and 2.4, the electrical production per m^2 increases until the peak values: 42.53 W/m^2, 47.07 W/m^2, and 48.55 W/m^2 for *TES* = 2 h, then it decreases as explained in Figure 4. The configuration of the plant in these values requires a large area, which is not profitable. Therefore, the electrical production at the start of the system for $SM = 1.8$ is larger, lowers slightly to $TES = 2$ h, and then coincides with the curve corresponding to $SM = 1.6$.

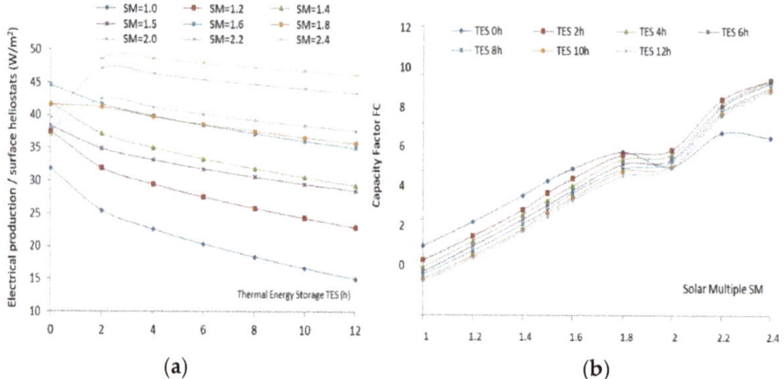

(a) (b)

Figure 4. (**a**) Effect of Thermal Energy Storage (*TES*) on electrical production/m^2 (surface heliostats) under different values of solar multiple (Béchar). (**b**) Solar multiple effect on capacity factor (*CF*) under different *TES* values (Béchar).

3.2. Region of El Oued

The capacity factor CF evolves proportionally with the *SM*. The graphs converge towards close values as shown in Figure 5. Except for the graph corresponding to $TES = 0$ h and $SM = 2.2$, which begins to descend slightly, this decrease is due to the loss of excess of the non-stored energy received by the receiver. From above, it can be concluded that the optimal point of operation of the system corresponds to the following coordinates: $SM = 1.8$, $TES = 2$ h, and the electrical production is 11.44 GWh/year.

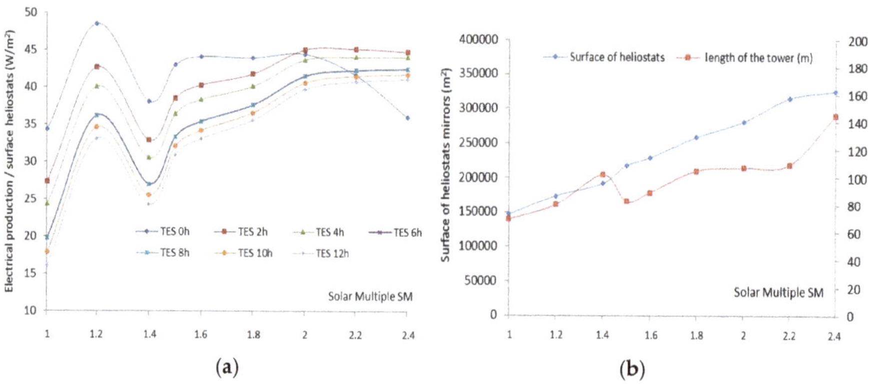

(a) (b)

Figure 5. (**a**) Solar multiple effect (*SM*) on electrical production per surface heliostats under different values of *TES*. (**b**) Solar multiple effect on the surface of heliostats mirrors (m^2) and length of the tower (m).

In Figure 5a, it can be seen that the electrical production per square meter of heliostats increase proportionally with the *SM*, except that there is a decrease recorded in the interval (1.2–1.4), where the length of the tower has exceeded the surface of the corresponding heliostat mirrors and then decreases to the value of 80 m, to resume the increase in electrical production in the interval (1.4–1.5), as shown in Figure 5b. Beyond $SM = 1.5$, the curves of electrical production continue parallel to their growth but it keeps constant the value of the electrical output at $SM = 1.2$ and view the optimal surface of the solar field.

For $SM = 2.0$, 2.2, and 2.4, the electrical production per m^2 increases until to peak values: 45.07 W/m^2, 45.18 W/m^2, and 44.79 W/m^2 for $TES = 2$ h then it drops as explained in Figure 6a. The configuration of the plant in these values requires a large area of the heliostat field, which is not profitable. Consequently, the electrical production at the start of the system for $SM = 1.2$ is greater, drops slightly up to $TES = 2$ h and then coincides with the curve corresponding to $SM = 1.8$.

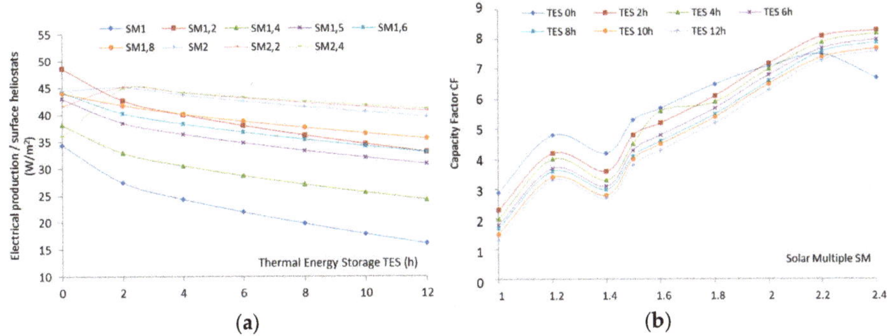

Figure 6. (**a**) The effect of Thermal Energy Storage (*TES*) on electrical production/m^2 (surface heliostats) under different values of solar multiple. (**b**) The solar multiple effect on capacity factor (*CF*) under different *TES* values.

The capacity factor (CF) evolves proportionally with the solar multiple (*SM*), the curves increase in parallel and tend towards close values as shown in Figure 6b, except for the curve corresponding to $TES = 0$ h and $SM = 2.2$ which begins to descend slightly; this decrease is due to the loss of excess of the non-stored energy received by the receiver.

From above, it can be concluded that the optimal point of operation of the system corresponds to the following coordinates $SM = 1.2$, $TES = 2$ h and the electrical production is 7.4 GWh per year.

3.3. Region of Djelfa

In Figure 7, the following variations can be distinguished. For $TES = 0$ h to 4 h, the electrical production per square meter of the heliostats increase respectively with the values $SM = 1.2$, $SM = 1.6$, and $SM = 2$, and then it decreases; for $TES = 6$ h to 12 h, the electrical production per square meter of the heliostats increase proportionally with to a converging value; for $SM = 2.0$, 2.2, and 2.4, the electricity production per m^2 of heliostat's increases proportionally with the *TES*. For the other *SM*, the electrical production increases until $TES = 2$ h, then it decreases. Except for $SM = 1.8$, the electrical production is interesting as *TES* exceeds 4 h. The configuration of the system becomes unprofitable (a large area, large dimensioning of the tower). On the other hand, the starting power of the plant is much better for $SM = 1.5$ and remains almost stable from $TES = 4$ h.

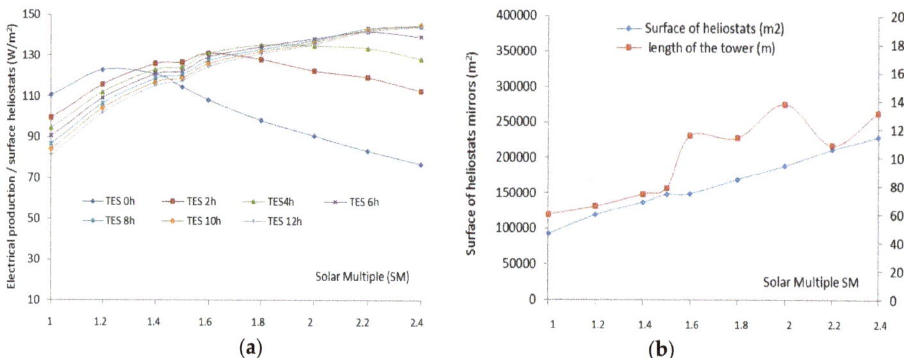

Figure 7. (**a**) Solar multiple effect (*SM*) on electrical production by surface heliostats under different values of *TES*. (**b**) Solar multiple effect on the surface of the heliostats mirrors (m^2) and the length of the tower (m).

The capacity factor (CF) evolves proportionally with the solar multiple (*SM*), the curves increase proportionally and tend towards close values except for *TES* = 0 h, 2 h and 4 h as shown in Figure 8.

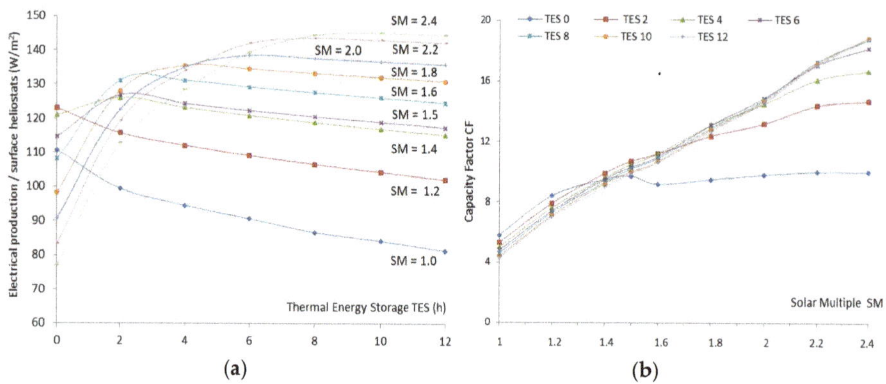

Figure 8. (**a**) Effect of Thermal Energy Storage (*TES*) on electrical production/m^2 (surface heliostats) under different values of solar multiple (Djelfa). (**b**) The solar multiple effect on capacity factor (*CF*) under different *TES* values (Djelfa).

From above, it can conclude that the optimal point of operation of the system corresponds to the following coordinates *SM* = 1.5, *TES* = 4 h, and the electrical production is 18.45 GWh/year. Table 3 shows the model validation parameters simulated in Djelfa and Batna.

Table 3. Parameters of the model validation.

Type of Parameter	Simulated Case, Scenario 1 [14]	Simulated Case, Scenario 2 [14]	Simulated Case, Study
Annual DNI (kWh/m^2)	1907.30	1907.30	2416.30
Hybridization (%)	0.00	15.00	0.00
Net energy production (GWh/year)	18.15	44.40	18.45
Net energy production difference (%)			1.60 (scenario 1.00 and study)
Annual capacity factor (%)	10.60	26.00	10.50

3.4. Optimization of the Field of Heliostats

From the above, for the optimal points of operation of CSP system of the three regions, it can conclude that the optimal heliostat field corresponds to the solar multiple and storage hours: $SM = 1.8$, $TES = 2$ h for Béchar region; $SM = 1.2$, $TES = 2$ h for El Oued region; and $SM = 1.5$, $TES = 4$ h for Djelfa region. The simulation results are shown in Figure 9.

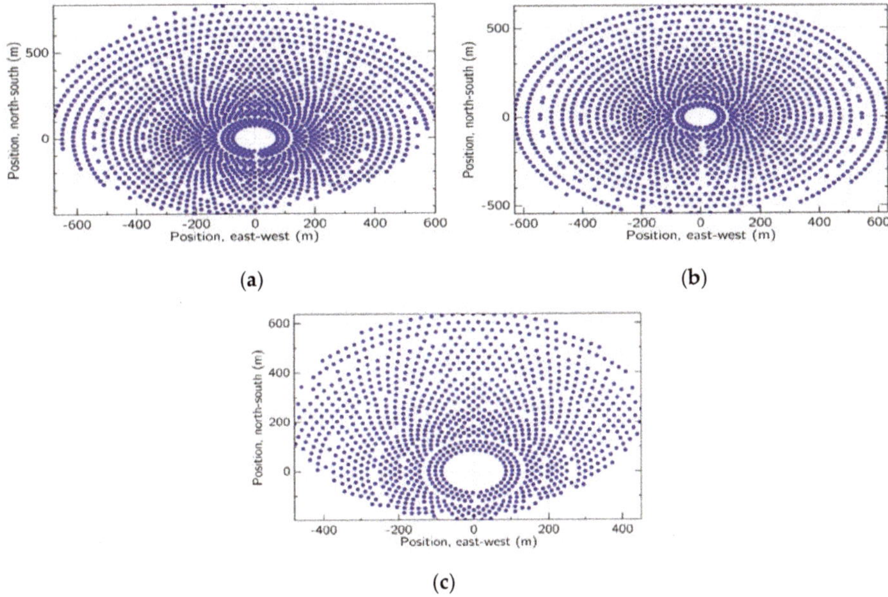

Figure 9. (**a**) Concentrated Solar Power (CSP) heliostats field configuration of Béchar region. (**b**) CSP heliostats field configuration of El Oued region. (**c**) CSP heliostats field configuration of Djelfa region.

Figure 9 shows how the largest STE was produced in the Djelfa region, although the highest *SM* was found in the CSP heliostat field configuration of Béchar region.

4. Discussion

Algeria is one of the most suitable countries for the cheap production of electricity from solar energy sources, especially from solar thermal concentration technology. Despite the great virtues of solar energy, the development of technologies that allow its use has been gradually slowed down by its disadvantages, including its high investment costs and the impossibility of generating energy at times when there is no solar radiation or it is intermittent due to the passage of clouds.

This study researches the influence of solar normal irradiation on solar power plants and its efficiency. It simulates the electricity production, capacity factors, and surface area required for solar field versus solar multiples considering the effect of *TES* at different capacities. The study focuses on optimization of CSP heliostat field configuration. Prediction of the field area and solar multiple can directly increase the efficiency of the solar power plant and power production rate. Also, the accumulation of energy through molten salt systems is an adequate solution to promote the use of solar energy.

According to study of the German Aerospace Centre (DLR), Algeria has with 1,787,000 km^2 of Sahara desert, the largest long-term land potential for concentrating solar thermal power plants. The insolation time over the quasi-totality of the national territory exceeds 2000 h annually, and may reach 3900 h (High Plains and Sahara) [35]. The daily obtained energy on a horizontal surface of 1 m^2 is of 5 kWh over the major part of the national territory, or about 1700 kWh/m^2 per year for the north and 2263 kWh/m^2 per year for the south of the country [36].

The solar thermal power plant is one of the promising renewable energy options to substitute the increasing demand of conventional energy [37]. The design of the solar towers allows the collector to reach a higher temperature than the cylindrical-parabolic ones. This higher temperature allows for more efficient conversion to electricity, as well as cheaper storage of thermal energy for later use.

The position of heating head is an important factor for power collection. If the sunlight can be concentrated to completely cover the heating head with small heat loss, it can obtain the maximum temperature of the heating head of the Stirling engine. Therefore, the temperature of heating head can be higher than 1000 °C on a sunny day [38].

The choice of the solar field is a difficult exercise, for example, a choice of solar multiples between 1.4 and 1.6 is suitable for high-optical performance with a lower surface of the field, and the starting arrangement must be very close to the optimum configuration, based on the experience obtained from the plants already in operation [39]. The thermal energy storage capacity is insufficient for the whole night, it covers between two and four hours hence the obligation to use an auxiliary energy source.

The starting layout must be very close to the optimum configuration (length of the tower, surface of the mirrors, dimension of the receiver), it is important therefore looking to design a cheaper and better performing heliostat concentrator.

In terms of direct solar irradiance, a measure of the gross energy received per unit area, Algeria is one of the most suitable countries for the cheap production of electricity from solar sources, especially from solar thermal concentration technology [40]. On the other hand, a reduction of taxes decreases leveled cost of electricity generated by CSP solar technologies [41].

However, it should be borne in mind that the type of geometry used in the construction of the reflective surface of a heliostat has a significant influence on the shape and size of the image generated in the plane of the receiver, and therefore on the energy density and the amount of energy intercepted by this element [42].

Concentrating solar power is clean and reliable, can be produced during high demand, and has the potential to meet a country's growing needs in the future. In addition, thermal storage systems prevent fluctuations in supply, allow production to continue in the absence of solar radiation, when direct generation is not possible, and allow production peaks to be transferred in accordance with demand requirements [43].

The research carried out in this work will be useful to optimize the performance of solar tower power plants of molten salt external receivers and to plan its design properly [44]. The results can hopefully help the Algerian government to decide on policies related to performance of solar tower power plants of molten salt external technologies. It has been proven that although energy production is lower in the northern regions of Algeria, it is profitable and allows for the efficient supply of electricity to regions that do not currently have electricity grids.

As prospects for future research, it will use the performance of the solar advisor model for integrating financial modeling into project models. It is also advisable to evaluate the incorporation of a wind barrier in the perimeter of the CSP in order to protect the components from high wind levels and the dust it carries. The barrier protects the heliostats from bursts and prevents the continuous movement of sand that may enter the solar field and deposit on the components.

5. Conclusions

A comparison study between three power stations located in the Algerian regions of Béchar, El Oued, and Djelfa was developed. Each plant is equipped with a molten salt storage mode, the receiver is of external type, and the implantation of the field of heliostats which have been defined for an annual production of 20 MW. It became evident that the regions in Northern Algeria are suitable for the production of concentrated solar energy,

It was found that the system advisor model software is a suitable tool to calculate normal direct irradiation on the performance of solar tower power plants of molten salt external receivers. The use of this software was very interesting in the study. It was observed that direct normal irradiation is a fundamental factor in order to choose an adequate region. High values correspond to a high performance of the solar power plant, resulting in high production and storage capacity.

The results showed that the optimal heliostat field corresponds to 1.8 *SM* and 2 *TES* hours in Béchar, 1.2 *SM* and 2 *TES* hours for El Oued, and 1.5 *SM* and 4 *TES* hours for Djelfa. Therefore, thermodynamic plants should be studied through their direct normal irradiation instead of global horizontal irradiation and diffuse horizontal irradiation.

Finally, this study shows that there is a strong and direct relationship between *SM*, power generation, and storage capacity hours. The higher value of *SM* corresponds to higher values of production and storage capacity. Since the storage hours do not cover the whole night, it is essential either to increase the number of heliostats, and therefore factor *SM*, or provide an auxiliary source of energy to guarantee permanent activity for 24 h.

As lessons learned, it became evident that satellite meteorological data give better results compared to actual data, as electricity production can be twice as high as actual data because the latter take into account atmospheric conditions (clouds, wind, pollution, etc.).

It was found that the system advisor model software is a suitable tool to calculate normal direct irradiation on the performance of solar tower power plants of molten salt external receivers. The use of this software was very interesting in the study. It is observed that direct normal irradiation is a fundamental factor in order to choose an adequate region. High values correspond to a high performance of the solar power plant, resulting in high production and storage capacity.

Author Contributions: Conceptualization, A.R.; Data curation, K.M.; Formal analysis, J.G.-S.-C.; Funding acquisition, D.B.; Investigation, A.R.; Methodology, D.B. and R.K.; Project administration, K.M.; Resources, R.K.; Supervision, A.A.-K. and J.G.-S.-C.; Validation, K.M.; Visualization, A.A.-K.; Writing-original draft, A.R.; Writing-review & editing, D.B.

Funding: This research received no external funding.

Acknowledgments: The authors of the present study would like to express their thankfulness to the financial support of Junta de Extremadura to a part of this study related to Research Projects GR-18029 and 18086, linked to the VI Regional Plan for Research, Technical Development, and Innovation from the Junta de Extremadura 2017-2020.

Conflicts of Interest: The authors declare no conflict of interest.

Nomenclature

STPP	Solar tower power plants
DNI	Direct normal irradiation
DHI	Diffuse horizontal irradiation
GHI	Global horizontal irradiation
SAM	System advisor module
SM	Solar multiple
TES	Thermal energy storage (h)
CSP	Concentrated solar power
ΔR	Radial distance between heliostats (m)
ΔA_Z	Azimuthal distance between heliostats (m)
θ	Receiver elevation angle from heliostat (°)
α	Heliostat loft angle in degree (°)
A_h	Surface heliostat (m^2)
I_d	Irradiation direct normal (W/m^2)
N_h	Number of heliostats.
η_h	Efficiency of the solar field
η_{cos}	Loss due to cosine effect
η_{omb}	Loss due to shading effect

N_t	Total number of the receiver tube
η_{bloc}	Loss due to blocking effect
η_{deb}	Loss due to overflow
η_{ref}	Loss due to reflection
η_{disp}	Losses due to dispersal
CF	Capacity factor
E_{gp}	Energy generated in part-time (W)
E_{gf}	Energy generated in full-time (W)
TMY	Typical Meteorological Year
$Q_{réc}$	Heat flow of the receiver (W)
Q_{inc}	Incident heat flow (W)
q_{ra}	Loss of radiant heat flux (W)
q_{conv}	Loss of convection heat flux (W)
q_{ref}	Loss of reflection flow (W)
q_{sf}	Energy generated by the field of heliostats (W)
q_{pb}	Energy required by the power block (W)
Q_h	Heat flow of molten salt (W)
S_i	Total surface/Surface total (m^2)
US_i	Heat transfer conductance coefficient (W/K)
m_{htf}	Molten salt flow rate (kg/s)
C_{htf}	Heat capacity of the molten salt fluid (kJ/kg·K)
T_{st}	Receiver temperature at the surface (K)
T_{htf}	Inlet temperature of the molten salt at x position (K)
$T_{ic\text{-}air}$	Temperature of the air in the inner cavity (K)
R_{conv}	Heat transfer resistance by convection (K/W)
R_{cond}	Heat transfer resistance by conduction (K/W)
D_{it}	Inner diameter of the tube (m)
D_{ot}	Outer diameter of the tube (m)
h_{htf}	Convection heat transfer of the molten salt (W/m^2·K)
h_{conv}	Convective heat losses from receiver tube (W/m^2·K)
K_t	Thermal conductivity of the receiver tube (W/m·K)
L_t	Length of the tube (m)
TRY	Test Reference Year
EPW	Energy Plus Weather
CSV	Comma Separated Value
WM	Heliostat width meters
HM	Heliostat height meters
HTF	Heat transfer fluid

References

1. Boudghene, S.; Khiat, Z.; Flazi, S.; Kitamura, Y. A review on the renewable energy development in Algeria: Current perspective, energy scenario and sustainability issues. *Renew. Sustain. Energy Rev.* **2012**, *16*, 4445–4460. [CrossRef]
2. Stambouli, A.B. Promotion of renewable energies in Algeria: Strategies and perspectives. *Renew. Sustain. Energy Rev.* **2011**, *15*, 1169–1181. [CrossRef]
3. Energies Nouvelles, Renouvelables et Maitrise de l'Energie. Available online: http://www.energy.gov.dz/francais/index.php?page=energies-nouvelles-renouvelables-et-maitrise-de-l-energie (accessed on 22 May 2018).
4. Medium Term RE Market Report, Agence Internationale de l'Energie. 2013. Available online: https://www.iea.org/publications/freepublications/publication/2013MTRMR.pdf (accessed on 22 May 2018).
5. Ghezloun, A.; Chergui, S.; Oucher, N. Algerian energy strategy in the context of sustainable development (Legal framework). *Energy Procedia* **2011**, *6*, 319–324. [CrossRef]

6. Portail Algérien des Energies Renouvelables. Available online: https://portail.cder.dz/spip.php?article4446 (accessed on 22 May 2018).

7. Imadojemu, H.E. Concentrating parabolic collectors: A patent survey. *Energy Convers. Manag.* **1995**, *36*, 225–237. [CrossRef]

8. Hocine Moussa, B. Thermal Solar Power Plants in Algeria and Desertec Mega project. In Proceedings of the Le 1erés Journées d'Etudes de Mécaniques, Batna, Algeria, 29–30 November 2011.

9. Rosen, M.A.; Dincer, I. Exergy–cost–energy–mass analysis of thermal systems and processes. *Energy Convers. Manag.* **2003**, *4*, 1633–1651. [CrossRef]

10. Adinberg, R. Simulation analysis of thermal storage for concentrating solar power. *Appl. Therm. Eng.* **2011**, *31*, 3588–3594. [CrossRef]

11. Wellmann, J.; Morosuk, T. Renewable Energy Supply and Demand for the City of El Gouna, Egypt. *Sustainability* **2016**, *8*, 314. [CrossRef]

12. Annuaire de la Filière Française du Solaire Thermodynamique. 2011. Available online: http://www.enr.fr/userfiles/files/Annuaires/2011171733_annuairethermo2011.pdf (accessed on 22 May 2018).

13. Ho, C.K.; Iverson, B.D. Review of high-temperature central receiver designs for concentrating solar power. *Renew. Sustain. Energy Rev.* **2014**, *29*, 835–846. [CrossRef]

14. Boudaoud, S.; Khellaf, A.; Mohammedi, K.; Behar, O. Thermal performance prediction and sensitivity analysis for future deployment of molten salt cavity receiver solar power plants in Algeria. *Energy Convers. Manag.* **2015**, *89*, 655–664. [CrossRef]

15. Behar, O.; Khellaf, A.; Mohammedi, K. A review of studies on central receiver solar thermal power plants. *Renew. Sustain. Energy Rev.* **2013**, *23*, 12–39. [CrossRef]

16. Mihoub, S.; Chermiti, A.; Beltagy, H. Methodology of determining the optimum performances of future concentrating solar thermal power plants in Algeria. *Energy* **2017**, *122*, 801–810. [CrossRef]

17. Quaschning, V. Technical and economical system comparison of photovoltaic and concentrating solar thermal power systems depending on annual global irradiation. *Sol. Energy* **2004**, *77*, 171–178. [CrossRef]

18. Zhu, Y.; Zhai, R.; Yang, Y.; Reyes-Belmonte, M.A. Techno-Economic Analysis of Solar Tower Aided Coal-Fired Power Generation System. *Energies* **2017**, *10*, 1392. [CrossRef]

19. Toro, C.; Rocco, M.V.; Colombo, E. Exergy and Thermoeconomic Analyses of Central Receiver Concentrated Solar Plants Using Air as Heat Transfer Fluid. *Energies* **2016**, *9*, 885. [CrossRef]

20. Boukelia, T.E.; Mecibah, M.S. Parabolic trough solar thermal power plant: Potential, and projects development in Algeria. *Renew. Sustain. Energy Rev.* **2013**, *21*, 288–297. [CrossRef]

21. Boudaoud, S.; Khellaf, A.; Mohammedi, K. Solar tower plant implementation in northern Algeria: Technico economic assessment. In Proceedings of the 2013 5th International Conference on Modeling, Simulation and Applied Optimization (ICMSAO), Hammamet, Tunisia, 28–30 April 2013; pp. 1–6.

22. Larbi, S.; Bouhdjar, A.; Chergui, T. Performance analysis of a solar chimney power plant in the southwestern region of Algeria. *Renew. Sustain. Energy Rev.* **2010**, *14*, 470–477. [CrossRef]

23. Viebahn, P.; Lechon, Y.; Trieb, F. The potential role of concentrated solar power (CSP) in Africa and Europe. A dynamic assessment of technology development, cost development and life cycle inventories until 2050. *Energy Policy* **2011**, *39*, 4420–4430. [CrossRef]

24. Abbas, M.; Merzouk, N.K. Techno economic study of solar thermal power plants for centralized electricity generation in Algeria. In Proceedings of the 2012 2nd International Symposium on Environment Friendly Energies and Applications, Newcastle upon Tyne, UK, 25–27 June 2012.

25. Freeman, J.M.; DiOrio, N.A.; Blair, N.J.; Neises, T.W.; Wagner, M.J.; Gilman, P.; Janzou, S. *System Advisor Model (SAM) General Description (Version 2017.9.5)*; Technical Report for National Renewable Energy Laboratory: Golden, CO, USA, May 2018.

26. Ouagued, M.; Khellaf, A.; Loukarfi, L. Estimation of the temperature, heat gain and heat loss by solar parabolic trough collector under Algerian climate using different thermal oils. *Energy Convers. Manag.* **2013**, *75*, 191–201. [CrossRef]

27. Gilman, P.; Blair, N.; Mehos, M.; Christensen, C.; Janzou, S.; Cameron, C. *Solar Advisor Model. User Guide for version 2.0*; Technical report NREL/TP-670-43704; August 2008. Available online: https://www.nrel.gov/docs/fy08osti/43704.pdf (accessed on 22 may 2018).

28. Kistler, B.L. *A User's Manual for DELSOL3. A Computer Code for Calculating the Optical Performance and Optimal System Design for Solar Thermal Central Receiver Plants.* SAND86-8018. Unlimited release UC-62a. 1986. Available online: http://ciTESeerx.ist.psu.edu/viewdoc/download?doi=10.1.1.467.8372&rep=rep1&type=pdf (accessed on 22 May 2018).

29. Ho, C.K. *Overview of Concentrating Solar Power Research at Sandia (Feb 2017);* Sandia National Lab.: Albuquerque, NM, USA, 2017.

30. Huang, W. Development of an analytical method and its quick algorithm to calculate the solar energy collected by a heliostat field in a year. *Energy Convers. Manag.* **2014**, *83*, 110–118. [CrossRef]

31. Wagner, M.J. Simulation and Predictive Performance Modeling of Utility Scale Central Receiver System Power Plants. Ph.D. Thesis, University of Wisconsin-Madison, Madison, WI, USA, January 2008.

32. Romero-Álvarez, M.; Zarza, E. Concentrating solar thermal power. In *Handbook of Energy Efficiency and Renewable Energy;* CRC Press: Boca Raton, FL, USA, 2007.

33. Modeling and Calculation of Heat Transfer Relationship for Concentrated Solar Power Receivers. Available online: https://search.library.wisc.edu/catalog/9910121576302121 (accessed on 30 May 2018).

34. ReGrid: Concentrated Solar Power. (RENAC) Renewables Academy. Schonhauser Alee 10–11. 10119. Berlin, Germany. Available online: https://www.renac.de/en/projects/ (accessed on 22 May 2018).

35. German Aerospace Centre (DLR). Available online: https://www.dlr.de/dlr/en/desktopdefault.aspx/tabid-10002/ (accessed on 30 May 2018).

36. Centre de Developpement des Energies Renouvelables (CEDER). Algeria (2013). Available online: http://www.cder.dz/ (accessed on 2 June 2018).

37. Siva Reddy, V.; Kaushik, S.C.; Ranjan, K.R.; Tyagi, S.K. State of the art of solar thermal power plants. A review. *Renew. Sustain. Energy Rev.* **2013**, *27*, 258–273. [CrossRef]

38. Cheng, T.C.; Yang, C.K.; Lin, I. Biaxial-Type Concentrated Solar Tracking System with a Fresnel Lens for Solar-Thermal Applications. *Appl. Sci.* **2016**, *6*, 115. [CrossRef]

39. Ohya, Y.; Wataka, M.; Watanabe, K.; Uchida, T. Laboratory Experiment and Numerical Analysis of a New Type of Solar Tower Efficiently Generating a Thermal Updraft. *Energies* **2016**, *9*, 1077. [CrossRef]

40. Gregorio-Muñiz, J.M. El Mercado de las Energías Renovables en Argelia. Oficina Económica y Comercial de la Embajada de España en Argel. 2004. Available online: https://www.icex.es/icex/ (accessed on 2 June 2018).

41. Owen, A.D. Renewable energy: Externality costs as market barriers. *Energy Policy* **2006**, *34*, 632–642. [CrossRef]

42. Pitz-Paal, R.; Botero, N.B.; Steinfeld, A. Heliostat field layout optimization for high-temperature solar thermochemical processing. *Sol. Energy* **2011**, *85*, 334–343. [CrossRef]

43. Zhang, H.L.; Baeyens, J.; Degrève, J.; Cacères, G. Concentrated solar power plants: Review and design methodology. *Renew. Sustain. Energy Rev.* **2013**, *22*, 466–481. [CrossRef]

44. Candelario-Garrido, A.; García-Sanz-Calcedo, J.; Reyes, A.M. A quantitative analysis on the feasibility of 4D planning graphic systems versus conventional systems in building projects. *Sustain. Cities Soc.* **2017**, *35*, 378–384. [CrossRef]

Article

A Hybrid Forecasting Method for Solar Output Power Based on Variational Mode Decomposition, Deep Belief Networks and Auto-Regressive Moving Average

Tuo Xie [1], Gang Zhang [1,*], Hongchi Liu [1], Fuchao Liu [2] and Peidong Du [1,2]

[1]   Institute of Water Resources and Hydro-Electric Engineering, Xi'an University of Technology, Xi'an 710048, China; xtxaut6863@aliyun.com (T.X.); 15594818128@163.com (H.L.); dpdong@126.com (P.D.)

[2]   State Grid Gansu Electric Power Company, Gansu Electric Power Research Institute, Lanzhou 730050, China; liufc8127@sina.com

*   Correspondence: zhanggang3463003@163.com

Received: 7 September 2018; Accepted: 8 October 2018; Published: 12 October 2018

Abstract: Due to the existing large-scale grid-connected photovoltaic (PV) power generation installations, accurate PV power forecasting is critical to the safe and economical operation of electric power systems. In this study, a hybrid short-term forecasting method based on the Variational Mode Decomposition (VMD) technique, the Deep Belief Network (DBN) and the Auto-Regressive Moving Average Model (ARMA) is proposed to deal with the problem of forecasting accuracy. The DBN model combines a forward unsupervised greedy layer-by-layer training algorithm with a reverse Back-Projection (BP) fine-tuning algorithm, making full use of feature extraction advantages of the deep architecture and showing good performance in generalized predictive analysis. To better analyze the time series of historical data, VMD decomposes time series data into an ensemble of components with different frequencies; this improves the shortcomings of decomposition from Empirical Mode Decomposition (EMD) and Ensemble Empirical Mode Decomposition (EEMD) processes. Classification is achieved via the spectrum characteristics of modal components, the high-frequency Intrinsic Mode Functions (IMFs) components are predicted using the DBN, and the low-frequency IMFs components are predicted using the ARMA. Eventually, the forecasting result is generated by reconstructing the predicted component values. To demonstrate the effectiveness of the proposed method, it is tested based on the practical information of PV power generation data from a real case study in Yunnan. The proposed approach is compared, respectively, with the single prediction models and the decomposition-combined prediction models. The evaluation of the forecasting performance is carried out with the normalized absolute average error, normalized root-mean-square error and Hill inequality coefficient; the results are subsequently compared with real-world scenarios. The proposed approach outperforms the single prediction models and the combined forecasting methods, demonstrating its favorable accuracy and reliability.

Keywords: solar output power forecasting; combined prediction model; variational model decomposition; deep belief networks; auto-regressive and moving average model

---

1. Introduction

To promote sustainable economic and social development, energy sources such as solar energy and wind power need to be leveraged to counteract the rapidly growing energy consumption and deteriorating environment caused by climate change. To promote increased solar energy utilization, photovoltaic (PV) power generation has been rapidly developed worldwide [1]. PV power generation is affected by solar radiation, temperature and other factors. It also has strong intermittency and volatility. Grid access by large-scale distributed PV power introduces significant obstacles to the planning, operation, scheduling and control of power systems. Accurate PV power prediction not only provides the basis for grid dispatch decision-making behavior, but also provides support for multiple power source space-time complementarity and coordinated control; this reduces pre-existing rotating reserve capacity and operating costs, which ensures the safety and stability of the system and promotes the optimal operation of the power grid [2].

According to the timescale, PV power forecasting can be divided into long-term, short-term, and ultra-short-term forecasts [3]. A medium-long-term forecast (i.e., with a prediction scale of several months) can provide support for power grid planning; Short-term prediction (i.e., with a prediction scale of one to four days in advance) can assist the dispatching department in formulating generator set start-stop plans. Super short-term forecast (i.e., with a prediction scale of 15 min in advance) can achieve a real-time rolling correction of the output plan curve and can provide early warning information to the dispatcher. The shorter the time scale, the more favorable the management of preventative situations and emergencies. Most of the existing literature describes short-term forecasting research with an hourly cycle. There are few reports on the ultra-short-term prediction of PV power generation [4–6]. In addition, in the previous research, PV power prediction methods mainly include the following: physical methods, statistical methods, machine learning methods, and hybrid integration methods.

(1) In physical methods, numerical weather prediction (NWP) is the most widely used method, which involves more input data such as solar radiation, temperature, and other meteorological information.

(2) As for the statistical approaches, their main purpose is to establish a long-term PV output prediction model. In literature [7–10], the auto-regressive, auto-regressive moving average, auto-regressive integral moving average and Kalman filtering model of short-term PV prediction are respectively established based on the time series, and obtain good prediction results. The above model is mainly based on a linear model, which only requires historical PV data and does not require more meteorological factors. In addition, the time series methods can only capture linear relationships and require stationary input data or stationary differencing data.

(3) Along with the rapid update of computer hardware and the development of data mining, prediction methods based on machine learning have been successfully applied in many fields. Machine learning models that have been widely applied in PV output prediction models are nonlinear regression models such as the Deep Neural Network (DNN), the Recurrent Neural Network (RNN), the Convolutional Neural Network (CNN), the Deep Belief Network (DBN) and so forth. Literature [11,12] establishes output prediction models based on the neural network, which can consider multiple input influencing factors at the same time. The only drawback is that the network structure and parameter settings will have a great impact on the performance of the models, which limits the application of neural networks. Literature [13,14] has analyzed various factors affecting PV power and established support vector machine (SVM) prediction models facing PV prediction. The results show that the SVM adopts the principle of structural risk minimization to replace the principle of empirical risk minimization of traditional neural networks; thus, it has a better generalization ability. To effectively enhance the reliability and accuracy of PV prediction results, related literature proposes the use of intelligent optimization algorithms to estimate model parameters; some examples of intelligent optimization algorithms include the gray wolf algorithm, the similar day analysis method and the particle swarm algorithm [15–17]. The example analysis illustrates the effectiveness of the optimization algorithm.

(4) In recent years, decomposition-refactoring prediction models based on signal analysis methods have attracted more and more attention from scholars. Relevant research shows that using signal analysis methods to preprocess data on PV power series can reduce the influence of non-stationary meteorological external factors on the prediction results and improve prediction accuracy. The decomposition methods of PV power output data mainly include wavelet analysis and

wavelet packet transform [18,19], empirical mode decomposition (EMD) [20], ensemble empirical mode decomposition (EEMD) [21] and local mean decomposition (LMD) [22]. Among them, wavelet analysis has good time-frequency localization characteristics, but the decomposition effect depends on the choice of basic function and the self-adaptability is poor [23]. EMD has strong self-adaptability, but there are problems such as end-effects and over-enveloping [24]. LMD has fewer iterations and lighter end-effects. However, judging the condition of purely FM signals requires trial and error. If the sliding span is not properly selected, the function will not converge, resulting in excessive smoothness, which affects algorithmic accuracy [25]. EEMD is the improved EMD method; the analysis of the signal is made via a noise-assisted, weakened influence of modal aliasing. However, this method has a large amount of computation and more modal components than the true value [26]. Variational mode decomposition (VMD) is a relatively new signal decomposition method. Compared to the recursive screening mode of EMD, EEMD, and LMD, by controlling the modal center frequency K, the VMD transforms the estimation of the sequence signal modality into a non-recursive variational modal problem to be solved, which can well express and separate the weak detail signal and the approximate signal in the signal. It is essentially a set of adaptive Wiener filters with a mature theoretical basis. In addition, the non-recursive method adopted does not transmit errors, and solves the modal aliasing phenomenon of EMD, EEDM and other methods appeared in the background of bad noise, and effectively weakens the degree of end-effect [27]. Literature [28] used this method for fault diagnoses and achieved good results.

Through the above literature research, we find that the previous prediction methods using traditional neural network models and single machine learning models cannot meet the performance requirements of local solar irradiance prediction scenarios with complex fluctuations. To further improve the prediction accuracy of PV output, this work proposes a new and innovative hybrid prediction method that can improve prediction performance. This method is a hybrid of variational mode decomposition (VMD), the deep belief network (DBN), and the auto-regressive moving average model (ARMA); it combines these prediction techniques adaptively. Different from the traditional PV output prediction model, the key features of the VMD-ARMA-DBN prediction model are the perfect combination of the following parts: (1) VMD-based solar radiation sequence decomposition; (2) ARMA-based low-frequency component sequence prediction model; and (3) DBN-based high-frequency component sequence prediction model. The original photovoltaic output sequences are decomposed into multiple controllable subsequences of different frequencies by using the VMD methods. Then, based on the frequency characteristics of each subsequence, the subsequence prediction is performed by using the advantages of ARMA and DBN, respectively. Finally, the subsequences are reorganized, and the final PV output prediction value is obtained. The main contributions of this article are as follows:

(1) To reduce the complexity and non-stationarity of the PV output data series, the VMD decomposition is used for the first time to preprocess the PV data sequence and decompose it into a series of IMF component sequences with good characteristics, achieving an effective extraction of advanced nonlinear features and hidden structures in PV output data sequences.

(2) An innovative method for predicting PV output based on VMD-DBM-ARMA is proposed. According to the characteristics of the IMF component sequence decomposed by the VMD, DBN and ARMA models are used to improve predictions of the high- and low-frequency component sequences, respectively. Based on this, the DBN is used for feature extraction and structural learning of the prediction results of each component sequence. Finally, the PV output predictive value is obtained.

(3) Taking the actual measured data of a PV power plant in China-Yunnan for application, the short-term PV output predictions of ARMA, DBN, EMD-ARMA-DBN, EEMD-ARMA-DBN, and VMD-ARMA-DBN were conducted and three prediction precisions were introduced, respectively. The evaluation indicators perform a statistical analysis on the prediction effect of each model. The results show that the proposed method can guarantee the stability of the prediction error and further improve the PV prediction accuracy.

The remainder of this paper is organized as follows: Section 2 describes our proposed approach: A Hybrid Forecasting Method for Solar Output Power Based on VMD-DBN-ARMA; experimental results are presented in Section 3; and the experimental comparison and conclusion are given in Sections 4 and 5, respectively.

2. Materials and Methods

2.1. Variational Mode Decomposition (VMD)

VMD is a new non-stationary, signal-adaptive decomposition estimation method. It was proposed by Konstantin Dragomiretskiy in 2014. The purpose of VMD is to decompose the original complex signal into K amplitude and frequency modulation sub-signals. Because K can be preset, and with a proper value of K, modal aliasing can be effectively suppressed [29]. VMD assumes that each "mode" has a finite bandwidth with unique center frequencies. The main process of this method is to: (1) use Wiener filtering to de-noise the signal; (2) obtain the K-estimated center angular frequency by initializing the finite bandwidth parameters and the central angular frequency; (3) use the alternating direction method of multipliers to update each modal function and its center frequency; (4) demodulate each modal function to the corresponding baseband; and (5) minimize the sum of each modal estimation bandwidth. The algorithm can be divided into the construction of a variational problem, subsequently obtaining the solution to the variation problem. The algorithm is described in detail in Sections 2.1.1–2.1.4.

2.1.1. The Construction of the Variational Problem

Assume that each mode has a finite bandwidth and a pre-defined center frequency. The variational problem is described as a problem that seeks K modal functions $u_k(t)(k = 1, 2, \ldots, K)$ such that the sum of the estimated bandwidths of each mode is minimized subject to the constraint that the sum of each mode is equal to the input signal f. The specific construction steps are as follows:

(1) Apply a Hilbert transform to the analytical signal of each modal function $u_k(t)$. Then, obtain its single-side spectrum

$$\left(\delta(t) + \frac{j}{\pi t}\right) * u_k(t)$$
$$\delta(t) = \begin{cases} 0 & t \neq 0 \\ \infty & t = 0 \end{cases}, \int_{-\infty}^{+\infty} \delta(t)dt = 1 \tag{1}$$

where $\delta(t)$ is the Dirac distribution.

(2) Modulate the spectrum of each mode to be the corresponding baseband based on the mixed-estimated center frequency $e^{-j\omega_k t}$ of various modal analysis signals

$$\left[\left(\delta(t) + \frac{j}{\pi t}\right) * u_k(t)\right]e^{-j\omega_k t}, \tag{2}$$

where $e^{-j\omega_k t}$ is the phasor description of the center frequency of the modal function in the complex plane and ω_K is the center frequency of each modal function.

(3) Calculate the square of the norm of the gradient of the analytical signal and estimate the bandwidth of each modal signal. The constrained variational problem is expressed as

$$\begin{cases} \min\limits_{\{u_k\},\{\omega_k\}} \left\{ \sum\limits_{k=1}^{K} \left\| \partial_t \left[\left(\delta(t) + \frac{j}{\pi t}\right) \otimes u_k(t) \right] e^{-j\omega_k t} \right\|_2^2 \right\} \\ \text{s.t. } \sum\limits_{k=1}^{K} u_k = f \end{cases}, \tag{3}$$

where $\{u_k\} = \{u_1, u_2, \cdots, u_K\}$ is the set of modal functions, $\{\omega_k\} = \{\omega_1, \omega_2, \cdots, \omega_K\}$ is the set of center frequencies that correspond to the modal functions, $\otimes$ is the convolution operation, and K is the total number of modal functions.

2.1.2. Solve the Variational Problem

(1) Introduce the second-level penalty factor C and the Lagrange multiplication operator $\theta(t)$ to change the constraint variational problem into a non-binding variational problem. Among them, C guarantees the reconstruction precision of the signal and $\theta(t)$ maintains the strictness of the constraint condition. The expanded Lagrange expression is characterized by

$$
\begin{aligned}
L(\{u_k\}, \{\omega_k\}, \theta) = C \sum_{k=1}^{K} \left\| \partial_t \left[\left(\delta(t) + \frac{j}{\pi t} \right) u_k(t) \right] e^{-j\omega_k t} \right\|_2^2 \\
+ \left\| f(t) - \sum_{k=1}^{K} u_k(t) \right\|_2^2 + \left\langle \theta(t), f(t) - \sum_{k=1}^{K} u_k(t) \right\rangle,
\end{aligned}
\tag{4}
$$

(2) VMD uses the alternating direction multiplication operator method to solve Equation (4) (the variational problem). In the expanded Lagrange expression, the "saddle point" is found by alternately updating u_k^{n+1}, ω_k^{n+1} and θ^{n+1}, where n denotes the number of iterations and where u_k^{n+1} can be transformed into the frequency domain using the Fourier isometric transformation

$$
\begin{aligned}
\hat{u}_k^{n+1} = \underset{\hat{u}_k, u_k \in X}{\mathrm{argmin}} \Big\{ C \| j\omega \{ [1 + \mathrm{sgn}(\omega + \omega_k)] \hat{u}_k(\omega + \omega_k) \} \|_2^2 \\
+ \left\| \hat{f}(\omega) - \sum_{k=1}^{K} \hat{u}_k(\omega) + \frac{\theta(\omega)}{2} \right\|_2^2 \Big\}
\end{aligned}
\tag{5}
$$

where ω is the random frequency and X contains all desirable sets of u_k. Replace ω with $\omega - \omega_k$, and the non-negative frequency interval integral form is

$$
\hat{u}_k^{n+1} = \underset{\hat{u}_k, u_k \in X}{\mathrm{argmin}} \left\{ \int_0^\infty \left[4C(\omega - \omega_k)^2 |\hat{u}_k(\omega)|^2 + 2 \left| \hat{f}(\omega) - \sum_{k=1}^{K} \hat{u}_k(\omega) + \frac{\theta(\omega)}{2} \right|^2 \right] d\omega \right\}.
\tag{6}
$$

Thus, the solution to the quadratic optimization problem is

$$
\hat{u}_k^{n+1}(\omega) = \frac{\hat{f}(\omega) - \sum_{k=1}^{K} \hat{u}_k(\omega) + \frac{\theta(\omega)}{2}}{1 + 2C(\omega - \omega_k)^2}.
\tag{7}
$$

According to the same process, the method for updating the center frequency is solved via

$$
\omega_k^{n+1} = \frac{\int_0^\infty \omega |\hat{u}_k(\omega)|^2 d\omega}{\int_0^\infty |\hat{u}_k(\omega)|^2 d\omega}.
\tag{8}
$$

Among them, $\hat{u}_k^{n+1}(\omega)$ is equivalent to the Wiener filter of the current residual quantity and $\hat{f}(\omega) - \sum_{k=1}^{K} \hat{u}_k(\omega)$; ω_k^{n+1} is the barycenter of the power spectrum of the current modal function. When an inverse Fourier transform is applied, we end up with $\hat{u}_k(\omega)$, where its real-part is $\{u_k(t)\}$.

2.1.3. VMD Algorithm Flow

- **Step 1**: Initialize parameters $\{u_k^1\}$, $\{\omega_k^1\}$, θ^1 and n. Set the number of iterations to be n to 1.
- **Step 2**: Update u_k and ω_k according to Equations (7) and (8).

- **Step 3**: Update θ via

$$\hat{\theta}^{n+1}(\omega) \leftarrow \hat{\theta}^{n}(\omega) + \tau \left[\hat{f}(\omega) - \sum_{k=1}^{K} \hat{u}_{k}^{n+1}(\omega) \right]. \tag{9}$$

- **Step 4**: If the discrimination precision $e > 0$ and $\sum_{k=1}^{K} \frac{\|\hat{u}_{k}^{n+1} - \hat{u}_{k}^{n}\|_{2}^{2}}{\|\hat{u}_{k}^{n}\|_{2}^{2}} < e$ are satisfied, the iteration ends and **Step 2** is returned.
- **Step 5**: Obtain the corresponding modal subsequences based on the given mode number.

2.1.4. Determine VMD Parameters

(1) Determine the number of modes

VMD needs to determine the number of K modalities before decomposing the signal. Study [29] found that if the K value is too small, multiple components of the signal in one modality may appear at the same time, or a component cannot be estimated. Conversely, the same component appears in multiple mode, and the modal center frequency obtained iteratively will overlap.

Considering these problems, this paper uses a simple and effective modal fluctuation method to determine the number of K modes. The algorithm flow is as follows [30]:

- **Step 1**: Estimate the initial value of the modal number k through the spectrum diagram of the signal;
- **Step 2**: The modal number is k, whether the modal center frequencies overlap;
- **Step 3**: If the center frequency overlaps, reduce the number of modalities for VMD decomposition until no center frequency overlap occurs, and output is K;
- **Step 4**: If there is no overlap in the center frequency, increase the number of modalities for VMD decomposition, until the center frequency overlaps and output is $K - 1$.

(2) Penalty factor

The introduction of the penalty factor changes the constraint variational problem into a non-binding variational problem. In the operation of the VMD program, only the modal bandwidth and convergence speed (after decomposition) are affected. To avoid modal aliasing and guaranteeing a certain convergence speed, the standard VMD has a strong adaptability with a penalty factor of 2000. In this work, the penalty factor adopts a default value of 2000 during the VMD decomposition [31].

2.2. Deep Belief Network (DBN)

The deep belief network (DBN) is an in-depth network efficient learning algorithm proposed by Hinton et al.; it processes high-dimensional and large-scale data problems [32] such as image feature extraction and collaborative filtering. DBN essentially consists of multiple Restricted Boltzmann Machine (RBM) networks and a supervised Back Propagation (BP) Network. The lower layer represents the details of the original data. The higher layer represents the data attribute categories or the characteristics, from a low-level to high-level layer-by-layer abstraction; it has the characteristics of gradually tapping deep data features. The specific structure is shown in Figure 1.

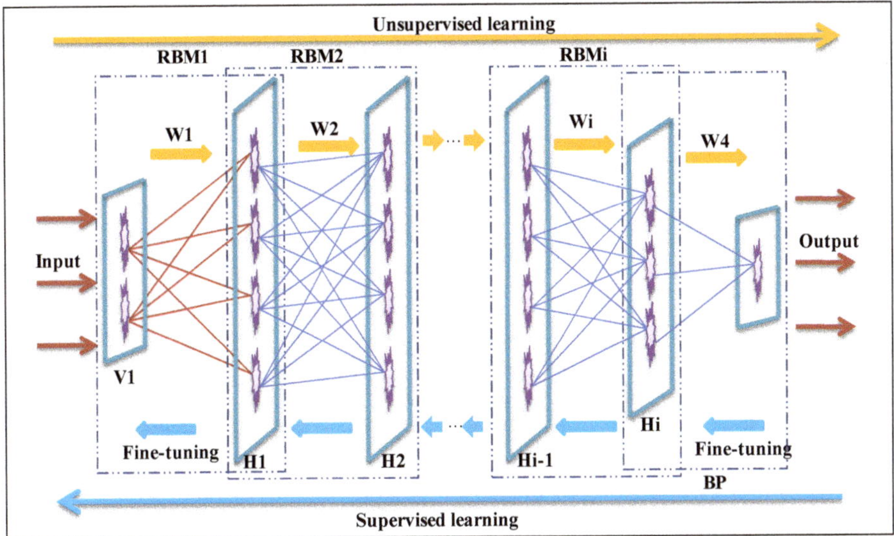

Figure 1. Deep belief network structure. RBM: Restricted Boltzmann Machine; BP: Back-Projection.

The DBN training process includes two stages: forward stack RBM pre-training and reverse BP fine-tuning. The pre-training phase initializes the parameters of the entire DBN model. The unsupervised learning of forward stacking is used to train each layer of RBM. The output of the previous RBM hidden layer could be used as the input of the next RBM visible layer. Since the RBM network can only ensure that the feature mapping in each layer of the DBN model is optimal, it cannot guarantee that the feature mapping can be optimized in the entire DBN model. Therefore, we need to enter the fine-tuning phase to optimize the parameters of the entire network. During the fine-tuning phase, supervised learning methods are used to further optimize and adjust the relevant parameters of the cyberspace. The errors resulting from the actual output and standard annotation information are propagated backward layer by layer, and the entire DBN weights and offsets are fine-tuned from the top to bottom.

2.2.1. Forward RBM Pre-Training

The Boltzmann machine (BM) is a probabilistic modeling method based on an energy function. It has a strong unsupervised learning ability. In theory, it can learn arbitrarily complex rules and apply the rules to the data. Inter-unit connections of the inner- and inter-layer are complex, so there are also disadvantages such as a long training time, a large number of calculations, and difficulty in obtaining the probability distribution [33].

RBM is an improvement over BM. It is a two-layer recursive neural network. Random binary input and random binary output are connected via symmetrical weights, as shown in Figure 2. The two-layer system consists of n dominant neurons (corresponding to the visible variable) and m recessive neurons (corresponding to the hidden variable h), where the v and h elements are binary variables whose state is 0 or 1. There is a weight value connection between the cell layer and the hidden cell layer, but there is no connection between the cells in the layer.

RBM is an energy-based model. For a given set of states (v, h), the joint configuration energy function [34] of the visible and hidden units is

$$E(v, h|\theta) = -\sum_{i=1}^{n}\sum_{j=1}^{m} w_{ij} h_j v_i - \sum_{j=1}^{m} c_j h_j - \sum_{i=1}^{n} b_i v_i. \tag{10}$$

where $\theta = (\omega, b, c)$ is the parameter of the RBM model; v_i, b_i are the respective states and offsets of the i-th visible unit; h_j, c_j are the respective states and offsets of the j-th hidden unit; ω_{ij} is the i-th visible unit and the connection weight between the j-th hidden units. The structure of the RBM model is very special. There are no connections within the layers and all the layers are fully connected. When the activation state of each visible layer unit is given, the activation states of the neurons in the hidden layer are independent of each other, and $\sigma(x) = \frac{1}{1+\exp(-x)}$ is a sigmoid activation function. Therefore, the activation probability of the j-th neuron in the hidden layer is

$$p(h_j = 1|v, \theta) = \sigma(b_j + \sum_i v_i w_{ij}) \tag{11}$$

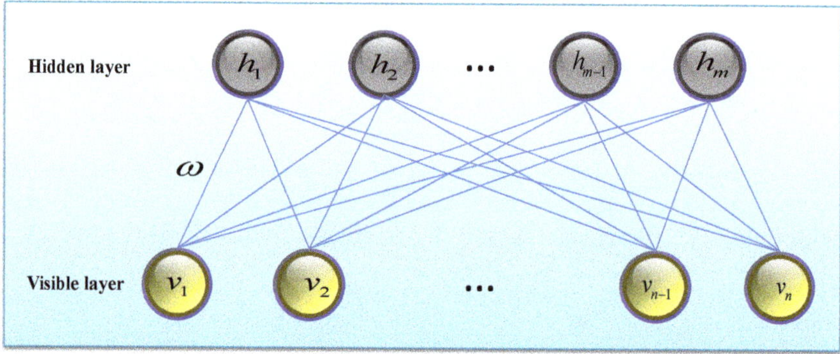

Figure 2. RBM structure diagram (V stands for the display element and h stands for the hidden element).

Similarly, when the state h of the hidden unit is given, the active states v_i of the visible units are also mutually independent, and the activation probability of the i-th visible unit is

$$p(v_i = 1|h, \theta) = \sigma(c_i + \sum_j h_j w_{ij}) \tag{12}$$

RBM is a stable network structure. By maximizing the log likelihood function $L(\theta)$ of the RBM on the input training set to obtain the model parameter θ, the training data set can be fitted. The hidden layer can subsequently be used as the characteristics of the visible layer data.

$$\hat{\theta} = \operatorname*{argmax}_\theta L(\theta) = \operatorname*{argmax}_\theta \sum_{t=1}^N \lg P(v^{(t)}|h, \theta) \tag{13}$$

To quickly train a log likelihood gradient of an RBM, the data distribution of Gibbs sampling [35] can be used as the RBM definition expectation, and then the weight and offset parameter update criteria can be obtained:

$$\begin{cases} w_{ij}^{k+1} = w_{ij}^k + \varepsilon(<v_i h_j>_{data} - <v_i h_j>_{recon}) \\ a_i^{k+1} = a_i^k + \varepsilon(<v_i>_{data} - <v_i>_{recon}) \\ b_j^{k+1} = b_j^k + \varepsilon(<h_j>_{data} - <h_j>_{recon}) \end{cases} \tag{14}$$

where ε is the learning rate, taking any value in the interval [0,1]; the data are the expectation of the distribution of the original observation data; recon is the desired distribution defined by the RBM model.

2.2.2. Reverse Back-Projection (BP) Trimming Phase

After pre-training, the DBN network is fine-tuned. This phase is achieved via reverse supervised learning. The BP network is set to be the last layer of the DBN, the output of the last layer of RBM is taken as the input of the BP, and supervised training is performed from top to bottom to optimize the parameters generated in the pre-training stage to optimize the prediction ability of the DBN. Unlike an unsupervised DBN training process that considers one RBM at a time, the reverse-trimmed DBN training process considers all DBN layers at the same time and uses the model output and target tag data to calculate training errors; it also updates DBN classifier model parameters to minimize training errors. In the process of backward BP propagation, the sensitivity of each layer needs to be calculated. The sensitivity calculation is described in [36].

2.3. Auto-Regressive and Moving Average Model (ARMA)

The auto-regressive moving average model is an important method for studying time series. It uses an auto-regressive model (referred to as the AR model) and a moving average model (referred to as the MA model) as a basis for "mixing". It uses modern statistics and information-processing techniques to investigate time series law, which is a group of powerful tools for solving practical problems. Time series laws have been widely used in many fields such as finance, economy, meteorology, hydrology, and signal processing. Based on the historical data of the sequence, this reveals the structure and regularity of the dynamic data, and quantitatively understands the linear correlation between observable data. Time laws use mathematical statistics to process and predict its future value. The ARMA model is used as a predictive model and its basic principles are as follows [37–39]:

Let $X(t) = \{X_t, t = 0, \pm 1, \cdots\}$ be a 0-mean stationary random sequence and satisfy for any t

$$X_t - \phi_1 X_{t-1} - \cdots - \phi_p X_{t-p} = \varepsilon_t + \theta_1 \varepsilon_{t-1} + \cdots + \theta_q \varepsilon_{t-q}, \tag{15}$$

where p, q is the ARMA model order; $\varepsilon = \{\varepsilon_t, t = 0, \pm 1, \cdots\}$ is the white noise sequence with variance σ^2; ϕ, θ is a non-zero undetermined coefficient; and $\{X_t\}$ is the ARMA (p, q) process with mean u [40].

2.3.1. Model Ordering

The ARMA model is the system's memory of its past self-state and noise of entering the system. Determining the order of the model and the value of the unknown parameter according to a set of observation data is the model order. Firstly, through the correlation analysis, the autocorrelation function (ACF) and partial autocorrelation function (PACF) of the sample are calculated, then the order of the model is preliminarily judged by the trailing nature or censored nature of the ACF and PACF, as shown in Table 1.

Table 1. Sequence characteristics table of ARMA (p, q) model. ARMA: Auto-Regressive Moving Average; ACF: autocorrelation function; PACF: partial autocorrelation function; AR: Auto-regressive; MA: Moving Average.

Model Type	AR(p)	MA(q)	ARMA(p, q)
ACF	trailing	censored	trailing
PACF	censored	trailing	trailing

It can be seen that the PACF of AR(p) is censored at point p, while the ACF of MA(q) is censored at point q. The p can also be determined by observing the oscillation period of the PACF. In this way, the order of AR(p) or MA(q) can be preliminary determined while the model is being identified.

The Akaike Information Criterion (AIC) [41] is used as the ARMA model ordering criterion. The AIC function is:

$$
\begin{aligned}
\text{AIC}(p,q) &= \min_{0 \le p,q \le L} (p,q) \\
&= \min_{0 \le p,q \le L} \left\{ \ln \hat{\sigma}^2 + \frac{2(p+q)}{N} \right\}
\end{aligned}
\qquad (16)
$$

Among them, $\hat{\sigma}^2$ is the estimation of the variance of the noise term, N is the known observation data sample size, and L is the highest order given in advance. The use of the AIC criterion to determine the order refers to the points p and q that seek to minimize the statistic AIC (p,q) within a certain range of p, q, which is used as an estimate of (p, q) [42]. Theoretically, the larger the L, the higher the prediction accuracy. However, considering the calculation time, and the increase in L, the prediction accuracy is not significantly improved. In general, the value is $N/10$, $\ln N$, or $\sqrt{N}$.

From Equation (16), when the model parameter N is gradually increased, the fitting error is improved significantly and the AIC (p,q) value will show a downward trend. As the model order increases, the fitting residual improves a little. The AIC value also increases gradually; when the AIC (p,q) obtains the minimum, the corresponding p, q is the ideal order of the model.

2.3.2. Parameter Estimation

There are many methods for estimating the parameters of the ARMA model. The Least Squares Estimation method is used in this work. See Reference [43] for the specific parameters. After each parameter is calculated, it is substituted into ARMA (p,q) to forecast each reconstructed component to obtain the predicted value $\hat{X}$ and the fitted value $\hat{c}_k$ of the modeling data. The residual $\gamma = c_k - \hat{c}_k$ between the measured data and the fitted value of the model data is calculated. Then, γ is used to describe the modeling data and to obtain the residual forecast $\hat{\gamma}$. After that, the revised forecast value is

$$
\widetilde{X}_t = \hat{\gamma} + \hat{X}_t
\qquad (17)
$$

In the formula, γ is the residual value of the observed value and the forecasted value; $c_k (k = 1, 2, \cdots)$ is the observation value of the modeling data; $\widetilde{X}_t$ is the prediction value after the prediction residual correction; and $\hat{X}_t$ and $\hat{\gamma}$ are the ARMA model prediction values and their corresponding residual values, respectively.

2.4. Combination Forecasting Model Based on VMD-ARMA-DBN

Considering the nonlinear, non-stationary, and periodic characteristics of PV output data, and considering that the time series ARMA (p,q) is a linear model, the prediction effect on non-stationary data is not good; however, the better-trained neural network has higher accuracy for non-stationary data prediction. Therefore, in this work, PV prediction based on the VMD-ARMA-DBN model is used to decompose the PV output time series into multiple IMFs with different frequencies. The predictive models for different IMFs sequences are established (respectively) to reduce the interaction between varying characteristics. Finally, DBN is used to reconstruct the prediction components to obtain the predicted value of the original sequence. The specific process is shown in Figure 3.

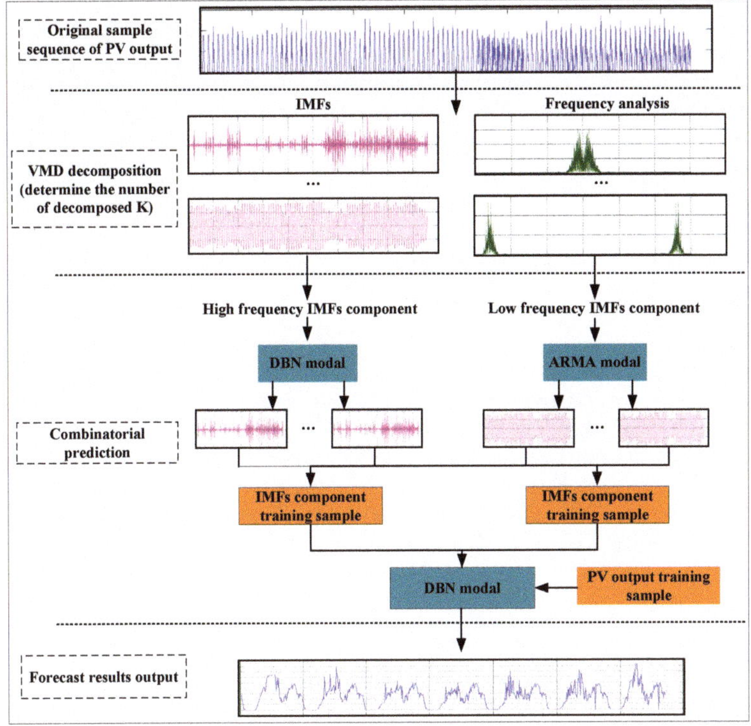

Figure 3. Combination Forecasting Model Based on VMD-ARMA-DBN. PV: photovoltaic; IMF: Intrinsic Mode Function; VMD: Variational Mode Decomposition; DBN: Deep Belief Network.

2.5. Data Model Accuracy Evaluation Index

To evaluate the predictive performance of the prediction model, the normalized mean absolute error e_{NMAE}, normalized mean-square-error e_{NRMSE}, and Theil coefficient (TIC) are used as the performance evaluation indicators of the prediction model.

$$e_{NMAE} = \frac{1}{P_r} \frac{1}{n} \sum_{i=1}^{n} |\hat{y}_i - y_i| \times 100\%, \tag{18}$$

$$e_{NRMSE} = \frac{1}{P_r} \sqrt{\frac{1}{n} \sum_{i=1}^{n} (\hat{y}_i - y_i)^2} \times 100\%, \tag{19}$$

$$Theil\ IC = \frac{\sqrt{\frac{1}{n} \sum_{i=1}^{n} (y_i - \hat{y}_i)^2}}{\sqrt{\frac{1}{n} \sum_{i=1}^{n} y_i^2} + \sqrt{\frac{1}{n} \sum_{i=1}^{n} \hat{y}_i^2}}. \tag{20}$$

where y_i is the actual observed value, $\hat{y}_i$ is the predicted value, $\overline{y}_i$ is the total average of the observed values, n is the number of samples, and P_r is the rated installed capacity of the PV power plant. The Hill unequal coefficient is always between 0 and 1. The smaller the value, the smaller the difference between the fitted value and the true value, which means the prediction accuracy is higher. When it is equal to 0, it means a 100% fit.

3. Results

Based on the multi-frequency combination forecasting model, this work selected the recorded data of a 50 MW PV power station monitoring platform in Yunnan Province in 2016 to conduct an empirical study of PV output forecasting. Since there are 96 load points in the output sample of the PV power plant every day, the data entries are numerous and the change is complex. Therefore, this work selected 35,040 output data entries from 1 January 2016 to 30 December 2016 and from 1 January to 23 December 2016 as research samples, which suggests that a total of 34,368 output data entries were used as sample points for model fitting and the basis for the selection parameters. This model was then used to make predictions for 768 loads for the period between 24 December 2016 and 30 December 2016.

3.1. Training Sample Construction Based on VMD

3.1.1. Initial Determination of VMD Mode

According to the decomposition principle of VMD in Section 2.1, the number of modalities is determined by studying the series of PV output samples. Figure 4a shows a sequence diagram of the PV output. Figure 4b shows the frequency spectrum after the PV output sequence is through the Fast Fourier Transform (FFT). Because there are many data, the full spectrum diagram is not easy to observe, while the spectrum diagram is symmetrical. Therefore, when analyzing, half of the spectrum diagram is to be taken for analysis.

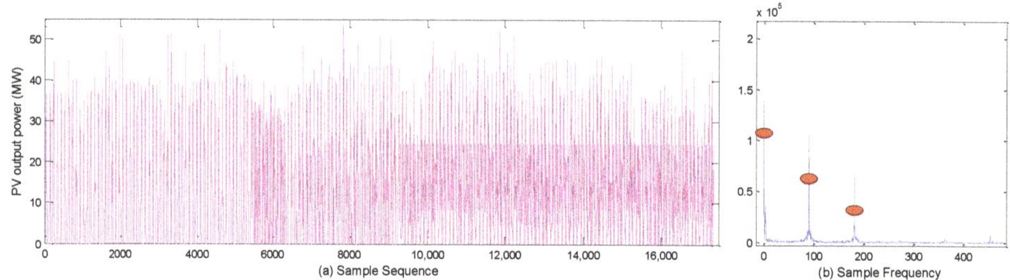

Figure 4. Photovoltaic output sample sequence.

As we can see from Figure 4b, the spectrum of the sample sequence contains three major frequency band components, and the symmetry of the spectrogram and the initial value of the modal number are taken as six. When K = 5, K = 6, and K = 7, the load data are separately decomposed via VMD to obtain an iterative curve of each modal center frequency under different K values, as shown in Figure 5.

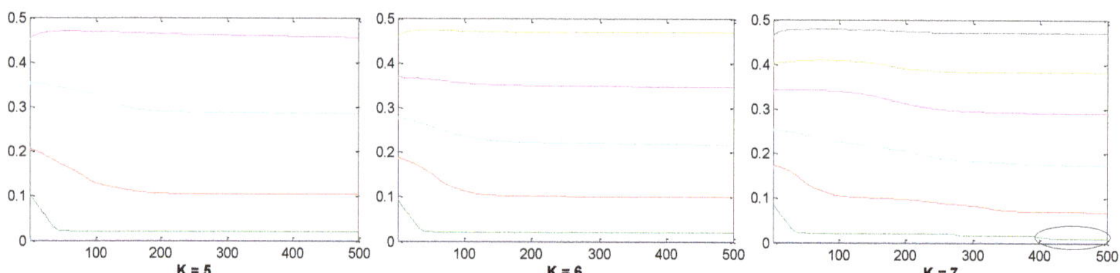

Figure 5. Sample center frequency iteration curves in different modes.

From the comparison of Figure 5, it can be found that when K = 7, the ends of the two iterative curves of the label are very close; in other words, central frequency aliasing appears. Therefore, the mode number was finally determined to be six.

3.1.2. Decomposition of Solar Output Power Data

The VMD decomposition method effectively improves modal aliasing and false components that appear in decomposition when the EMD and EEMD are decomposed. From a mathematical point-of-view, the phenomenon of mode mixing is that the components of each mode are intercoupling, which does not satisfy the orthogonality requirement. False components, as the name suggests, imply that they are mathematically calculated modal components. Modal aliasing leads directly to the appearance of false components. As demonstrated above, the number of modalities here is K = 6 and the original load is decomposed. The modal decomposition diagrams and spectrograms are shown in Figure 6.

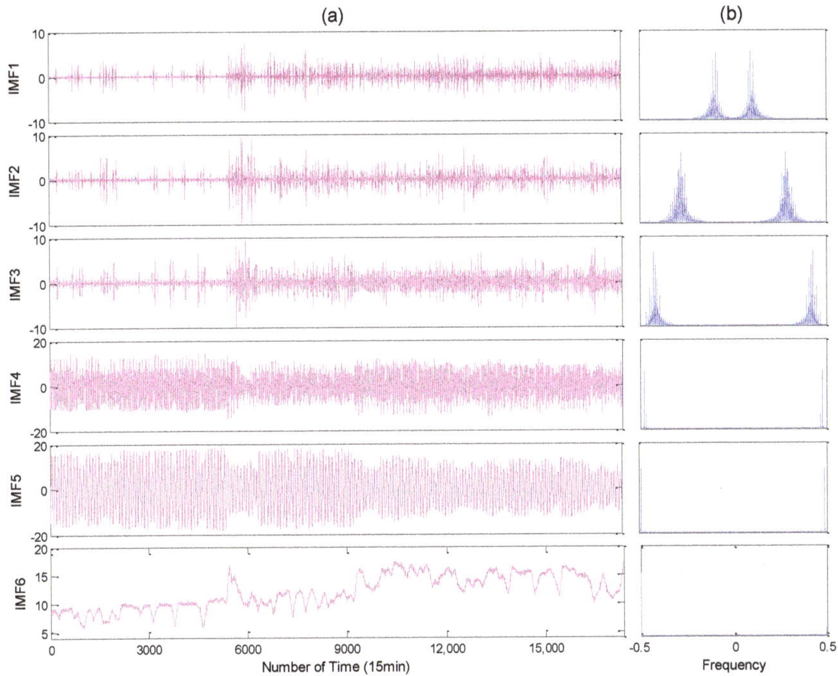

Figure 6. VMD decomposition diagram and spectrum diagram. (**a**) Modal components, (**b**) Frequency.

As in the spectrogram shown in Figure 6b, the spectral distributions of each modal component do not appear to be coupled with one another, satisfying the requirement of orthogonality. Therefore, no modal aliasing phenomenon occurs and the false component due to modal aliasing is also greatly improved (reduced). According to the spectrum period size of each modal component, the modes found via VMD decomposition are divided into two categories: the first three cycle short modal components IMF1, IMF2 and IMF3 are classified as high-frequency data, while the longer periods IMF4, IMF5 and IMF6 are used as low-frequency data.

3.2. Prediction of VMD Components

3.2.1. Rolling Prediction of High-Frequency Components Based on DBN Model

As described above, the DBN is used to predict high-frequency components. First, the DBN network is trained. Because the number of hidden layers of the DBN network and the number of cells in each hidden layer have a great influence on the prediction accuracy and computation time, this work focuses on the selection for these two parameters. The weight of the DBN model is initialized via a normal random distribution. The threshold of the DBN model is initialized at 0, the maximum number of iterations of the RBM is 100, the learning rate is 0.1, the momentum parameter is 0.9, and the model parameters are set as in Reference [44,45]. In the training model, the rolling prediction model with eight inputs and one output is adopted, i.e., the number of input layer nodes are the data from the first 2 h before the predicted time (eight nodes). The output layer consists of the predicted time for one node. The enumeration method is used to select the number of hidden units, layer by layer, to verify the influence of the deep network structure on the prediction effect and time consumption. First, we determine the optimal number of hidden units in the RBM1 layer and fix them; then, we add a hidden layer to determine the optimal value of the number of hidden units in the RBM2 layer. This continues until the prediction accuracy is no longer improved. The output error and time spent are obtained by changing the number of hidden layers and the number of nodes. Since the corresponding weight and threshold are automatically initialized during each training, the number of hidden cells in each layer is set to four to 32 (the interval is four), which is a total of eight levels. The number of layers of the hidden layer is sequentially set to the RBM1, RBM2, and RBM3 layers. IMF1 is taken as an illustration example, and the others will not be described again. The specific DBN parameters are shown in Table 2.

Table 2. Training times and output errors of IMF1 hidden layer nodes. IMF1: Intrinsic Mode Function 1.

Hidden Layer RBM1								
Number of hidden nodes	4	8	12	16	20	24	28	32
Output error/%	2.7427	2.1418	1.6717	1.2419	1.0242	1.3237	1.5517	1.8152
Training time/s	13.27	13.36	13.21	13.53	13.47	14.18	19.97	24.41
Hidden Layer RBM2								
Number of hidden nodes	4	8	12	16	20	24	28	32
Output error/%	1.5024	1.3024	1.2124	1.9122	2.0038	2.1522	2.0155	2.0014
Training time/s	15.18	15.94	16.47	17.08	21.78	28.15	30.32	37.77
Hidden Layer RBM3								
Number of hidden nodes	4	8	12	16	20	24	28	32
Output error/%	4.5174	4.2415	3.7791	2.8172	3.4014	3.1179	3.1563	4.7791
Training time/s	22.47	25.78	30.11	33.91	55.78	80.14	117.35	135.88

Table 2 shows that when the number of neurons in the hidden layer RBM1 is 20, the output error reaches a minimum of 1.0242%, which takes 13.47 s; the average output error of its two neighboring neurons is 1.28281. When the number of neurons in the hidden layer RBM2 is 12, the output error reaches a minimum of 1.21242%, which takes 15.94 s; the output error is smaller than the average output errors of neurons that are adjacent to the optimal neurons in the RBM1 layer. When the number of neurons in the hidden layer RBM3 is 16, the output error reaches the minimum value of 2.81824%, which takes 33.91 s; both the output error and the training time of RBM3 are higher than the average output errors of neurons that are adjacent to the optimal neurons in the RBM2 layer. For the IMF1 component data set, the DBN prediction model has better effects when it adopts a four-layer structure of "8-20-12-1" (that is, the numbers of hidden units of RBM1 and RBM2 are 20 and 12, respectively, which are marked in red in the table). By using this method, the DBN model structure of the high-frequency components of IMF2 and IMF3 and the number of nodes of each hidden layer are obtained, as shown in Table 3.

Table 3. Variational Mode Decomposition (VMD) decomposition high-frequency components' Deep Belief Network (DBN) structure.

VMD Decomposition High-Frequency	IMF1	IMF2	IMF3
DBN structure	8-20-12-1	8-16-12-4-1	8-12-8-1

Through the analysis above, the IMF1-IMF3 components are predicted using the trained model, as shown in Figure 7.

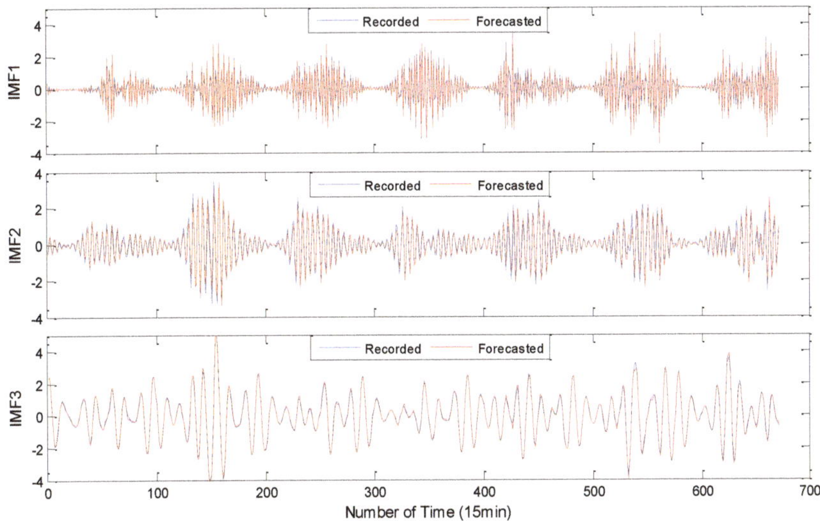

Figure 7. Prediction of IMF1-IMF3 component.

Figure 7 shows that the high-frequency components IMF1 and IMF2 have strong volatility, their prediction errors are larger than other components, the IMF3 components fluctuate less than the previous ones, and the error is reduced. Overall, the high self-learning and adaptive capabilities of the DBN model are suitable for predicting high-frequency components with strong volatility and short periods.

3.2.2. Prediction of Low-Frequency Components Based on ARMA Model

Thus, the low-frequency components are predicted via the ARMA model. First, the sample's autocorrelation function (ACF) and partial autocorrelation function (PACF) are obtained to determine the initial order of the model. This work uses IMF4 as the example to illustrate, and other low-frequency components are not described again. The ACF and PACF of IMF4, respectively, are shown in Figure 8.

According to the AIC order code determined in Reference [41] and the above figure, both the autocorrelation and the partial autocorrelation plots of the IMF4 component have tailing characteristics. Also, the autocorrelation coefficient is not zero when the lag order is 3, and the trailing characteristic is apparent when the lag order is greater than 3. The partial autocorrelation coefficient is not zero when the lag order is 6, but the trailing characteristic is obvious after the lag order is greater than 6. Thus, we can initially determine that $p = 6$ and $q = 3$. To make the model more accurate, the values of p and q can be relaxed. Using the AIC criterion, the minimum value is taken as the optimal model. The AIC values under the various model orders are shown in Table 4.

Table 4. The Akaike Information Criterion (AIC) values of different order models.

Order	ARMA(6, 3)	ARMA(6, 4)	ARMA(7, 3)	ARMA(7, 4)	ARMA(8, 3)	ARMA(8, 4)
The value of AIC	−2.902	−2.869	−2.609	−2.823	−2.807	−2.817

From the above table, the optimal model of IMF4 is ARMA(7, 3). Then, using the above method, the components IMF5 and IMF6 are ordered. The specific situation is shown in Table 5.

Table 5. Orders of VMD low-frequency components.

Component	IMF4	IMF5	IMF6
Order	ARMA(7, 3)	ARMA(8, 4)	ARMA(6, 4)

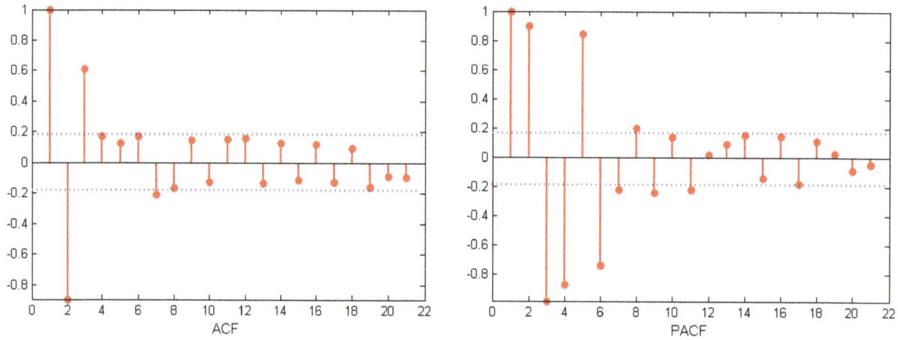

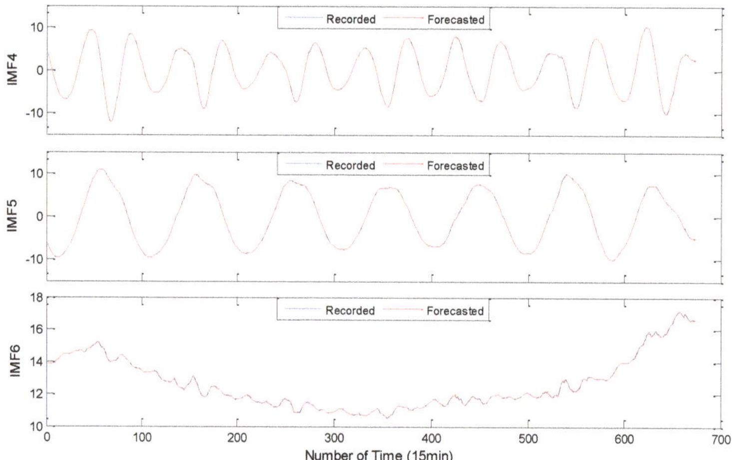

Figure 8. The autocorrelation function (ACF) and partial autocorrelation function (PACF) of IMF4 decomposed.

Through the above analysis, the components IMF4 to IMF6 were predicted using the trained model, which is shown in Figure 9.

Figure 9. IMF4–IMF6 component prediction.

From Figure 9, we can see that using the ARMA model to predict low-frequency components, with relatively gentle fluctuations, results in a small error. ARMA has strong nonlinear fluctuation data learning ability, which is suitable for low-frequency component prediction.

3.2.3. Combination Prediction Based on VMD-ARMA-DBN Model

In this work, DBN is used for the combined reconstruction of the prediction value of each component (IMF1 to IMF6). Taking each IMF sample data as input, the actual PV output sample value is used as an output to train the model. Then, the prediction value of each component is taken as input, and the prediction value of each component, that is, the final load prediction value, is obtained.

Among them, the DBN model is a five-layer implicit structure "12-24-16-8-1". The final PV output forecasting chart is shown in Figure 10.

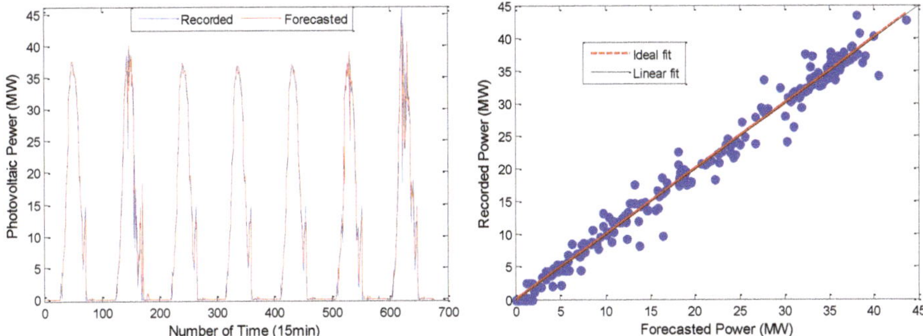

Figure 10. VMD-ARMA-DBN combination forecast results.

4. Discussions and Comparison

To test the prediction effect of the model proposed in this paper, we compared the results of the following prediction models: (1) the single prediction models (ARMA, DBN) used in this paper; (2) the common neural network prediction model, RNN and Gradient Boost Decision Tree (GBDT) in literature [46,47], used on a representative basis; (3) the combined prediction model, Discrete Wavelet Transformation (DWT) in literature [48] and traditional EMD and EEMD are used on a representative basis. The prediction results for each model are shown in Figure 11.

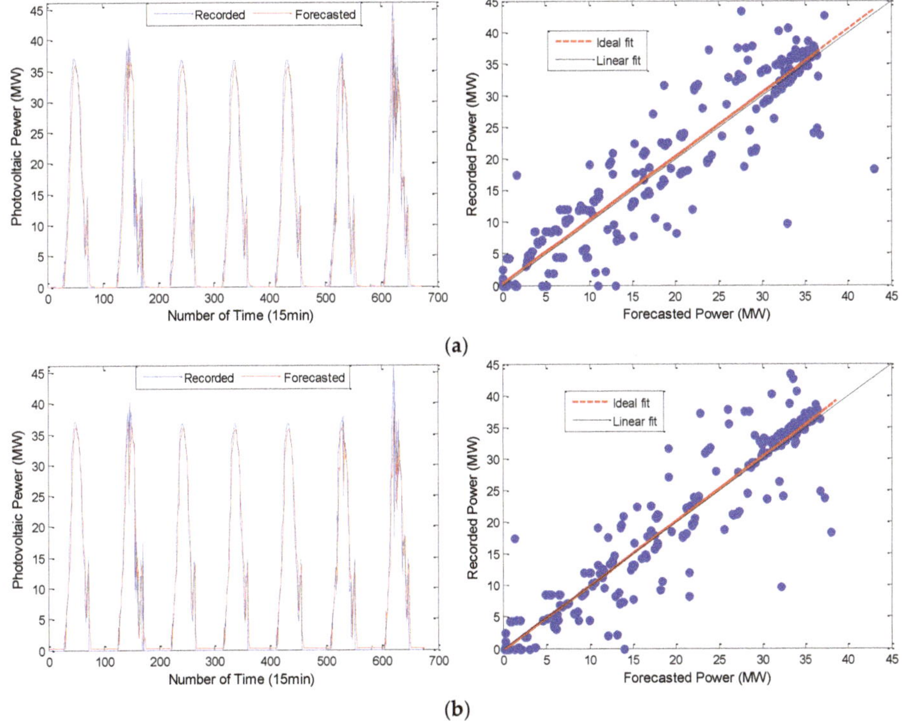

Figure 11. *Cont.*

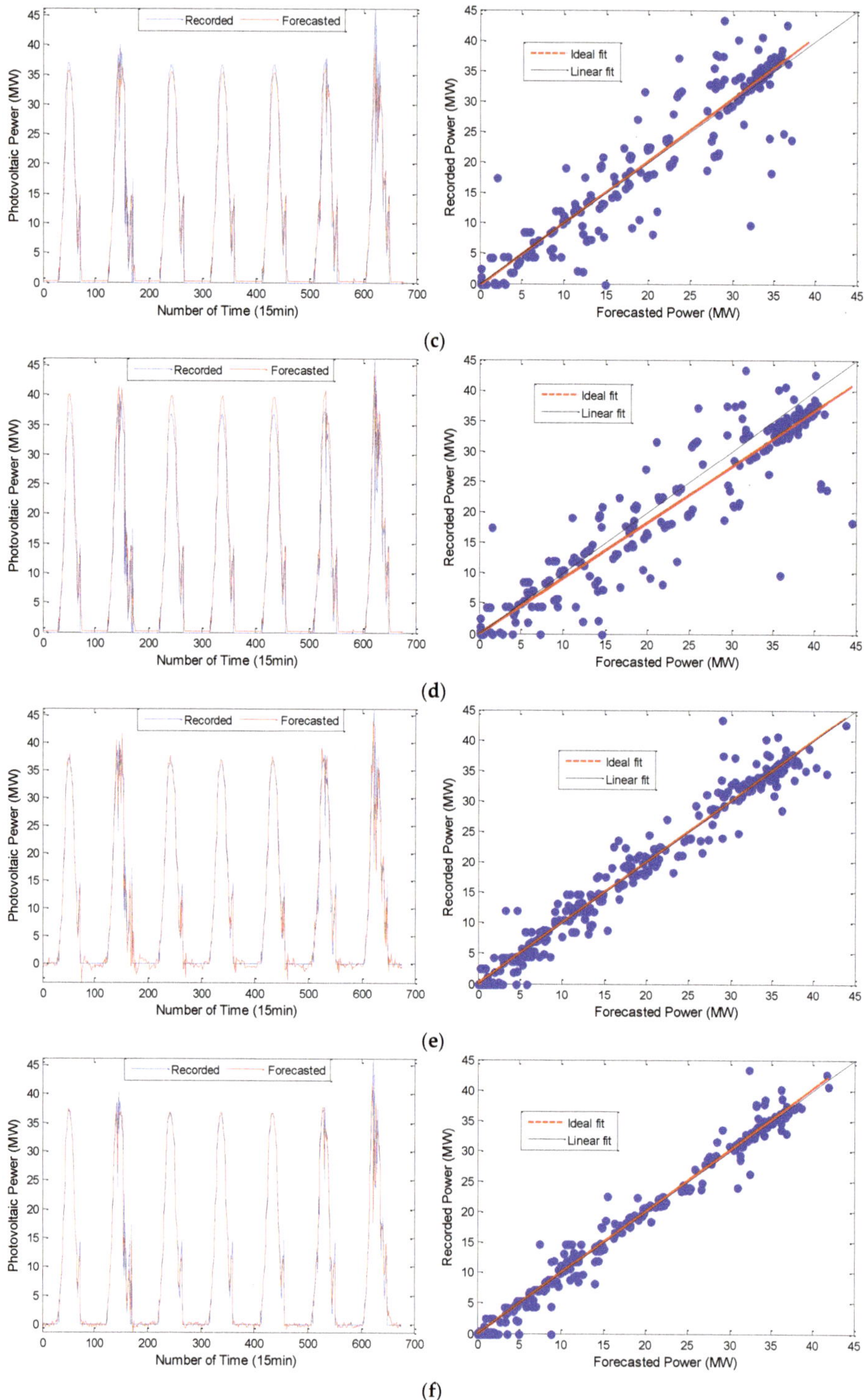

(c)

(d)

(e)

(f)

Figure 11. *Cont.*

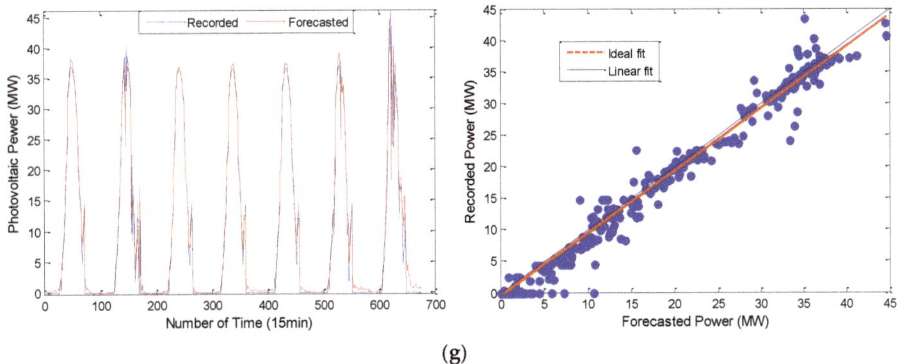

(g)

Figure 11. Prediction results for multiple models. (**a**) ARMA model prediction results; (**b**) DBN model prediction results; (**c**) RNN model prediction results; (**d**) GBDT model prediction results; (**e**) EMD-ARMA-DBN model prediction results; (**f**) EEMD-ARMA-DBN model prediction results; (**g**) DWT-RNN-LSTM model prediction results. RNN: Recurrent Neural Network; GBDT: Gradient Boost Decision Tree; EEMD: Ensemble Empirical Mode Decomposition; DWT: Discrete Wavelet Transformation; LSTM: Long Short-Term Memory.

From the simulation results shown in Figures 10 and 11, the VMD-ARMA-DBN combined models have a better tracking and fitting ability for the PV output curve. Compared to the single models, the combined model prediction accuracy (after using the modal decomposition technique) shows different degrees of improvement. Figure 12 is a bar graph demonstrating the absolute error of prediction in each model.

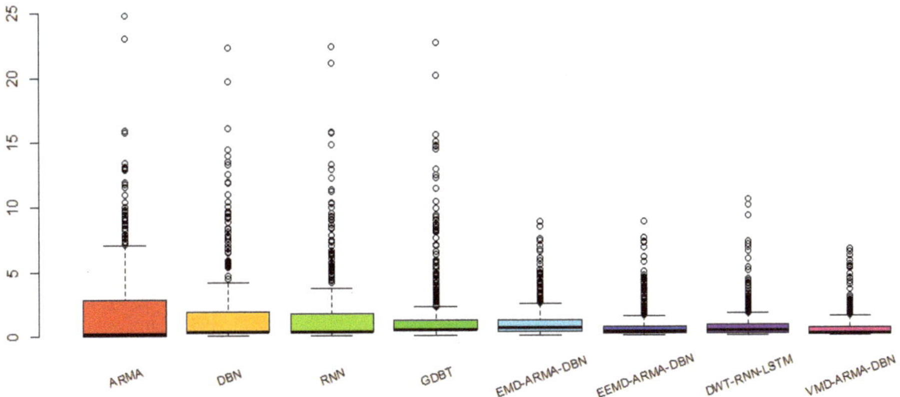

Figure 12. Simulation results of each model: absolute error box plot.

From the perspective of the absolute error distribution of the prediction results, the stability of the prediction accuracy for the single models is poor, and the error distribution interval is large. Among them, the absolute error distribution interval of each model is [0, 24.7821], [0.0051, 22.2464], [0.0082, 22.3289], [0.0363, 22.6686], [0.0005, 8.8306], [0.0018, 8.7955], [0.0008, 10.5322], and [0.0017, 6.6526], respectively, and the prediction error median is 0.1692, 0.2791, 0.2708, 0.4523, 0.5926, 0.3078, 0.0351, and 0.1414, respectively. The absolute error distribution of the VMD-ARMA-DBN model is more concentrated and the median of the error is the smallest, which is the most ideal of the eight groups of prediction models. In summary, the VMD-based multi-frequency combined forecasting model presented in this paper is superior to other models. To compare the prediction effects of each model more intuitively, we used quantitative evaluation indicators. Table 6 shows the evaluation results for each model.

Table 6. Forecast results for each model.

Evaluation Index	e_{NMAE}	e_{NRMSE}	*TIC*
ARMA	3.74146	7.17435	0.11359
DBN	3.19754	6.43502	0.10139
RNN	3.21145	6.548871	0.10328
GBDT	3.27612	6.434691	0.10812
EMD-ARMA-DBN	2.11062	3.49985	0.05456
EEMD-ARMA-DBN	1.35409	2.67181	0.04171
DWT-RNN-LSTM	1.48241	2.74976	0.04265
VMD-ARMA-DBN	1.03374	2.05776	0.03201

First, it can be concluded from Figure 12 and Table 6 that, compared with the single prediction models containing ARMA, DBN, RNN, and GBDT, the introduction of the modal decomposition method has a great influence on the accuracy of the prediction results. The modal decomposition method is used to effectively decompose the original PV output, and the prediction method is selected according to the characteristics of different modal vectors, which can make the prediction result more accurate and stable, and the result can be anticipated. The PV system power output has high volatility, variability, and randomness; through modal decomposition it can effectively eliminate the unrelated noise components to make each component easier to predict. In the single prediction models, the error of the ARMA prediction model is the largest, which is not suitable for effectively tracking the undecomposed solar PV output; DBN, RNN and GBDT belong to machine learning, and the prediction error is essentially the same. However, the parameters of the RNN prediction model are more difficult to choose and more easily fall into the local optimum, and GBDT is easy to over-fit for complex models. However, combined prediction methods can effectively avoid these problems.

Second, the proposed VMD-ARMA-DBN model prediction results are always better than those of other combined prediction models (such as EMD, EEMD, and DWT). This is mainly because different decomposition methods have different ways of controlling the modal number, affecting the size of the prediction error. The center frequency of the VMD modal decomposition is controllable, which can effectively avoid modal aliasing compared with other modal decomposition models. The original sequences are decomposed according to the frequency components, and different prediction models are used for fitting purposes. According to the prediction results in Table 6, this kind of combined prediction method significantly improves the prediction accuracy.

Finally, it should be noted that the prediction results of the VMD-ARMA-DBN models are different under different modal center frequencies K. Specifically, when K is too large, it is easy to cause excessive frequency decomposition, which increases the degree of complexity of the model prediction; when K is too small, it will cause modal overlapping, and the single frequency components cannot be effectively predicted, so only the appropriate K can be used to make an effective prediction. Moreover, this will be an important problem that must be overcome in the next research stage.

5. Conclusions

The short-term prediction accuracy of the nonlinear PV power time series in this work proposes a multi-frequency combined prediction model based on VMD mode decomposition. Specifically, the following was observed:

(1) For the first time, this paper introduces the VMD method into PV power plant output forecasting, decomposes the unstable PV output sequence, and conducts in-depth research on the characteristics of VMD. When traditional decomposition methods deal with sequences that contain both trend and wave terms, accurately extracting the shortcomings of the trend items is impossible. A combination method based on VMD-ARMA-DBN is proposed, which not only reflects

the development trend of the size of PV output, but also decomposes the fluctuation series into a set of less complex and some strong periodical parameters, which greatly reduced the difficulty of prediction.

(2) If the VMD cannot restore the original sequence completely, and cannot determine the number of decomposition layers automatically, we propose a method to determine the number of VMD decomposition layers via a spectrum analysis, which can restore the original sequence to a large extent and ensure the stability of the component. First, according to the spectrum diagram of the sample data, we determined the number of modal components. If the overlapping phenomenon occurred in the center frequency iteration curve, the number of decompositions was selected and divided into high- and low-frequency components according to the characteristics of the different components. ARMA and DBN were used to simulate and predict high-frequency and low-frequency components. Then, the predictive value of each component was determined using DBN. Each one had a strong nonlinear mapping ability, and high self-learning ability and self-adaptive capability. The sample data of each component was taken as input, and the actual PV sample value was used as an output to train the model. Then, the predicted value of each component was used as input for the prediction. Finally, the PV output predicted value was obtained.

(3) To test the prediction effect of the VMD combinatorial model, the normalized absolute mean error, normalized root-mean-square error, and the Hill inequality coefficient were used to compare the single prediction models with the combined prediction models. The simulation results show that the different decomposition methods have been improved to varying degrees in terms of forecast accuracy. Thus, the VMD-ARMA-DBN model proposed in this work offers better accuracy and stability than the single prediction methods and the combined prediction models.

In the prediction process, we found that although the VMD improved the phenomenon of modal aliasing and false components, it was not eliminated. In addition, the DBN's component–RBM needs to be further improved. The weight and offset of each layer of RBM are initialized during training. Therefore, even if the number of hidden layer nodes is compared and selected, the optimal model cannot be obtained, and the final prediction result will show diversification; that is, the same model yields different results, and to obtain optimal results, it is necessary to train the model multiple times, which makes the workload cumbersome. The above deficiencies inevitably increase the errors in the prediction process of each component and affect the final prediction results.

Author Contributions: T.X. and G.Z. conceived and designed the experiments; T.X. and H.L. performed the experiments; G.Z. and F.L. analyzed the data; H.L. and P.D. contributed reagents/materials/analysis tools; T.X. and P.D. wrote the paper.

Funding: This research was funded by the National Natural Science Foundation of China under grant 51507141, 51679188 and the National Key Research and Development Program of China, grant number 2016YFC0401409, and the Key Research and Development Plan of Shaanxi Province, grant number 2018-ZDCXL-GY-10-04.

Conflicts of Interest: The authors declare no conflict of interest.

Nomenclature

ACF	autocorrelation function
AIC	Akaike Information Criterion
AR	Auto-Regressive Model
ARMA	Auto-Regressive Moving Average Model
BM	Boltzmann machine
BP	Back-Projection
CNN	Convolutional Neural Network
DBN	Deep Belief Network
DNN	Deep Neural Network
DWT	Discrete Wavelet Transformation
EMD	Empirical Mode Decomposition
EEMD	Ensemble Empirical Mode Decomposition
FFT	Fast Fourier Transform

GBDT	Gradient Boost Decision Tree
IMFs	Intrinsic Mode Functions
LMD	Local mean decomposition
LSTM	Long Short-Term Memory
MA	Moving-Average Model
NWP	numerical weather prediction
PACF	Partial autocorrelation function
PV	Photovoltaic
RBM	Restricted Boltzmann Machine
RNN	Recurrent Neural Network
SVM	Support vector machine
VMD	Variational Mode Decomposition

References

1. Gobind, P.; Husain, A. Techno-economic potential of largescale photovoltaics in Bahrain. *Sustain. Energy Technol. Assess.* **2018**, *27*, 40–45. [CrossRef]
2. Eltawil, M.; Zhao, Z. Grid-connected photovoltaic power systems: Technical and potential problems-a review. *Renew. Sustain. Energy Rev.* **2010**, *14*, 112–129. [CrossRef]
3. Atsushi, Y.; Tomonobu, S.Y. Determination Method of Insolation Prediction with Fuzzy and Applying Neural Network for Long-Term Ahead PV Power Output Correction. *IEEE Trans. Sustain. Energy* **2013**, *4*, 527–533. [CrossRef]
4. Shi, J.; Lee, W.J. Forecasting power output of photovoltaic systems based on weather classification and support vector machines. *IEEE Trans. Ind. Appl.* **2012**, *48*, 1064–1069. [CrossRef]
5. Selmin, E.; Rusen, A. Coupling satellite images with surface measurements of bright sunshine hours to estimate daily solar irradiation on horizontal surface. *Renew. Energy* **2013**, *55*, 212–219. [CrossRef]
6. Ricardo, M.; Hugo, T.C. Hybrid solar forecasting method uses satellite imaging and ground telemetry as inputs to ANNs. *Sol. Energy* **2013**, *92*, 176–188. [CrossRef]
7. Wang, J.Z.; Jiang, H. Forecasting solar radiation using an optimized hybrid model by Cuckoo Search algorithm. *Energy* **2015**, *81*, 627–644. [CrossRef]
8. Bone, V.; Pidgeon, J. Intra-hour direct normal irradiance forecasting through adaptive clear-sky modelling and cloud tracking. *Sol. Energy* **2018**, *159*, 852–867. [CrossRef]
9. Costa, E.B.; Serra, G.L. Self-Tuning Robust Fuzzy Controller Design Based on Multi-Objective Particle Swarm Optimization Adaptation Mechanism. *J. Dyn. Syst. Meas. Control ASME* **2017**, *139*, 071009. [CrossRef]
10. Atrsaei, A.; Alarieh, H. Human Arm Motion Tracking by Inertial/Magnetic Sensors Using Unscented Kalman Filter and Relative Motion Constraint. *J. Intell. Robot. Syst.* **2018**, *90*, 161–170. [CrossRef]
11. Xia, L.; Ma, Z.J. A model-based design optimization strategy for ground source heat pump systems with integrated photovoltaic thermal collectors. *Appl. Energy* **2018**, *214*, 178–190. [CrossRef]
12. Al-Waeli, A.H.; Sopian, K. Comparison of prediction methods of PV/T nanofluid and nano-PCM system using a measured dataset and artificial neural network. *Sol. Energy* **2018**, *162*, 378–396. [CrossRef]
13. Yousif, J.H.; Kazem, H.A. Predictive Models for Photovoltaic Electricity Production in Hot Weather Conditions. *Energies* **2017**, *10*, 971. [CrossRef]
14. Wang, F.; Zhen, Z. Comparative Study on KNN and SVM Based Weather Classification Models for Day Ahead Short Term Solar PV Power Forecasting. *Appl. Sci.-Basel* **2018**, *8*, 28. [CrossRef]
15. Eseye, A.T.; Zhang, J.H. Short-term photovoltaic solar power forecasting using a hybrid Wavelet-PSO-SVM model based on SCADA and Meteorological information. *Renew. Energy* **2018**, *118*, 357–367. [CrossRef]
16. Shan, Y.H.; Fu, Q. Combined Forecasting of Photovoltaic Power Generation in Microgrid Based on the Improved BP-SVM-ELM and SOM-LSF With Particlization. *Proc. CSEE* **2016**, *36*, 3334–3342.
17. Semero, Y.K.; Zheng, D.H. A PSO-ANFIS based Hybrid Approach for Short Term PV Power Prediction in Microgrids. *Electr. Power Compon. Syst.* **2018**, *46*, 95–103. [CrossRef]
18. Liu, D.; Niu, D.X. Short-term wind speed forecasting using wavelet transform and support vector machines optimized by genetic algorithm. *Renew. Energy* **2014**, *62*, 592–597. [CrossRef]

19. Farhan, M.A.; Swarup, K.S. Mathematical morphology-based islanding detection for distributed generation. *IET Gener. Transm. Dis.* **2017**, *11*, 3449–3457. [CrossRef]

20. Liu, H.; Chen, C. A hybrid model for wind speed prediction using empirical mode decomposition and artificial neural networks. *Renew. Energy* **2012**, *48*, 545–556. [CrossRef]

21. Dai, S.Y.; Niu, D.X. Daily Peak Load Forecasting Based on Complete Ensemble Empirical Mode Decomposition with Adaptive Noise and Support Vector Machine Optimized by Modified Grey Wolf Optimization Algorithm. *Energies* **2018**, *11*, 163. [CrossRef]

22. Sun, B.; Yao, H. Short-term wind speed forecasting based on local mean decomposition and multi-kernel support vector machine. *Acta Energiae Solaris Sin.* **2013**, *34*, 1567–1573.

23. Zhu, T.; Wei, H. Clear-sky model for wavelet forecast of direct normal irradiance. *Renew. Energy* **2017**, *104*, 1–8. [CrossRef]

24. Li, F.; Wang, S. Long term rolling prediction model for solar radiation combining empirical mode decomposition (EMD) and artificial neural network (ANN) techniques. *J. Renew. Sustain. Energy* **2018**, *10*, 013704. [CrossRef]

25. Wang, L.; Liu, Z.W. Complete ensemble local mean decomposition with adaptive noise and its application to fault diagnosis for rolling bearings. *Mech. Syst. Signal Process.* **2018**, *106*, 24–39. [CrossRef]

26. Wang, C.; Zhang, H.L. A new chaotic time series hybrid prediction method of wind power based on EEMD-SE and full-parameters continued fraction. *Energy* **2017**, *138*, 977–990. [CrossRef]

27. Zhang, Y.C.; Liu, K.P. Short-Term Wind Power Multi-Leveled Combined Forecasting Model Based on Variational Mode Decomposition-Sample Entropy and Machine Learning Algorithms. *Power Syst. Technol.* **2016**, *40*, 1334–1340. [CrossRef]

28. Liu, C.L.; Wu, Y.J. Rolling bearing fault diagnosis based on variational model decomposition and fuzzy C means clustering. *Proc. CSEE* **2015**, *35*, 3358–3365.

29. Dragomiretskiy, K.; Zosso, D. Variational mode decomposition. *IEEE Trans. Signal Process.* **2014**, *62*, 531–544. [CrossRef]

30. Naik, J.; Dash, S. Short term wind power forecasting using hybrid variational mode decomposition and multi-kernel regularized pseudo inverse neural network. *Renew. Energy* **2018**, *118*, 180–212. [CrossRef]

31. Li, C.S.; Xiao, Z.G. A hybrid model based on synchronous optimisation for multi-step short-term wind speed forecasting. *Appl. Energy* **2018**, *215*, 131–144. [CrossRef]

32. Hinton, G.; Salakhutdinov, R. Reducing the dimensionality of data with neural networks. *Science* **2006**, *313*, 504–507. [CrossRef] [PubMed]

33. Zhu, Q.M.; Dang, J. A Method for Power System Transient Stability Assessment Based on Deep Belief Networks. *Proc. CSEE* **2018**, *38*, 735–743.

34. Hinton, G.; Deng, L. Deep neural networks for acoustic modeling in speech recognition: The shared views of four research groups. *IEEE Signal Process. Mag.* **2012**, *29*, 82–97. [CrossRef]

35. Hinton, G. Training products of experts by minimizing contrastive divergence. *Neural Comput.* **2002**, *14*, 1771–1800. [CrossRef] [PubMed]

36. Wang, Y.J.; Na, X.D. State Recognition Method of a Rolling Bearing Based on EEMD-Hilbert Envelope Spectrum and DBN Under Variable Load. *Proc. CSEE* **2017**, *37*, 6943–6950.

37. Tian, B.; Piao, Z.L. Wind power ultra short-term model based on improved EEMD-SE-ARMA. *Power Syst. Protect. Control* **2017**, *45*, 72–79.

38. Yang, Z.; Ce, L. Electricity price forecasting by a hybrid model, combining wavelet transform, ARMA and kernel-based extreme learning machine methods. *Appl. Energy* **2017**, *190*, 291–305. [CrossRef]

39. Li, C.B.; Zheng, X.S. Predicting Short-Term Electricity Demand by Combining the Advantages of ARMA and XG Boost in Fog Computing Environment. *Wirel. Commun. Mob. Comput.* **2018**, 5018053. [CrossRef]

40. Lu, J.L.; Wang, B. Two-Tier Reactive Power and Voltage Control Strategy Based on ARMA Renewable Power Forecasting Models. *Energies* **2017**, *10*, 1518. [CrossRef]

41. Li, L.; Zhang, S.F. Ionospheric total electron content prediction based on ARMA model. *J. Basic Sci. Eng.* **2013**, *21*, 814–822.

42. Wu, J.; Chan, C.K. Prediction of hourly solar radiation using a novel hybrid model of ARMA and TDN. *Sol. Energy* **2011**, *85*, 808–817. [CrossRef]

43. Voyant, C.; Muselli, M. Numerical weather prediction (NWP) and hybrid ARMA/ANN model to predict global radiation. *Energy* **2012**, *39*, 341–355. [CrossRef]

44. Alshamaa, D.; Chehade, F.M. A hierarchical classification method using belief functions. *Signal Process.* **2018**, *148*, 68–77. [CrossRef]
45. Fu, G.Y. Deep belief network based ensemble approach for cooling load forecasting of air-conditioning system. *Energy* **2018**, *148*, 269–282. [CrossRef]
46. Zhang, B.; Wu, J.L.; Chang, P.C. A multiple time series-based recurrent neural network for short-term load forecasting. *Soft Comput.* **2018**, *22*, 4099–4112. [CrossRef]
47. Wang, F.; Yu, Y.L.; Zhang, Z.Y. Wavelet Decomposition and Convolutional LSTM Networks Based Improved Deep Learning Model for Solar Irradiance Forecasting. *Appl. Sci.-Basel* **2018**, *8*, 1286. [CrossRef]
48. Wang, J.D.; Li, P.; Ran, R. A Short-Term Photovoltaic Power Prediction Model Based on the Gradient Boost Decision Tree. *Appl. Sci.-Base7l* **2018**, *8*, 689. [CrossRef]

Article

An Ultrashort-Term Net Load Forecasting Model Based on Phase Space Reconstruction and Deep Neural Network

Fei Mei [1,*], Qingliang Wu [1], Tian Shi [2], Jixiang Lu [3], Yi Pan [2] and Jianyong Zheng [2]

1   College of Energy and Electrical Engineering, Hohai University, Nanjing 211100, China; 15150682019@163.com
2   School of Electrical Engineering, Southeast University, Nanjing 210096, China; 15851832207@163.com (T.S.); 230159517@seu.edu.cn (Y.P.); jy_zheng@seu.edu.cn (J.Z.)
3   State Key Laboratory of Smart Grid Protection and Control, NARI Group Corporation, Nanjing 211000, China; lujixiang@sgepri.sgcc.com.cn
*   Correspondence: meifei@hhu.edu.cn; Tel.: +86-152-9552-9785

Received: 28 February 2019; Accepted: 2 April 2019; Published: 9 April 2019

Abstract: Recently, a large number of distributed photovoltaic (PV) power generations have been connected to the power grid, which resulted in an increased fluctuation of the net load. Therefore, load forecasting has become more difficult. Considering the characteristics of the net load, an ultrashort-term forecasting model based on phase space reconstruction and deep neural network (DNN) is proposed, which can be divided into two steps. First, the phase space reconstruction of the net load time series data is performed using the C-C method. Second, the reconstructed data is fitted by the DNN to obtain the predicted value of the net load. The performance of this model is verified using real data. The accuracy is high in forecasting the net load under high PV penetration rate and different weather conditions.

Keywords: net load forecasting; phase space reconstruction; deep neural network

---

1. Introduction

In recent years, an increasing number of photovoltaic (PV) power generations have been connected to the distribution network. The use of new energy brings huge benefits to human beings. However, it also has negative impacts on the power grid while improving the environment. The PV power is greatly influenced by weather conditions and fluctuates with the change in irradiance. To ensure the quality of the power supply and safe operation of the power system, it is necessary to maintain equal amounts of power generation and power consumption in the power system. The power load forecasting is an indispensable mean of maintaining this dynamic balance. In addition, power load forecasting is also of great significance for the planning and scheduling of power systems and the planning of power maintenance. However, the fluctuations in PV power can cause fluctuations in the load when many PV systems are connected. Therefore, load forecasting will become more difficult.

Generally, load forecasting can be divided into long-term forecasting, medium-term forecasting, short-term forecasting, and ultrashort-term forecasting [1]. Recently, many systematic and fruitful studies on traditional load forecasting have been conducted. The load forecasting methods mainly include the similar day prediction method [2,3], time series prediction method [4,5], expert system [6,7], and regression analysis method [8,9]. Artificial intelligence and machine learning algorithm are types of prediction methods that have rapidly developed in recent years. Support vector regression (SVR) [10], relevance vector regression (RVR) [11], artificial neural network (ANN) [12], deep neural network (DNN) [13], and their improved hybrid algorithms [14] have been applied in the field of load forecasting. In Moon et al. [15], the random forest and multilayer perceptron are combined to predict the daily electrical load. In Nazar et al. [16], the wavelet and Kalman machines, Kohonen self-organizing map (SOM), multi-layer perceptron artificial neural network (MLP-ANN) and adaptive neuro-fuzzy inference system (ANFIS) are used to establish a hybrid three-stage forecasting model. In Zhang et al. [17], a short-term power load forecasting method with wavelet neural network (WNN) and an adaptive mutation bat optimization algorithm (AMBA) are proposed. Liang et al. [18] propose a hybrid model that combines the empirical mode decomposition (EMD), minimal redundancy maximal relevance (mRMR), general regression neural network (GRNN), and fruit fly optimization algorithm (FOA). In Dai et al. [19], a load forecasting model is proposed based on the complete ensemble empirical mode decomposition (EEMD) with adaptive noise and support vector machine (SVM), which is optimized by the modified grey wolf optimization (MGWO) algorithm.

Since traditional load data are typically in hours as the smallest unit, short-term and ultrashort-term forecasts are often not strictly distinguished. With the large-scale access of distributed PV, the short-term load forecasting usually in hours cannot satisfy the requirements of the real-time safety analysis. For the real-time safety analysis of power systems and the reliable operation of economic dispatch, a more detailed ultrashort-term prediction is required. Considering the volatility of the distributed PV power generation and the real-time requirements of the ultrashort-term prediction, the PV power should also be considered a load and merged with the traditional load to form a net load [20–22]. Because the net load is a set of nonlinear time series with large volatility, in this paper, the phase space reconstruction of net load data is first performed to project the data into the moving point with certain regularity of the trajectory in the phase space. Then, the excellent nonlinear fitting ability of the deep neural network is used to fit the moving point trajectory to obtain the final prediction value. Finally, the actually measured load data is applied to verify the prediction effectiveness and prediction effect of the model under different weather conditions. The phase space is a tool to feature a dynamic system that is reconstructed from a univariate or multivariate time series [23], which is widely used in forecasting models. The DNN is the development of the traditional ANN and suitable for net load forecasting because the nonlinear fitting ability is strengthened [24]. The high accuracy in forecasting the net load under high PV penetration rate and different weather conditions is verified using real data. The contribution of this paper can be summarized as follows:

- The bus load prediction model is established considering distributed PV power supply. The prediction results are necessary guidance for the power grid dispatching, which is conducive to the improvement of PV consumption.
- The phase space reconstruction is used to process bus load data, and one-dimensional time series data are inversely constructed into the phase space structure of the original system, which can better describe the dynamic characteristics and adapt to the strong fluctuation of the bus load.
- Levenberg-Marquardt back propagation (LMBP) algorithm is used to train DNN, which accelerates the training speed. Compared with the single hidden layer neural network, DNN can fit the historical data better and significantly improve the accuracy of ultrashort-term load forecasting.

2. Fundamental of Phase Space Reconstruction and Deep Neural Network

2.1. Phase Space Reconstruction

Phase space reconstruction is a method proposed by Takens to analyze the time series. The basic idea of the phase space reconstruction is to regard the time series as a component produced by a certain nonlinear dynamic system. The equivalent high-dimensional phase space of the power system can be reconstructed by the variation law of the component. Among them, the key to reconstruction is to determine the optimal embedding dimension m_{opt} and delay t_{opt}.

In this paper, the optimal embedding dimension m_{opt} and delay t_{opt} are simultaneously obtained by the C-C method. If a set of time series is $x = \{x_1, x_2 \cdots x_N\}$, the embedding dimension is m, and the time delay is t, then the set of points in the reconstructed phase space can be expressed by Formula (1), where $M = N - (m-1)t$ [25].

$$
\begin{bmatrix} X_1 \\ X_2 \\ \vdots \\ X_M \end{bmatrix} = \begin{bmatrix} x_1 & x_{1+t} & \cdots & x_{1+(m-1)t} \\ x_2 & x_{2+t} & \cdots & x_{2+(m-1)t} \\ \vdots & \vdots & \cdots & \vdots \\ x_M & x_{M+t} & \cdots & x_N \end{bmatrix}. \tag{1}
$$

At this time, the correlation integral is Formula (2), where $\theta(x) = \begin{cases} 0 & x < 0 \\ 1 & x \geq 0 \end{cases}$. According to the statistical conclusion of BDS, the range of M and r_k can be obtained when $N > 3000$; $m \in \{2,3,4,5\}$, and $r_k = k \times 0.5\sigma$ which is a real number representing a given range of distances. σ is the standard deviation of time series, and $k \in \{1,2,3,4\}$. Correlation integral indicates the probability that the distance between any two points in the phase space is less than r_k.

$$
C(m, N, r_k, t) = \frac{2}{M(M-1)} \sum_{1 \leq i < j \leq M} \theta\left(r_k - \|X_i - X_j\|\right). \tag{2}
$$

We define the test statistic S and ΔS, and use the block averaging strategy, as shown in Formula (3).

$$
\begin{cases} S(m, N, r_k, t) = \frac{1}{t} \sum\limits_{i=1}^{M} C_i\left(m, \frac{N}{t}, r_k, t\right) - C_i^m\left(m, \frac{N}{t}, r_k, t\right) \\ \Delta S(m, N, t) = \max[S(m, N, r_k, t)] - \min[S(m, N, r_k, t)] \end{cases}. \tag{3}
$$

Formula (4) is the average of S and ΔS. Rounding the t value of the first zero of $\overline{S}$ or the first minimum of ΔS is the optimal delay t_{opt}.

$$
\begin{cases} \overline{S}(t) = \frac{1}{4 \times 4} \sum\limits_{m=2}^{5} \sum\limits_{k=1}^{4} S(m, N, r_k, t) \\ \overline{\Delta S}(t) = \frac{1}{4} \sum\limits_{m=2}^{5} \Delta S(m, N, t) \end{cases}. \tag{4}
$$

Formula (5) is the test statistic. The global minimum of $S_{cor}(t)$ is the optimal embedded window t_ω.

$$
S_{cor}(t) = \overline{\Delta S}(t) + \left|\overline{S}(t)\right|. \tag{5}
$$

Then:

$$
t_\omega = (m_{opt} - 1)t_{opt}. \tag{6}
$$

Therefore, the optimal delay t_{opt} determined by Equation (4) and the optimal embedded window t_ω determined by Equation (5) can be substituted into Formula (6) and rounded to obtain the optimal embedding dimension m_{opt}.

2.2. Deep Neural Network

A neural network generally consists of three layers: An input layer, a hidden layer, and an output layer. As shown in Figure 1, it is a simple neural network with three inputs and two outputs and four neurons in a single hidden layer. The neurons between layers are connected by weight ω. The DNN contains multiple hidden layers, which has a significant improvement in the nonlinear fitting ability of the DNN compared with the single hidden layer. However, too many hidden layers are likely to cause over-fitting. The learning process of the network is the process of adjusting and determining the connection weights ω of each neuron through training samples.

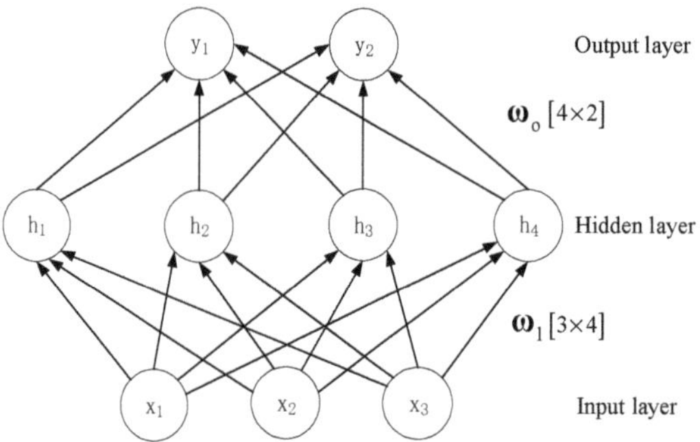

Figure 1. Single-hidden-layer neural network.

In this paper, the LMBP algorithm is used to train DNN. Compared with the traditional back propagation (BP) algorithm, the LMBP algorithm has faster convergence speed and higher convergence reliability. It is more suitable for training the DNN and can also satisfy the requirements of real-time ultrashort-term prediction. Unlike the traditional BP algorithm, which uses a gradient descent, the LMBP algorithm is based on the Gauss-Newton method of the least squares solution and takes the square of error v as the objective function.

$$E(\omega) = v^{\mathrm{T}}(\omega)v(\omega). \tag{7}$$

The second-order Taylor expansion and derivation of the objective function of Equations (2)–(7) can obtain the change of weight ω as follows:

$$\Delta\omega = -\left[\nabla^2 E(\omega)\right]^{-1}\nabla E(\omega). \tag{8}$$

where:

$$\begin{cases} \nabla E(\omega) = 2J^{\mathrm{T}}(\omega)v(\omega) \\ \nabla^2 E(\omega) = 2J^{\mathrm{T}}(\omega)J(\omega) + 2\nabla^2 v^{\mathrm{T}}(\omega)v(\omega) \end{cases}. \tag{9}$$

$J(\omega)$ is the Jacobian matrix of $v(\omega)$. If $v(\omega)$ consists of a elements, $J(\omega)$ can be written as follows:

$$J(\omega) = \begin{bmatrix} \frac{\partial v_1}{\partial \omega_1} & \frac{\partial v_1}{\partial \omega_2} & \cdots & \frac{\partial v_1}{\partial \omega_a} \\ \frac{\partial v_2}{\partial \omega_1} & \frac{\partial v_2}{\partial \omega_2} & \cdots & \frac{\partial v_2}{\partial \omega_a} \\ \vdots & \vdots & \ddots & \vdots \\ \frac{\partial v_a}{\partial \omega_1} & \frac{\partial v_a}{\partial \omega_2} & \cdots & \frac{\partial v_a}{\partial \omega_a} \end{bmatrix}. \tag{10}$$

Since $2\nabla^2 v^{\mathrm{T}}(\omega)v(\omega)$ is usually negligible, Formula (8) can be rewritten as follows:

$$\Delta\omega = -\left[J^{\mathrm{T}}(\omega)J(\omega)\right]^{-1}J^{\mathrm{T}}(\omega)v(\omega). \tag{11}$$

Considering that $J^{\mathrm{T}}(\omega)J(\omega)$ may be irreversible, Formula (11) is modified by adding the correction coefficient μ, where I is the unit matrix.

$$\Delta\omega = -\left[J^{\mathrm{T}}(\omega)J(\omega) + \mu I\right]^{-1}J^{\mathrm{T}}(\omega)v(\omega). \tag{12}$$

Similar to the BP algorithm, the modification of weight $\omega^{(k+1)}$ in the kth iteration is shown in Formula (13). $\Delta\omega^{(k)}$ can be obtained by Formula (12). When $E\left(\omega^{(k+1)}\right) < \varepsilon$, the algorithm has converged, where ε is the given error limit.

$$\omega^{(k+1)} = \omega^{(k)} + \Delta\omega^{(k)}. \tag{13}$$

The initial value of μ generally takes a small positive number such as 0.001. If the objective function $E\left(\omega^{(k)}\right)$ becomes lower in the kth iteration, $\mu^{(k)}$ is divided by a factor θ as $\mu^{(k+1)}$ of the next iteration. If the objective function $E\left(\omega^{(k)}\right)$ becomes higher in the kth iteration, the iteration will be restarted and multiply $\mu^{(k)}$ by the factor xxx as θ of this iteration. θ generally takes a number greater than 1, such as 4.

3. Ultrashort-Term Load Forecasting Model Based on Phase Space Reconstruction and Deep Neural Network

The traditional load fluctuation is mainly caused by user fluctuations in power usage. Although the electricity consumption of the user is uncertain, there are certain rules in general, and the fluctuation range is not large. For ultrashort-term prediction, linear extrapolation, time series prediction, and other methods can usually achieve the required accuracy. With the massive access of distributed energy sources such as PV power plants, the net load can be expressed by Formula (14). p_t is the actual net load, p' is the user's electricity load, and p_t^{PV} is the opposite number of PV power generation.

$$p_t = p'_t + p_t^{PV}. \tag{14}$$

Since the amount of PV power generation is as uncertain as the power load and different from the traditional power supply with a known power output, the PV power generation can be considered a load, which reduces the dispatching burden of the system. As a result, the uncertainty of load increases, and the range of fluctuation enlarges, even the situation of power reversal will occur at noon on sunny days. If the traditional forecasting method is also used, it will produce larger errors and cannot accurately predict the load.

Since the load is a non-linear time series with large fluctuations after the distributed energy access, it is difficult to directly predict. Therefore, the complex short-term prediction model may not satisfy the real-time requirements. In this paper, the phase space reconstruction is used to project the load time series into a time-varying and short-term regularity point in the high-dimensional phase space. Then, the non-linear fitting ability and fast convergence speed of LMBP DNN are used to fit and predict the locus of the points in the phase space to realize the ultrashort-term prediction of the load considering the distributed energy.

3.1. Modelling Steps of Prediction Model

For a series of net load time series $p = \{p_1, p_2 \cdots p_N\}$ considering the PV power generation, the modelling, and forecasting steps of ultrashort-term forecasting model based on phase space reconstruction and DNN are as follows:

- Step 1: The load time series is linearly normalized to facilitate the training of DNN. The maximum and minimum values of the data are saved for the reverse normalization of the load forecasting value to restore the actual value.
- Step 2: The C-C method is used to process the load time series, and the optimal embedded dimension m_{opt} and optimal delay t_{opt} of the time series are obtained.

- Step 3: The load time series are reconstructed according to the embedding dimension m and delay t obtained in Step 2. The phase space matrix of the reconstructed load time series is as follows. In Formula (15), $M = N - (m-1)t$.

$$\begin{bmatrix} p_1 \\ p_2 \\ \vdots \\ p_M \end{bmatrix} = \begin{bmatrix} p_1 & p_{1+t} & \cdots & p_{1+(m-1)t} \\ p_2 & p_{2+t} & \cdots & p_{2+(m-1)t} \\ \vdots & \vdots & \cdots & \vdots \\ p_M & p_{M+t} & \cdots & p_N \end{bmatrix}. \tag{15}$$

- Step 4: The p neural network is constructed, and the phase space matrix of load time series reconstructed in Step 3 is used as the training set to train the DNN. The trained DNN is used to predict the load value immediately after training.
- Step 5: Using the maximum and minimum values stored in Step 1, the load prediction values returned by the DNN are inverse-normalized to obtain the actual load prediction values.
- The model workflow chart is shown in Figure 2.

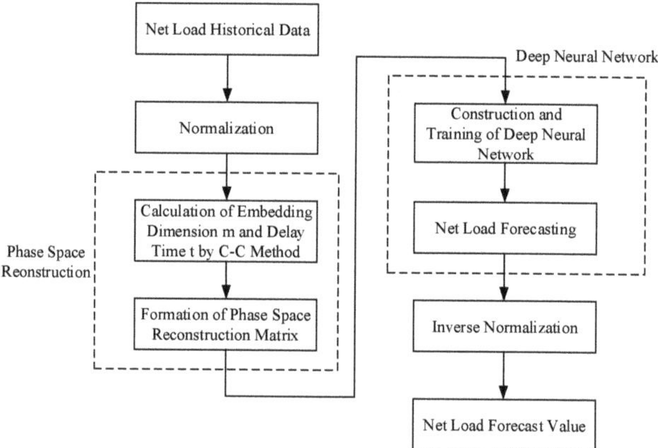

Figure 2. Forecasting model flow chart.

3.2. Determination of the Structure of Deep Neural Networks

The determination of the structure of the DNN is a link of the neural network hyper-parameter adjustment. An unreasonable structure can make the prediction results of the DNN seriously deviate. If the training time is too long, the work is half the effort. The specific method of determination is as follows:

- Input layer

If the DNN is directly trained using the original load data, the determination of the number of neurons in the input layer can be very difficult and requires a lot of debugging to obtain the optimal value. Moreover, when the training set data changes, previous optimal values may no longer be applicable, and the structure of the input layer must be re-debugged. In this model, the input data of the DNN is the phase space reconstructed matrix. Therefore, the number of neurons in the input layer is directly determined by the embedding dimension m obtained by the C-C method without artificial designation or after debugging to select the optimal value.

- Hidden layer

The number of hidden layers can be heuristically determined. When there are few hidden layers, the model will have an under-fitting and cause a large deviation in the predicted value.

Conversely, too many hidden layers can cause a model overfitting. The number of hidden layers can be gradually increased during the trial until the predicted value shows a significant over-fitting. Then, we gradually reduce the number of hidden layers so that the predicted value and true value of the model are as similar as possible on the verification set to determine the optimal number of hidden layers. The number of neurons in each hidden layer can be taken as 75% of the number of neurons in the upper layer but generally more than the number of neurons in the output layer. The activation function of the hidden layer neurons usually uses the tanh function and rectified linear unit (ReLU) function. The tanh function is used in this paper.

- Output Layer

The prediction model in this paper adopts the one-step prediction. Only the load value at the next moment is predicted at a given time. Therefore, after the load time series is projected to the moving point in the phase space, the model must output the position vector of the point in the phase space of the next time. In fact, if the input of the model is $p_i(1 \leq i \leq M)$ in the phase space reconstruction matrix of Formula (15), only $p_{i+1+(m-1)t}$ is unknown in position vector p_{i+1} at the next moment. Therefore, the output layer must only output the predicted value of load $\hat{p}_{i+1+(m-1)t}$. If $i+1 > M$, the phase space reconstruction matrix of Formula (15) must be extended downward, and p_{i+1} is added as a new line. The expression of p_{i+1} is shown in Formula (16), where $p_{i+1+(m-1)t}$ is the true value of the newly measured load.

$$p_{i+1} = \begin{bmatrix} p_{i+1} & p_{i+1+t} & \cdots & p_{i+1+(m-1)t} \end{bmatrix} \tag{16}$$

p_{i+1} is used as the input of the model to obtain the predicted value of $p_{i+2+(m-1)t}$; then, the matrix is augmented, and the predicted value of $p_{i+3+(m-1)t}$ is obtained, and the process continues until the end of the prediction. The activation function of the output layer is a linear function. The DNN structure is shown in Figure 3.

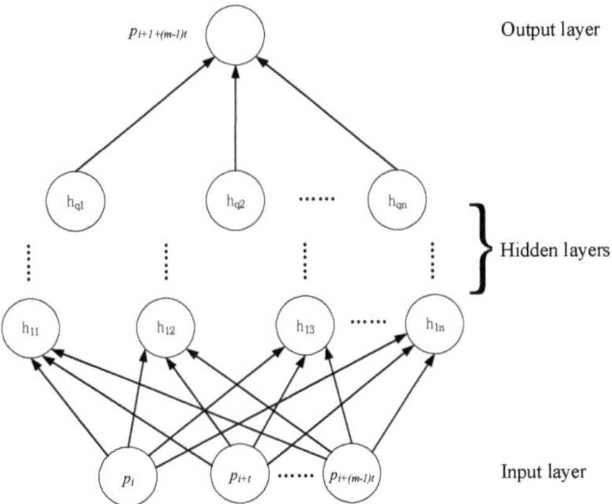

Figure 3. Deep neural network structure.

DNN can be widely used in different areas. In addition to load forecasting, DNN can also solve the problems including image processing, speech recognition, and fault diagnosis. The training method and network structure are basically the same. The difference is the training data. For the prediction of time series load data, the output layer is the actual load data. For the image processing and fault diagnosis, the output layer is the data label. The structure of DNN is basically the same as the traditional ANN, but the training method is improved which make it have better performance.

4. Analysis of Prediction Results

To verify the validity of the model, the proposed model was validated in MATLAB R2018a. In this paper, the net load data of the upper bus of a city's PV substation in the first 15 days of May 2017 were used. The sampling interval of the net load data is 5 min, and the installed capacity of the distributed PV power station is approximately 50 MW. The data from 1–11 May were selected as the training set to model the prediction model, and the parameters of the model were adjusted by cross-validation. The data from 12–15 May were selected as the samples of the prediction test, where 12 May was sunny, and 13–15 May were cloudy.

4.1. Result of the Phase Space Reconstruction by the C-C Method

The net load data from 1–11 May were processed by the C-C method. The corresponding statistics of $\overline{\Delta S}(t)$ and $S_{cor}(t)$ are shown in Figure 4. The first extremum point of $\overline{\Delta S}(t)$ was $t = 7$. $S_{cor}(t)$ had no obvious minimum point, and the optimal embedded window t_ω was not obtained. According to the BDS statistics, when $N > 3000$, $m \in \{2,3,4,5\}$, so the maximum value of m could only take 5. According to Formula (6), the final optimal embedding dimension $m_{opt} = 5$ and optimal delay $t_{opt} = 7$ were obtained.

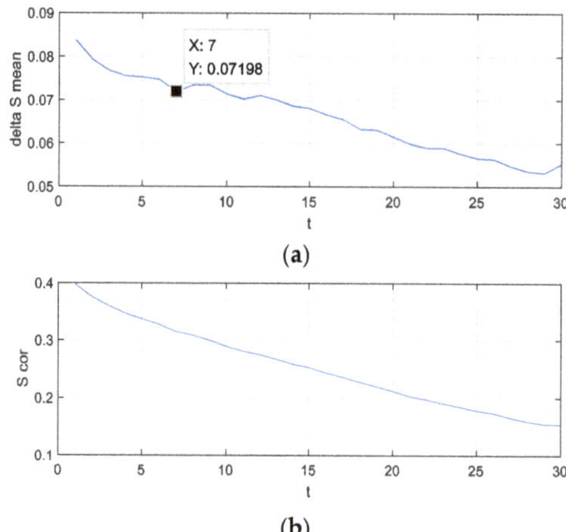

Figure 4. Curves of $\overline{\Delta S}(t)$ and $S_{cor}(t)$. (**a**) Curve of statistic $\overline{\Delta S}(t)$; (**b**) Curve of statistic $S_{cor}(t)$.

4.2. Prediction Results of the Deep Neural Network

Since the embedding dimension is $m = 5$ as determined by the C-C method, there were five input neurons in the DNN. The cross-validation shows that when the number of layers in the hidden layer was 5, the predicted value showed a significant over-fitting. Therefore, the number of layers of the hidden layer was taken as 4, and the neurons of each hidden layer were taken as 5, 4, 3, and 2. The DNN used the single-step prediction and only predicted the next 5 min load value at a time. The predicted result is shown in Figure 5. The black solid line is the actual net load, the red solid line is the ultrashort-term prediction value based on the phase space reconstruction and DNN, and the blue dotted line is the ultrashort-term prediction value based on the traditional BP neural network.

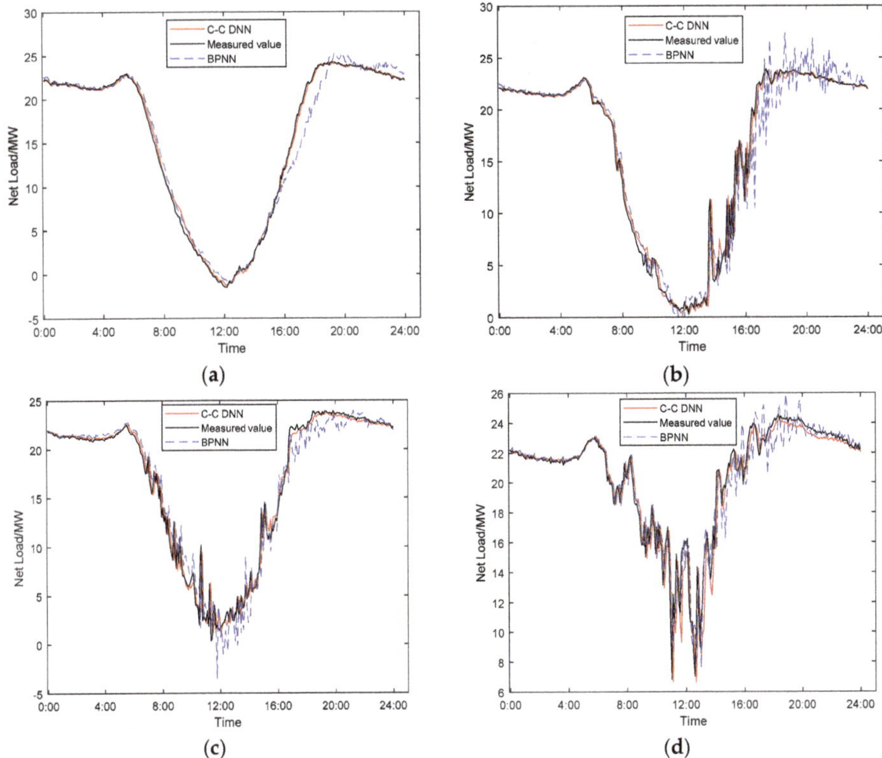

Figure 5. Load forecasting results based on different models: (**a**) 12 May; (**b**) 13 May; (**c**) 14 May; (**d**) 15 May.

In Figure 5, at approximately 12:00 noon on a clear day (12th), a negative power was present in the payload due to an increase in the amount of PV power generation. The prediction results based on the phase space reconstruction and DNN were closer to the actual net load value, which was obviously better than the prediction results using the traditional BP neural network.

On cloudy days (13–15 May), when the net load sharply fluctuated due to the fluctuation of the PV output, although the actual net load value was stable, the predicted value of the traditional BP neural network still had large fluctuations, which resulted in a large deviation. However, the prediction results based on the phase space and DNN did not strongly deviate and basically conformed to the actual trend of the net load.

To accurately evaluate the accuracy of the model prediction and the accuracy of prediction, the mean absolute percentage error (MAPE) and root mean square error (RMSE) were used as evaluation indicators. In Formulas (17) and (18), n is the number of predicted samples, p_i is the actual value of the net load at time i, and $\hat{p}_i$ is the predicted value of the net load at time i.

$$\text{MAPE} = \frac{1}{n} \sum_{i=1}^{n} \left| \frac{p_i - \hat{p}_i}{p_i} \right| \times 100\% \tag{17}$$

$$\text{RMSE} = \sqrt{\frac{1}{n} \sum_{i=1}^{n} (p_i - \hat{p}_i)^2} \tag{18}$$

The prediction accuracy is shown in Table 1. Compared with the prediction model based on the traditional BP neural network, the forecasting scheme proposed in this paper improved the accuracy of net load forecasting under different weather conditions. On a cloudy day (15 May), the power of PV power generation was very small. The accuracy of the prediction model based on the phase space reconstruction and the deep neural network was basically identical to that based on the traditional BP neural network. However, on a sunny day (12 May), the PV power was relatively high, and the difference of MAPE between the two models was nearly 10%. The predictive models based on the phase space reconstruction and DNN still had higher prediction accuracy even in the case of large distributed PV power access and large power fluctuation.

Table 1. Prediction accuracy of different models.

Date	MAPE (%)		RMSE	
	C-C DNN	**BPNN**	**C-C DNN**	**BPNN**
12 May (Sunny)	8.8838	18.3725	0.8973	1.3327
13 May (Cloudy)	9.9978	17.6827	1.3970	1.8242
14 May (Cloudy)	15.4156	17.6415	1.1193	1.3648
15 May (Sunny)	4.0648	4.6388	1.0972	1.1916

5. Conclusions

- A large amount of access to the distributed PV power generation results in the increasing fluctuation of the net load power, which challenges the ultrashort-term load forecasting. Considering this phenomenon and the characteristics of ultrashort-term load forecasting, this paper presents a model of ultrashort-term load forecasting based on the phase space reconstruction and DNN. Based on the phase space reconstruction, the time series is projected into a moving point in the phase space, and the DNN is subsequently used to fit the trajectory to realize the load forecasting.

- The prediction of the actual load data and a comparison experiment with the BP neural network prediction model have verified that the proposed model has higher prediction accuracy even in the case of large distributed PV power fluctuations and different weather conditions. In particular, when there is a large amount of PV power access and high penetration, there is also an ideal predictive performance.

Author Contributions: F.M. analyzed the data and wrote the paper. Q.W. performed the simulation and modeling; T.S. analyzed the results and reviewed the modeling and text; J.L. did funding acquisition; Y.P. prepared the data; J.Z. contributed analysis tools.

Funding: This research was funded by "State Key Laboratory of Smart Grid Protection and Control, SKL of SGPC".

Conflicts of Interest: The authors declare no conflicts of interest.

References

1. Li, L.L.; Sun, J.; Wang, C.H.; Zhou, Y.T.; Lin, K.P. Enhanced Gaussian process mixture model for short-term electric load forecasting. *Inf. Sci.* **2019**, *477*, 386–398. [CrossRef]
2. Sun, W.; Zhang, C.C. A Hybrid BA-ELM Model Based on Factor Analysis and Similar-Day Approach for Short-Term Load Forecasting. *Energies* **2018**, *11*, 1282. [CrossRef]
3. Barman, M.; Choudhury, N.B.D.; Sutradhar, S. A regional hybrid GOA-SVM model based on similar day approach for short-term load forecasting in Assam, India. *Energy* **2018**, *145*, 710–720. [CrossRef]
4. Zhang, J.L.; Wei, Y.M.; Li, D.Z.; Tan, Z.F.; Zhou, J.H. Short term electricity load forecasting using a hybrid model. *Energy* **2018**, *158*, 774–781. [CrossRef]
5. Li, Y.Y.; Han, D.; Yan, Z. Long-term system load forecasting based on data-driven linear clustering method. *J. Mod. Power Syst. Clean Energy* **2018**, *6*, 306–316. [CrossRef]
6. Bennett, C.J.; Stewart, R.A.; Lu, J.W. Forecasting low voltage distribution network demand profiles using a pattern recognition based expert system. *Energy* **2014**, *67*, 200–212. [CrossRef]

7. Kazemi, S.M.R.; Hoseini, M.M.S.; Abbasian-Naghneh, S.; Rahmati, S.H.A. An evolutionary-based adaptive neuro-fuzzy inference system for intelligent short-term load forecasting. *Int. Trans. Oper. Res.* **2014**, *21*, 311–326. [CrossRef]

8. Ahmad, T.; Chen, H.X. Nonlinear autoregressive and random forest approaches to forecasting electricity load for utility energy management systems. *Sustain. Cities Soc.* **2019**, *45*, 460–473. [CrossRef]

9. Massidda, L.; Marrocu, M. Smart Meter Forecasting from One Minute to One Year Horizons. *Energies* **2018**, *11*, 3520. [CrossRef]

10. Alamaniotis, M.; Ikonomopoulos, A.; Tsoukalas, L.H. Evolutionary Multiobjective Optimization of Kernel-Based Very-Short-Term Load Forecasting. *IEEE Trans. Power Syst.* **2012**, *27*, 1477–1484. [CrossRef]

11. Alamaniotis, M.; Bargiotas, D.; Tsoukalas, L.H. Towards smart energy systems: Application of kernel machine regression for medium term electricity load forecasting. *Springerplus* **2016**, *5*, 58. [CrossRef] [PubMed]

12. Amjady, N. Evolutionary Multiobjective Short-term bus load forecasting of power systems by a new hybrid method. *IEEE Trans. Power Syst.* **2007**, *22*, 333–341. [CrossRef]

13. Hossen, T.; Plathottam, S.J.; Angamuthu, R.K.; Ranganathan, P.; Salehfar, H. Short-Term Load Forecasting Using Deep Neural Networks (DNN). In Proceedings of the 2017 North American Power Symposium (NAPS), Morgantown, WV, USA, 17–19 September 2017.

14. Chen, Y.; Luh, P.B.; Guan, C.; Zhao, Y.; Michel, L.D.; Coolbeth, M.A.; Friedland, P.B.; Rourke, S.J. Short-Term Load Forecasting: Similar Day-Based Wavelet Neural Networks. *IEEE Trans. Power Syst.* **2010**, *25*, 322–330. [CrossRef]

15. Moon, J.; Kim, Y.; Son, M.; Hwang, E. Hybrid Short-Term Load Forecasting Scheme Using Random Forest and Multilayer Perceptron. *Energies* **2018**, *11*, 3283. [CrossRef]

16. Nazar, M.S.; Fard, A.E.; Heidari, A.; Shafie-Khah, M.; Catalao, J.P.S. Hybrid model using three-stage algorithm for simultaneous load and price forecasting. *Electr. Power Syst. Res.* **2018**, *165*, 214–228. [CrossRef]

17. Zhang, B.; Liu, W.; Li, S.T.; Wang, W.G.; Zou, H.; Dou, Z.H. Short-term load forecasting based on wavelet neural network with adaptive mutation bat optimization algorithm. *IEEJ Trans. Electr. Electron. Eng.* **2019**, *14*, 376–382. [CrossRef]

18. Liang, Y.; Niu, D.X.; Hong, W.C. Short term load forecasting based on feature extraction and improved general regression neural network model. *Energy* **2019**, *166*, 653–663. [CrossRef]

19. Dai, S.Y.; Niu, D.X.; Li, Y. Daily Peak Load Forecasting Based on Complete Ensemble Empirical Mode Decomposition with Adaptive Noise and Support Vector Machine Optimized by Modified Grey Wolf Optimization Algorithm. *Energies* **2018**, *11*, 163. [CrossRef]

20. Heymann, F.; Silva, J.; Miranda, V.; Melo, J.; Soares, F.J.; Padilha-Feltrin, A. Distribution network planning considering technology diffusion dynamics and spatial net-load behavior. *Int. J. Electr. Power Energy Syst.* **2019**, *106*, 254–265. [CrossRef]

21. Elghitani, F.; El-Saadany, E. Smoothing Net Load Demand Variations Using Residential Demand Management. *IEEE Trans. Ind. Inform.* **2019**, *15*, 390–398. [CrossRef]

22. Kuppannagari, S.R.; Kannan, R.; Prasanna, V.K. Optimal Discrete Net-Load Balancing in Smart Grids with High PV Penetration. *ACM Trans. Sens. Netw.* **2018**, *14*, 24. [CrossRef]

23. Zhou, X.; Zi, X.H.; Liang, L.Q.; Fan, Z.B.; Yan, J.W.; Pan, D.M. Forecasting performance comparison of two hybrid machine learning models for cooling load of a large-scale commercial building. *J. Build. Eng.* **2018**, *21*, 64–73. [CrossRef]

24. Kong, W.C.; Dong, Z.Y.; Jia, Y.W.; Hill, D.J.; Xu, Y.; Zhang, Y. Short-Term Residential Load Forecasting Based on LSTM Recurrent Neural Network. *IEEE Trans. Smart Grid* **2019**, *10*, 841–851. [CrossRef]

25. Kim, H.S.; Eykholt, R.; Salas, J.D. Nonlinear dynamics, delay times, and embedding windows. *Physica D* **1999**, *127*, 48–60. [CrossRef]

 *applied sciences*

Article

Net-Metering and Self-Consumption Analysis for Direct PV Groundwater Pumping in Agriculture: A Spanish Case Study

Alvaro Rubio-Aliaga [1], Angel Molina-Garcia [1,*], M. Socorro Garcia-Cascales [2] and Juan Miguel Sanchez-Lozano [3]

[1] Department of Electrical Engineering, Universidad Politécnica de Cartagena, 30202 Cartagena, Spain; alvaro.rual@gmail.com
[2] Deptartment of Electronics, Computer Technology and Projects, Universidad Politécnica de Cartagena, 30202 Cartagena, Spain; socorro.garcia@upct.es
[3] Centro Universitario de la Defensa, Universidad Politécnica de Cartagena, 30720 Murcia, Spain; juanmi.sanchez@cud.upct.es
* Correspondence: angel.molina@upct.es; Tel.: +34-968-32-5462

Received: 28 February 2019; Accepted: 9 April 2019; Published: 20 April 2019

Abstract: International policies mainly that are focused on energy-dependence reduction and climate change objectives have been widely proposed by most developed countries over the last years. These actions aim to promote the integration of renewables and the reduction of emissions in all sectors. Among the different sectors, agriculture emerges as a remarkable opportunity to integrate these proposals. Indeed, this sector accounts for 10% of the total greenhouse gas (GHG) emissions in the EU, representing 1.5% of gross domestic product (GDP) in 2016. Within the agriculture sector, current solutions for groundwater pumping purposes are mainly based on diesel technologies, leading to a remarkable fossil fuel dependence and emissions that must be reduced to fulfill both energy and environmental requirements. Relevant actions must be proposed that are focused on sustainable strategies and initiatives. Under this scenario, the integration of photovoltaic (PV) power plants into groundwater pumping installations has recently been considered as a suitable solution. However, this approach requires a more extended analysis, including different risks and impacts related to sustainability from the economic and energy points of view, and by considering other relevant aspects such as environmental consequences. In addition, PV solar power systems connected to the grid for groundwater pumping purposes provide a relevant opportunity to optimize the power supplied by these installations in terms of self-consumption and net-metering advantages. Actually, the excess PV power might be injected to the grid, with potential profits and benefits for the agriculture sector. Under this scenario, the present paper gives a multidimensional analysis of PV solar power systems connected to the grid for groundwater pumping solutions, including net-metering conditions and benefit estimations that are focused on a Spanish case study. Extensive results based on a real aquifer (Aquifer 23) located in Castilla La Mancha (Spain) are included and discussed in detail.

Keywords: economic–energy–environment (3E) analysis; solar pumping; renewable energy source (RES) integration; net-metering; sustainable rural development

1. Introduction

Presently, the sustainability of the globalized society is at potential risk because of climate change, involving an important level of atmospheric pollution. These environmental effects have been evidenced in the climate and in the availability of natural resources, mainly water. With respect to this resource, the growing water demand requires government support to avoid undesired overexploitation. In addition, climate change can affect all sectors of society. In fact, certain effects are beginning to cause concern in the agricultural sector, such as minor rainfalls and increasing temperatures. These impacts also affect the sustainability of this sector as well as other dimensions, such as energy and productivity, and finally end up affecting the social and economic global structure, especially in rural areas and areas with water scarcity. To overcome these negative impacts, international organizations have promoted several agreements as global strategies, such as the Kyoto Protocol and the COP21 Conference of the Parties on Climate Change held in Paris in 2015 [1], aiming to reduce the impacts of those climate changes. At this last event, it was agreed to contain the increase in global average temperature below 2 °C at the end of the current century. To fulfill this objective, different actions were proposed, mainly focused on (*i*) reducing dependence on fossil fuels; (*ii*) increasing the integration of renewable energy sources [2]; and (*iii*) decreasing CO_2 emissions into the atmosphere. The change in the energy model toward a major use of renewable energy resources within a framework of sustainable development in all economic sectors of society implies the need for a firm Research and Development and Innovation (R+D+i) strategy. Although solutions to achieve these targets in the domestic, industrial, and transport sectors have been widely studied, there is a lack of contributions for the agriculture sector, which requires a more detailed and multidimensional analysis. Actually, from an agro-energy perspective, this issue must be studied widely in a global analysis on the environmental, hydrological, and socioeconomic effects that have certain influences on pumping irrigation. Therefore, the energy demand and proposals for renewable energy alternatives must be considered in all applications of agriculture, and specifically in pumping facilities. Villamayor-Tomas affirms that remaining institutional challenges must include an important water rights reform, including the promotion of a distributed energy network and irrigation modernization within Spain and at the European level [3]. The change in the energy model of pumping agriculture thus represents a strategy to reduce dependence on fossil fuels, creates wealth in rural areas, settles employment, and allows participation in the reduction of CO_2 emissions. Photovoltaic (PV) solutions for agriculture pumping present a viable and profitable alternative to replacing diesel generators in isolated and individual installations [4–6]. This has mainly been motivated by high fuel costs and easily amortized investment costs. In fact, Cuadros et al. defines 'photoirrigation' as a procedure to estimate PV installations for irrigation pumping purposes [7]. Some significant agriculture–energy synergy studies have been conducted by different authors [8–10]. However, most contributions in the agricultural sector are focused on standalone solutions without considering distributed generation purposes. In this way, battery and water tanks are proposed in [11] to store energy obtained from solar panels increasing the system stability. Mohana Rao et al evaluate PV-based water pumping system for agricultural sector under standalone conditions [12]. Similar analysis can be found in [13,14], where standalone PV water pumping systems described and evaluated. Binshad et al. investigates the operation and analysis of the photovoltaic water pumping system without considering grid connection requirements [15]. A grid-connected hybrid renewable energy system example is described in [16], consisting of PV and wind power technologies applied on rural township in the Mediterranean climate region of central Catalonia (Spain). Therefore, there is a lack of contributions focusing on grid-connected PV pumping systems for water supplies and human consumption where self-consumption and net-metering schemes are evaluated. This lack of contributions thus implies that (*i*) analysis of global irrigation pumping is not available in the specific literature; (*ii*) these solutions depend on different variables that must be evaluated accordingly; and (*iii*) PV pumping solutions need to be analyzed annually to include the problem of low use of these PV installations depending on the crops. In fact, optimal use and exploitation of the facility should be properly evaluated. Moreover, it is necessary to analyze energy generated in periods when irrigation is not demanded by crops and periods when an excess of PV generation power is provided by the installation. Some studies confirm that PV installations are usually oversized for individual PV solar pumping solutions [17], which are used for irrigation purposes only 180–200 h per year. In most cases, for the rest of the potential PV solar hours, when energy is available from their locations, PV power plants are disconnected from the grid and this additional energy is not used as a potential resource. For this reason, the use of this surplus energy, which in some cases could reach 80–90% of the annual potential energy generated by the PV system, must be analyzed in detail. In this way, Langarita et al. affirm that in irrigated agriculture, a producer-consumer can be systematically exposed to energy shortfalls and surpluses [18]. An example of hybrid power plant with wind turbines, photovoltaic panels, and compressed air energy storage is described in [19], where positive income due to sale of surplus energy to the national power grid is analyzed.

Presently, the idea of systems organized in agro-smart grids or rural smart grids, conceived as distributed generation in rural areas, has been widely studied [20–23]. This organizational structure represents an alternative way of carrying out energy development integration, energy storage, automation, measurement systems, information, and communication related to power generation/demand. In addition, it provides not only better and more efficient distribution/production energy management [24], but also an optimal localized use of resources [25,26]. This concept also includes efficient water management, automation, and precision agriculture, and generation/demand balance in rural areas. Figure 1 summarizes schematically the integration of the agricultural sector into a smart grid. However, one of the main limitations of these solutions is the power line construction cost and the auxiliary elements to inject the power from those PV power plants to the grid. Moreover, Bassi affirms that it is difficult to connect millions of scattered wells, fitted with solar pumps (earlier operating with diesel pumps), to the power grid [27]. Another important drawback of these systems in general—including other sectors such as the residential sector [28]—is the current legislation and requirements on distributed generation and net-metering policies. Christoforidis et al. affirm that there is a lack of a universal policy harmonizing the respective legislations of the EU member countries in terms of net-metering schemes [29]. Nevertheless, there is a favorable legislative framework for this type of facility in some countries such as Belgium and Denmark, in other countries, such as Spain, there is currently no advantageous regulation for net-metering implementation [30]. For the Spanish case, and after a long series of changes in the regulatory and legal framework of renewable energy installations in Spain (RD1699/2011, RDL 1/2012, L15/2012, OM1491/2013, RD413/2014), the regulation of self-consumption and net-metering facilities through RD900/2015 [31] implies a series of taxes that must be paid by the facilities connected to the grid when they inject power into the grid. Further information focused on self-supply and net-balance Spanish policies can be found in [32].

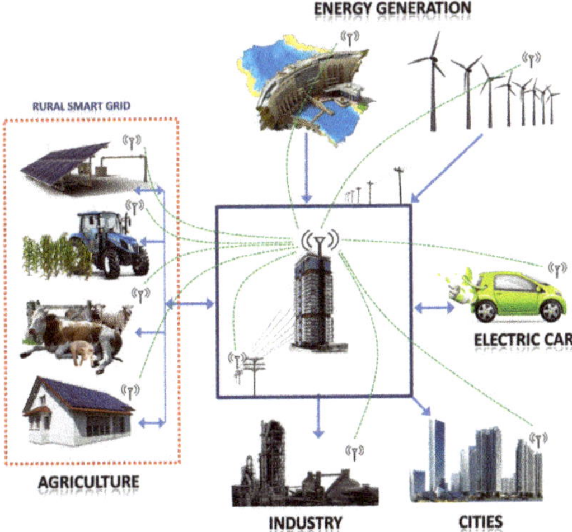

Figure 1. Integration of the agricultural sector into a smart grid: supply and demand-side active roles.

By considering previous contributions and the lack of analysis from a multidimensional perspective regarding PV power plants connected to the grid for groundwater pumping solutions, this paper aims to:

- Analyze and identify, from a socioeconomic, environmental, and energy perspective, the problems of current agriculture groundwater pumping systems based on fossil fuel technologies.
- Propose and analyze PV solar pumping alternatives connected to the grid by including surplus energy and its injection into the grid, evaluating the possible economic profits from the sector.
- Evaluate a case study based on including this alternative in a real environment and crops located in the southeast of Spain (Castilla La Mancha Region).

The rest of the paper is structured as follows: Section 2 discusses the methodology, focused on a global analysis of the problem in agriculture, describing the problems and their most important impacts, as well as the process of determining the surplus energy and the possible economic return from the sale of such additional energy. Section 3 describes the case study. Results are given in Section 4, including estimations of the surplus energy and the potential economic benefits of the sale of energy. In addition, benefits provided to the agriculture sector with the integration of this solar resource are also included. Finally, conclusions are given in Section 5.

2. Multifocused Analysis Methodology

The proposed methodology can be divided into two parts. The first part is a preliminary approach focused on analyzing, from a multidimensional perspective, the energy problem of groundwater pumping for agriculture. In this way, a study that considers a relevant number of specific factors, derived in part from the current use of fossil fuel–based solutions usually implemented for irrigation purposes, is conducted by the authors. We analyze how future changes related to an upcoming energy model, through the implementation of renewable resources (mainly PV technology as proposed this work), can address relevant positive impacts on the agriculture sector. The second part of the methodology describes a process for characterizing the energy alternative of PV pumping installations connected to the grid, identifying and quantifying the benefits provided by this solution [33]. Figure 2 schematically summarizes the proposed methodology.

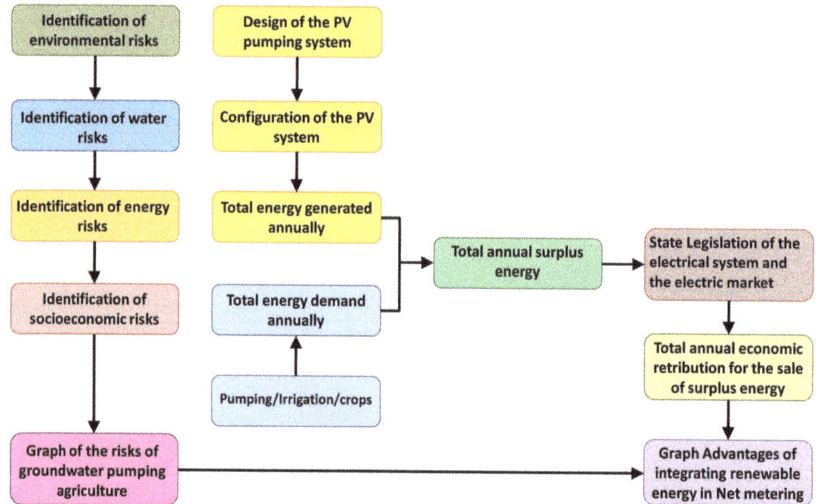

Figure 2. Description of the proposed methodology.

2.1. Groundwater Agriculture Problem: Multifocused Analysis

A multidimensional analysis is proposed by the authors to characterize groundwater pumping in agriculture. With this aim, the problem is analyzed considering: (*i*) an environmental problem (climate change, rainfall, temperature); (*ii*) a water scarcity problem (decreased aquifer phreatic level, among others); (*iii*) an energy problem; and (*iv*) a socioeconomic problem. Figure 3 shows this multidimensional analysis and the relationships among the different points of view. This methodology is in line with other contributions. Moreover, the proposed methodology considers some aspects that have been neglected or not considered in other works. Actually, the problem of sustainability related to water and aquifer resource exploitation as well as PV solar installation analysis has been previously considered in [34–37]. Figure 3 summarizes the dependencies and influences among the different approaches, which are discussed in detail below.

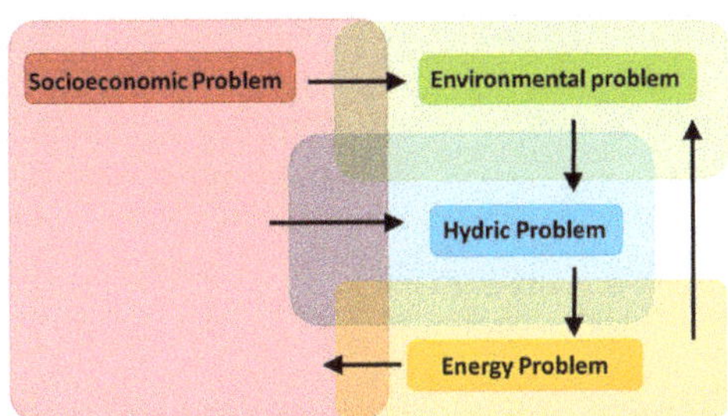

Figure 3. Multifocused analysis for agriculture groundwater pumping purposes.

2.1.1. Environmental Analysis

The environmental problem emerges as one of the most crucial impacts. In fact, this issue can involve important problems for the agricultural sector and its irrigation requirements, especially for groundwater irrigation proposals. In an arid climate, climate change can lead to a decrease in precipitation, consequently reducing the natural recharge of aquifers. These negative conditions are a limiting factor for agricultural development in those areas of the world [38]. Reduced rainfall as a result of climate change is not the only environmental impact, but also rising temperatures and other collateral effects. In fact, some analyses and studies focused on climate forecasting suggest a gradual temperature increase, with warmer and drier summer periods. Therefore, water reservoirs and lakes exposed to solar radiation can lose more water by evaporation, and crops will demand greater amounts of water. From a climate perspective, analysis of these data shows a clear tendency from semiarid areas to severely arid conditions, almost becoming desertified areas.

Water resource management in the agriculture sector emerges as a crucial and relevant factor, affecting irrigation and slightly increasing crop water requirements. Desertification and soil erosion are then collateral problems as a result of poor agricultural practices and inefficient use of irrigation strategies, leading to the loss of soil moisture in semiarid areas. Presently, the climate change problem is leading to unsustainability due to the overexploitation of some aquifers. Other physicochemical problems have also been identified as a consequence of such overexploitation. Leachates, pesticides, and inorganic fertilizers can come into contact with the aquifer and produce a contaminated environment. As an additional drawback to the overexploitation problem, groundwater salinization is becoming extreme. Consequently, water is unusable for either agriculture or human consumption unless highly expensive pretreatment is carried out. Finally, CO_2 emissions produced by pumping irrigation in agriculture must be also estimated and analyzed. Traditional agriculture has mostly been based on diesel equipment. Obviously, these systems do not contribute to mitigating climate change effects, but increase emissions. Therefore, alternative solutions based on clean and renewable energy technologies in the agriculture sector can promote the reduction of emissions.

2.1.2. Water Analysis

As previously discussed, climate change has an important impact on pumping solutions and it must be considered for irrigation groundwater purposes. In fact, well-irrigated areas have increased rapidly over the last century thanks to a large investment in pumping technology. Thanks to advances in irrigation technology, farmers have changed their agricultural models from rainfed lands (with a low agricultural productivity ratio) to high-yield irrigated crops. Indeed, some of these crops have a very high water demand, such as maize, beets, and rice. Therefore, these advances have given farmers a remarkable opportunity to diversify their crops for greater economic value but also higher water demand value, by increasing pressure on aquifers [39].

Concerns about groundwater sustainability became relevant when many aquifers reached overexploitation and encountered emergency situations [40]. Overexploitation of aquifers also involves direct environmental impacts on discharges or sources [41,42]. Indeed, sources and rivers dependent on aquifers have considerably reduced their flows, generating problems downstream of the aquifers, for both irrigation and human consumption purposes. At the same time, lake and fluvial aquatic ecosystems have been degraded, depending on maintenance of the phreatic level. Examples of these situations can be found in the Tables of Daimiel (Spain), the Saiss plain (Morocco), the flow losses of the Mikkes River, and other sources [43]. Overexploitation and other water problems are transformed into greater energy demands. Irrigation methods present different efficiency values, with significant discrepancies among them. For example, methods based on gravity, furrows, or flooding are the most inefficient irrigation solutions (50%). Apart from water inefficiency, their use can bring serious consequences to underground aquifers [44]. To achieve suitable crop maintenance, a more efficient use of water resources should be proposed, such as localized irrigation (90% efficiency) or irrigation by spraying (70–80%) [45]. In addition to the previous problems, which are easily discernible by their immediate impact on the agricultural economy, other problems associated with the continuous depletion of aquifer resources can be identified, such as the problematic subsidence of the terrain due to different pressures of water storage inside [46].

2.1.3. Energy Analysis

The environmental issues not only affect the water balance of the river basins, but are also involved in one of the main water problems in agriculture: aquifer overexploitation with high energy consumption [47]. Once a well is built, the energy required to raise water to the surface is the most relevant annual cost for these systems. This cost depends mainly on: (*i*) the unit price of energy, (*ii*) the depth of the phreatic level, (*iii*) the generator-pump system efficiency, and (*iv*) the hydrogeological characteristics of the aquifer. The high energy dependence of fossil fuels poses an international problem for any sector, and it usually involves high costs. The agriculture sector also suffers from these consequences, as it is a demanding sector of fossil fuels. To solve this, geopolitical and economic factors must be considered to find a suitable solution. For a specific crop, a decreasing phreatic level is closely related to the corresponding energy requirements. Indeed, proper hydric maintenance requires pumping from a deeper source of water and thus more energy is required in the process. Reducing the phreatic level requires a large amount of energy to raise, transport, and distribute water to crops. Increased energy demand involves major production costs for farmers, regardless of their country. This is a difficult problem to be borne by farmers, since it implies more economic effort to pay for fuel for pumping. Different contributions have been devoted to solving this energy problem [48]. The different solutions depend on proper water management to meet high energy demands at low cost [49]. Other inefficiencies, such as poorly performing irrigation methods, a lack of maintenance, or oversized facilities, can mean an excess of energy demand and high economic costs. To solve these problems, some countries have developed different energy policies for agriculture, aiming to reduce the cost of energy production. In some cases, national policies advocate the subsidization of fossil fuels for any sector or exclusively focused on agricultural use. Other policies are based on fiscal subsidies of hydrocarbon taxes on farmers and ranchers, whereas in other countries, the price of fuel is totally regulated by the government.

2.1.4. Socioeconomic Analysis

The problems mentioned above usually imply an increasing price of fuels to meet the relevant energy demand. Because of this, small plots with wells are disappearing and the current tendency is to aggregate larger areas able to decrease the costs associated with pumping maintenance at very deep groundwater levels. This effect is the complete opposite of maintaining traditional agriculture and land democratization [50]. The typical way of dealing with this problem is to raise food prices by farmers; usually prices at the farmer level are then increased to improve their profit margin. This option reduces the competitiveness of their products compared to similar and cheaper products from other countries where the costs of production are considerably lower [50]. A more drastic option is to give up crops or plots, which results in poor economic benefits and does not allow this solution to continue over time. This last option generates depopulation in agricultural and rural areas, where the opportunities and jobs could decrease drastically [50]. Both options have a great impact on the agricultural sector, involving a loss of economic value in the sector, a loss of competitiveness for national products, a reduction in investment, and a subsequent loss of plot value in rural areas. In addition, there is a loss of social value, such as loss of employment in the countryside, loss of traditional agriculture, and depopulation of rural areas. These situations mean that governments, including international associations such as the European Union and the United Nations Organization, must offer alternative actions, strategies, and energy policies to provide solutions to these problems. These strategies are intended to help or subsidize the agricultural sector, such as the Community Agricultural Policy (CAP), which subsidizes, with nuances, such loss of competitiveness of European products directly to farmers. In other cases, there is protectionism toward national agriculture, such as an agrarian policy.

2.2. PV System Configuration and Surplus Energy Estimation

Presently, customers of electricity that have installed energy sources at their households are transformed into 'prosumers' [51]. As was previously discussed, different countries use diverse schemes of support for 'prosumers' [52]. In fact, diverse mechanisms supporting the self-consumption of electricity in key countries all over the world and to highlight the challenges and opportunities associated with their developments have been recently discussed by the IEA [53]. Under this framework, the present section characterizes the sale of energy from PV installations, which supplies energy for agricultural irrigation by groundwater. This characterization process starts with an initial database, where the energy demanded by the irrigated area and the energy-demanding facilities are estimated. Subsequently, a preliminary configuration of the PV facility is determined by including the type of technology (Mono-Si, Te-Cd…), solar tracking options, connection to the grid, and injection of surplus energy to the grid. Other parameters such as depth of the aquifer, plot grouping, and water needs of crops are also taken into account [9]. It is then possible to estimate the rate power of the PV installation under different groundwater pumping scenarios, which depend on the depth of the aquifer level, the averaged crop water demand, and the hydraulic system pressure. For the purpose of comparing different alternatives, the rate power required by the pump is first estimated (P_p) [54]:

$$P_p = \frac{H_t \cdot Q_{mx} \cdot \rho \cdot g}{v_{MP}} \rightarrow P_d = \frac{P_p \cdot K_d}{v_d} \rightarrow P_g = P_d \cdot K_g \tag{1}$$

where H_t is the total dynamic head (m), Q_{mx} is the maximum flow rate (m^3/s), r is the water density (kg/m^3), g is the earth's gravitational acceleration (m/s^2), and v_{MP} is the pump efficiency (%). For PV solar power estimations (P_{PV}), the following expression is proposed [55]:

$$P_{PV} = \frac{E_{dem}}{E_{(\alpha,\beta)} \cdot PR} \tag{2}$$

where E_{dem} is the expected averaged energy consumption (kWh/day) by considering the crop water need, $E_{(\alpha,\beta)}$ is the expected averaged energy production of a PV power plant from an average monthly value of a typical daily irradiation on the horizontal surface (kWh/m²·day) and PR is the performance ratio of the PV installation. The surplus energy from the PV pumping system can be then determined from the global PV-generated power and the global crop water need:

$$Se_{(x,y,z)} = \sum_{k=1}^{k=8760} \frac{E_{gen}(k) - E_{dem}(k)}{1000} \tag{3}$$

where $Se_{(x,y,z)}$ is the surplus annual energy (MWh/year), x is the aggregated areas (ha), y is the phreatic level of groundwater depth (m), z the global crop water need (m³), k is hours in a year, E_{gen} is the energy produced in a specific k-hour (kWh), and E_{dem} is the energy demanded in a specific k-hour (kWh).

The next step is to apply the rules and requirements to enable pouring surplus energy into the electricity grid, determining the economic values to consider possible economic retribution. At this point, as discussed in the introduction, the net-metering is differentiated by applying the prices of the electricity market and self-consumption defined and regulated by the corresponding national authorities. The following section describes the case study, which focuses on current Spanish legislation. Nevertheless, the proposed methodology can also be applied to other legislative frameworks under different national authority requirements.

3. Case Study

3.1. Preliminaries

Recently, Barbel affirms that in Spain, irrigated agriculture accounts for 20% of the total agricultural area, consumes 75% of total water resources, and generates 60% of the total agricultural production and 80% of agricultural exports [56]. Under these circumstances, Aquifer 23 located in Castilla La Mancha, Spain, is considered for the case study. Figure 4 shows the location of this aquifer and the agricultural area that depends on this water resource. The area is basically a sedimentary basin immersed in a karstic system. This aquifer varies in depth between 10 and 70 m, occupying an area of 5500 km². Recently, it has been declared an overexploited aquifer as a consequence of not only poor management and a lack of environmental and water control, but also a lack of planning of water resources. Indeed, it has reached drops of 2.3 m/year over several years of severe extraction. Over the last decade, it has been considered as a remarkable resource recovery example, increasing the groundwater level of the aquifer, as depicted in Figure 5. Presently, this aquifer is still considered overexploited, mainly due to high influence of recent periods of low rainfall. The recovery process is a consequence of the awareness of this situation and farmers' economic dependence on the aquifer [57]. Irrigation is one of the main economic drivers and sources of sustenance of the rural society in this area [58]. Regarding the climate in the area, it can be classified as continental Mediterranean with dry and hot summers with high solar irradiance levels, and cold winters with certain frost periods. Spring and autumn are characterized by soft and humid periods. Annual rainfall is a determining factor, which in the study area presents relevant oscillations between wetter periods and drier periods, accounting for 350–400 mm per year. However, with the conditions imposed by climate change in recent decades, average annual temperatures are slightly rising while rainfall is being partially reduced, posing a serious risk of desertification. Solar resource has high average potential during sunshine hours, with more than 4900 sunshine hours per year. Figure 6 depicts solar irradiance levels and aquifer depth for the case study.

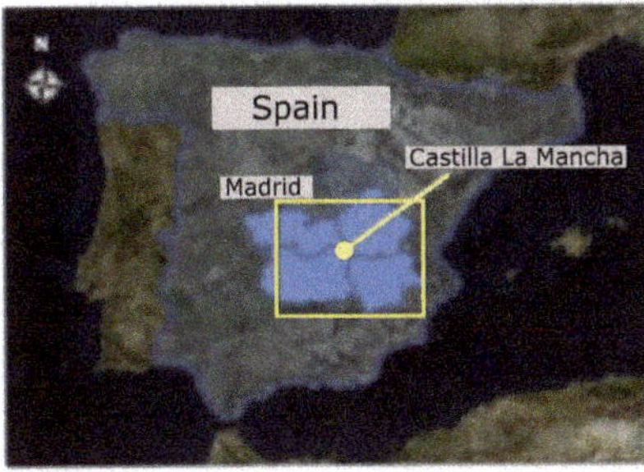
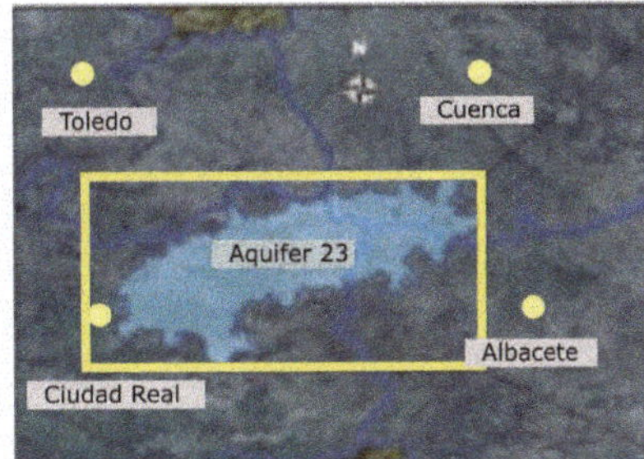

Figure 4. Location of study area and aquifer.

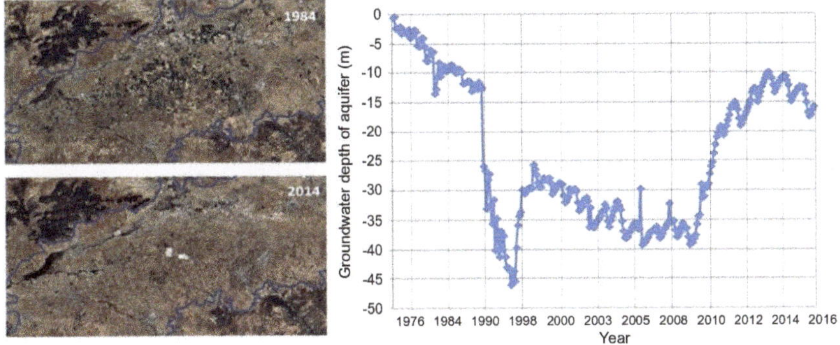

Figure 5. Satellite image of regime of exploitation of water resources and chronological graph of groundwater aquifer level. Source: Authors' elaboration through Google Earth images and CHG data.

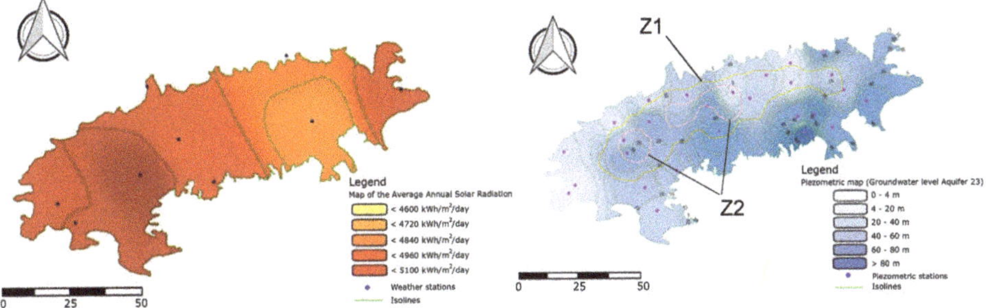

Figure 6. Irradiance solar resource and groundwater depth level: case study. Description of Z1 and Z2.

3.2. Crop Water Need

According to Figure 6, the central band of the aquifer (labelled as Z1) has the highest concentration of irrigation plots: 79% of the irrigation surface of the entire aquifer. This part accounts for 257,456 ha and 58% of this surface (149,647 ha) has irrigated crops. In other areas, labelled as Z2, groups of plots have an irrigated vs. unirrigated ratio of around 75%, accounting for around 15% of the agricultural surface on the global aquifer. In addition, the average depth of the aquifer estimated in 2016 was 29.41 m, according to reference piezometers used by the Guadiana Hydrographic Confederation (Spain) and data from the SIG maps. Due to the initial conditions related to the present case study, two levels of crop water need are considered: 1500 m^3/ha per year and 3000 m^3/ha per year. Both values are the result of water constraints on crop irrigation with the aim of preserving the aquifer. Although there are crops that have higher water requirements, most crops in the study area (mainly vineyards) currently have an average water requirement within the selected range, by considering usual and real grouping plots. Subsequently, both crop water need values (1500 m^3/ha and 3000 m^3/ha per year) are representative of averaged crop irrigation necessities.

3.3. PV Power Plant Configuration

As was previously described in Section 2.2, the rate power of the PV installation can be estimated by considering the depth of the aquifer level, the averaged crop water demand and the aggregated crop area. Under these requirements, the specific PV power plant configuration is not in line with usual individual installations, mainly promoted in the agricultural sector and based on isolated pumping systems. In our case, we propose an aggregated PV pumping solution without accumulation, directly connected to the grid and excluding any water reservoir facility. The proposed pumping solution thus requires more power, but lower annual maintenance. Therefore, PV power plant solutions with PV modules based on mono-silicon PV technology in fixed installations is considered for the analysis, being the rate PV power estimated to cover the average daily demand according to a specific crop water need. From the aquifer characteristics, ranges to be considered for the study can be then summarized as follows: groups from 1 to 2000 ha, groundwater pumping levels between 10 and 55 m of aquifer depth; and two representative crop water need: 1500 m^3/ha and 3000 m^3/ha—discussed in Section 3.2. Different PV power plant solutions are determined based on the different configurations assumed in the case study. In this way, Figure 7 summarizes the PV solutions (in kWp power capacity) for the different scenarios. In all cases, PV power plant is determined to supply the averaged power demand according to the crop water need, the aggregated area (ha) and groundwater pumping level (m). Therefore, the PV power plants to be installed (in kWp) would provide power enough to supply the corresponding pumping groundwater requirements.

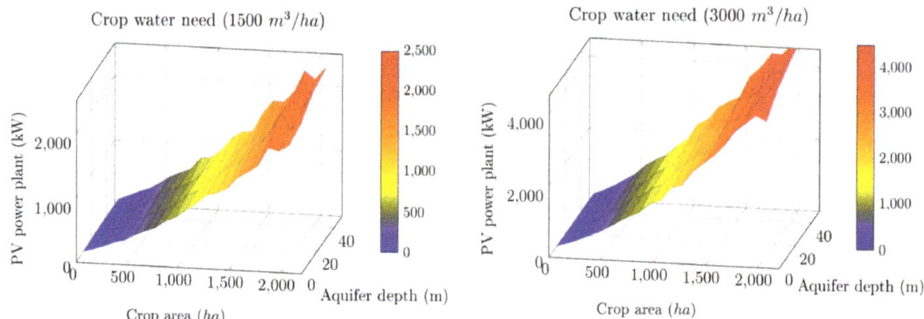

Figure 7. PV power plant capacities for self-consumption scenarios: 1500 m^3/ha and 3000 m^3/ha.

3.4. Self-Consumption: Spanish Legislative Framework

As previously discussed, the proposed methodology can be applied to any legislative framework and according to the corresponding different national authority requirements. In our case, the aquifer is in Spain (Aquifer 23), and thus, the Spanish legislation based on RD 900/2015 is applied [31]. Through this directive, two types of self-consumption are defined: (*i*) Type 1, lower than 100 kW of rate power; most individual pumping irrigation facilities can be classified as Type 1; and (*ii*) Type 2, self-consumption with more than 100 kW rate power, injected into the grid at a price established by the electricity market pool. In line with the PV power plant capacities estimated and summarized in Figure 7, most communities of solar pumping irrigators would be considered as Type 2. In the case of Spain, taxes and fees that would reduce the final remuneration for the sale of energy must be imposed. These conditions represent a burden at a time of encouraging the implementation of solar solutions—mainly in this case that PV power plants connected to the grid cannot be amortized in a relatively short period of time. Indeed, the costs include a variable charges component associated with the system costs and determined from the variable terms, and a capacity payment component to compensate for system support, compensation to the market and system operators, and interrupted service and adjusted service. It is also necessary to add the recent 7% tax on electricity generation (in September 2018 this was removed by the Spanish government) and the value added tax (VAT) of 21%. Tables 1–3 summarize the representative Spanish fixed, variable, and additional costs to be currently considered for the sale of energy.

Table 1. Spanish fixed fees and taxes applied to the sale of surplus energy through self-consumption: requirements according to RD900/2015 [31].

Access Cost	Annual Fixed Tax (Euro/kWh)					
	Period 1 P1	Period 2 P2	Period 3 P3	Period 4 P4	Period 5 P5	Period 6 P6
2.0 (P ≤ 10 kW)	8.682019					
2.1 (10 ≤ P ≤ 15 kW)	15.083303					
3.0 A (P > 15 kW)	32.083923	6.212601	14.245468			
3.1 A (1 kV to 36 kV)	35.952537	6.717794	4.985851			
6.1 A (1 kV to 30 kV)	22.169359	7.844864	9.790954	11.926548	14.278122	4.882162
6.1 B (30 kV to 36 kV)	14.050921	3.782129	6.817708	8.953302	11.304876	3.525577
6.2 (36 kV to 72.5 kV)	9.082012	1.409534	4.372144	6.352856	8.073738	2.442188
6.3 (72.5 kV to 145 kV)	9.279523	2.525841	3.909548	5.479569	6.893947	1.911493
6.4 (≥145 kV)	2.815509	0.000000	1.718359	3.457606	4.990376	0.970612

Table 2. Spanish variable fees and taxes applied to the sale of surplus energy through self-consumption: requirements according to RD900/2015 [31].

Access Cost	Annual Variable Tax (Euro/kWh)					
	Period 1 P1	Period 2 P2	Period 3 P3	Period 4 P4	Period 5 P5	Period 6 P6
2.0 A (P ≤ 10 kW)	0.043187					
2.0 DHA (P ≤ 10 kW)	0.057144	0.006148				
2.0 DHS (P ≤ 10 kW)	0.057938	0.006430	0.006112			
2.1 A (10 ≤ P ≤ 15 kW)	0.054883					
2.1 DHA (10 ≤ P ≤ 15 kW)	0.068081	0.015450				
2.1 DHS (10 ≤ P ≤ 15 kW)	0.068875	0.018220	0.011370			
3.0 A (P > 15 kW)	0.020568	0.013696	0.008951			
3.1 A (1 kV to 36 kV)	0.015301	0.009998	0.012035			
6.1 A (1 kV to 30 kV)	0.011775	0.011336	0.007602	0.009164	0.009986	0.006720
6.1 B (30 kV to 36 kV)	0.011775	0.008312	0.007322	0.008260	0.009403	0.006349
6.2 (36 kV to 72.5 kV)	0.012669	0.011554	0.007881	0.008377	0.008716	0.006245
6.3 (72.5 kV to 145 kV)	0.015106	0.012816	0.008530	0.008510	0.008673	0.006278
6.4 (≥145 kV)	0.011775	0.008531	0.007322	0.007788	0.008257	0.006104

Table 3. Spanish additional fees and taxes applied to the sale of surplus energy through self-consumption: requirements according to RD900/2015 [31].

Annual Additional Tax (Euro/kWh)	
Electricity market operation	0.000025
Power system operation	0.000109
Interruptibility service	0.002000
Provision of adjustment services	0.003210

In Spain, the times of reduced power are usually distributed in three periods. However, for power higher than 450 kW, the Spanish electricity market offer six time periods (P1, P2, P3, P4, P5, P6). Figure 8 shows the electricity rates for the different time periods under the Spanish electricity system legislation. As an example, and for the systems described in this case study (direct PV solar pumping installations) and the selected crop water-need values—1500 m^3/ha and 3000 m^3/ha, the typical periods for this type of facility are the following: P5 in May, P3 and P4 in June, P1 in the rest of June and July, and P6 in August. Irrigation is not usual in September for the considered crops, but it would be framed in periods P3 and P4. To clarify the Spanish electricity market, in terms of selling the excess energy to the grid at a price determined by such electricity market, Figure 9 shows an example for an 870 kWp PV installation, 1000 ha aggregated crop area, 40 m aquifer depth and 1500 m^3/ha crop water need. Energy demanded by the crop, surplus of energy and estimated benefits—excluding and including Spanish taxes—are determined by the different months. Time periods to be applied according to the Spanish electricity market, see Figure 8, are also included.

Hours	0-8	8	9	10	11	12	13	14	15	16	17	18	19	20	21	22	23
January	P6	P2	P2	P1	P1	P1	P2	P2	P2	P2	P2	P1	P1	P1	P2	P2	P2
February	P6	P2	P2	P1	P1	P1	P2	P2	P2	P2	P2	P1	P1	P1	P2	P2	P2
March	P6	P4	P4	P4	P4	P4	P4	P4	P4	P3	P3	P3	P3	P3	P3	P4	P4
April	P6	P5	P5	P5	P5	P5	P5	P5	P5	P5	P5	P5	P5	P5	P5	P5	P5
May	P6	P5	P5	P5	P5	P5	P5	P5	P5	P5	P5	P5	P5	P5	P5	P5	P5
June (1-15)	P6	P4	P3	P3	P3	P3	P3	P3	P4	P4	P4	P4	P4	P4	P4	P4	P4
June (15-30)	P6	P2	P2	P2	P1	P1	P1	P1	P1	P1	P1	P1	P2	P2	P2	P2	P2
July	P6	P2	P2	P2	P1	P1	P1	P1	P1	P1	P1	P1	P2	P2	P2	P2	P2
Augost	P6	P6	P6	P6	P6	P6	P6	P6	P6	P6	P6	P6	P6	P6	P6	P6	P6
September	P6	P4	P3	P3	P3	P3	P3	P3	P4	P4	P4	P4	P4	P4	P4	P4	P4
October	P6	P5	P5	P5	P5	P5	P5	P5	P5	P5	P5	P5	P5	P5	P5	P5	P5
November	P6	P4	P4	P4	P4	P4	P4	P4	P4	P3	P3	P3	P3	P3	P3	P4	P4
December	P6	P2	P2	P1	P1	P1	P2	P2	P2	P2	P2	P1	P1	P1	P2	P2	P2

Figure 8. Description of electricity rates for different time periods in the Spanish electricity system.

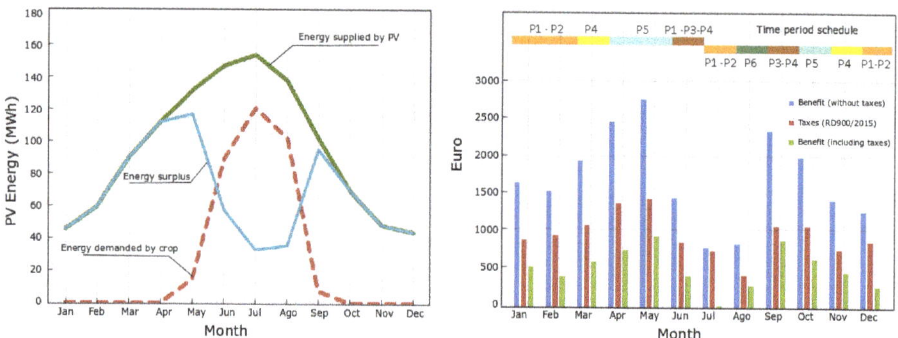

Figure 9. Example of PV generation and surplus of energy. Costs and benefits for the Spanish electricity system.

4. Results and Discussion

According to the proposed methodology described in Section 2, the case study is analyzed from a multidimensional perspective with the goals of reducing the intense dependence on fossil fuels, increasing the integration of solar solutions and preserving the aquifer to avoid future lower phreatic levels that would require more energy resources and thus relevant economic efforts. Furthermore, PV power plants connected to the grid can give farmers additional benefits through net-metering scenarios and annual energy surpluses.

4.1. PV Installations Connected to the Grid: Surplus of Annual Energy

Depending on the agronomic management of irrigation, the amount of water demanded by certain crops, the climatic conditions, and the state of the soil, the energy required by crops can vary considerably. As previously discussed, most crops require irrigation during specific periods of the year and their demand can be considered as seasonal. For example, for the case study, the months are limited to May, June, July, August, and September. Therefore, an important part of the potentially generated annual power is initially wasted. Under this hypothesis, the energy generated for the case study has been quantified-based on the selected PV configuration and the corresponding 1500 and 3000 m^3/ha crop water needs, which represents vineyard crops and a mosaic of vegetation and vineyard for typical areas of the case study. Figure 10 shows the surplus energy gradient based on the surface in hectares and the aquifer depth for the different PV configurations summarized in Figure 7. As shown in these results, greater depth aquifer and greater crop water need would imply more power required by the system, and consequently, the potential annual generated energy will be higher. This is due to the fact that both parameters have a relevant influence on the preliminary estimation of the PV solar pumping installation. Nevertheless, the investments are related to size of the PV power plant, and subsequently, the higher the PV system the higher the annual profits. A more detailed economic analysis, including investments, should be conducted to estimate the best solution. A recent detailed economic analysis carried out by the authors can be found in [59].

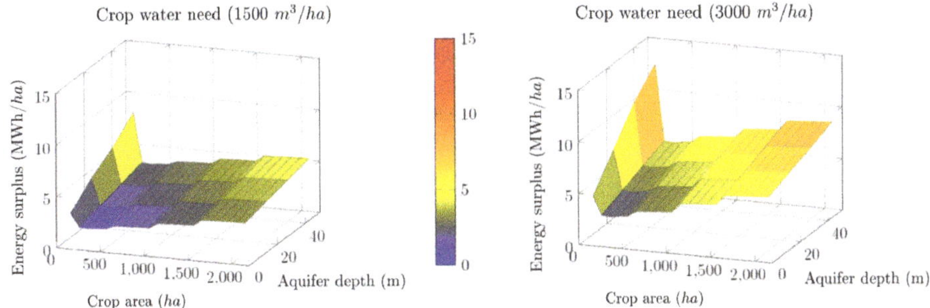

Figure 10. PV installation connected to the grid: annual surplus energy estimation examples (1500 m^3/ha and 3000 m^3/ha).

4.2. Grid Injection from PV Systems into the Grid: Net-Metering Schemes

Firstly, and from the annual surplus energy estimation examples depicted in Figure 10, a preliminary estimation of annual benefits can be determined by considering the current Spanish legislation—aimed at promoting net-metering policies—but excluding current taxes on electrical generation summarized in Tables 1–3. With this aim, Figure 11 shows annual estimated economic returns provided by the corresponding PV installations previously determined and summarized in Figure 7. It is important to point out that these benefits are highly dependent on the irrigation profiles required by the crops, and subsequently, they could be different when considering other crops and water needs. Nevertheless, the proposed can be applied to other legislative scenarios.

From these preliminary analyses, the following results estimate the economic compensation of PV facilities under the current Spanish legislation. Figure 12 gives the benefits under the legislation defined in RD900/2015 and the application of the corresponding taxes and charges. The final economic compensation, compared to Figure 11, is reduced for both 1500 m^3/ha and 3000 m^3/ha cases. The analysis of the results and the comparison between economic return on surplus energy for 1500 m^3/ha and 3000 m^3/ha, with a law aimed at developing renewable energy and Spanish legislation defined by RD 900/2015, means that only between 40% and 60% of economic compensation for the sale of energy is obtained with application of this legislation regarding a net-metering scheme excluding taxes and fees. Subsequently, a PV solar configuration for 3000 m^3/ha allows us to provide between 1.6 and 1.8 times more surplus energy than the 1500 m^3/ha-based solution. For example, an area of 1000 hectares with an aquifer depth of 30 m and a vineyard crop of 1500 m^3/ha of annual water requirements is estimated to cost 180 Euro/ha (per year) for the sale of energy. The same solution under current Spanish self-consumption legislation would be significantly reduced by up to 81 Euro/ha (per year).

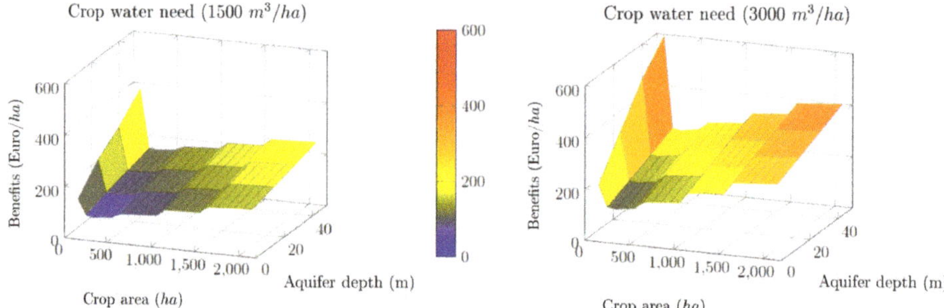

Figure 11. PV installation connected to the grid: annual benefit estimation examples excluding taxes and fees (1500 m^3/ha and 3000 m^3/ha).

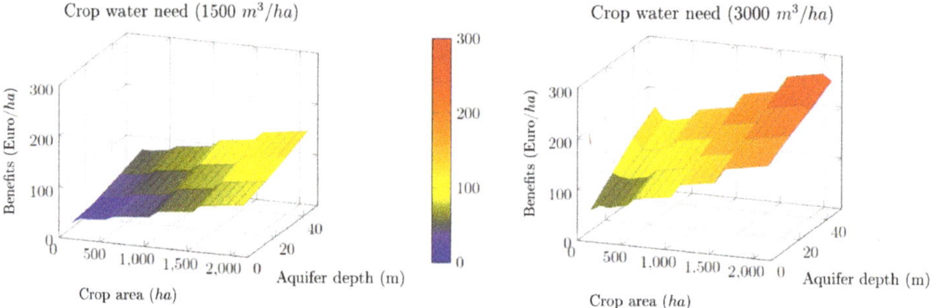

Figure 12. PV installation connected to the grid: annual benefit estimation examples including taxes and fees (1500 m^3/ha and 3000 m^3/ha).

4.3. PV Integration into Net-Metering Schemes: Aquifer 23 Discussion

By considering that PV solar installations for pumping groundwater purposes can be used more efficiently under net-metering schemes, significant economic benefits and environmental profits can be provided by these facilities. The proposed PV configurations allow reduction in the energy costs and subsequently the production costs, becoming more competitive without changing the profit margin. In addition, these solutions give rural areas an opportunity to maintain their population and, at the same time, reduce their economic dependency mainly based on subsidies. A remarkable reduction of emissions in the agricultural sector can also be achieved. According to the annual benefit estimation for the self-consumption and net-metering schemes previously described are summarized in Figure 12, it is possible to extrapolate the data to the rest of the aquifer (Aquifer 23). In this way, we consider the point where the concentration of wells is larger: within zones Z1 and Z2, accounting for 58% of the wells. If communities of irrigators of 800 ha, such as an existing one of this size, were connected to the grid, considering the average aquifer depth in those zones, emissions would be reduced in a more than relevant way. As previously discussed, after implementing a PV power plant connected to the grid in a community of irrigators, the economic benefits are highly dependent on the specific crop water need and the aquifer depth, which corresponds to 50,000 to 90,000 Euro in Z1 and 100,000 to 140,000 Euro in Z2, based on a net-metering scheme excluding taxes and fees; and from 28,000 to 50,000 euros in Z1 and 40,000 to 90,000 euros in Z2 according to current legislation in Spain (RD900/2015). Extrapolating the economic benefits for the entire aquifer, direct benefits to farmers of between 8 and 13 million Euro could be achieved in Z1, and between 3 and 4 million Euro in Z2, in accordance with a preliminary net-metering scheme without taxes and fees. However, with the current legislation in Spain regarding self-consumption and net-metering, between 4 and 8 million Euro in Z1 and between 1 and 2.5 million

Euro in Z2 would be estimated annually. Tables 4 and 5 summarizes the economic benefits from the corresponding net-metering schemes by Z1 and Z2 zones, respectively.

Finally, Figure 13 summarizes the analyzed approaches considered in this work and the corresponding advantages from these different perspectives. Presently, to undertake projects and design aid for the promotion of new renewable technologies and energy efficiency in agriculture, the European Union promotes several programs along this line under the European Agricultural Fund for Rural Development (EAFRD), to which is added the Green Fund for Climate [60] for other countries.

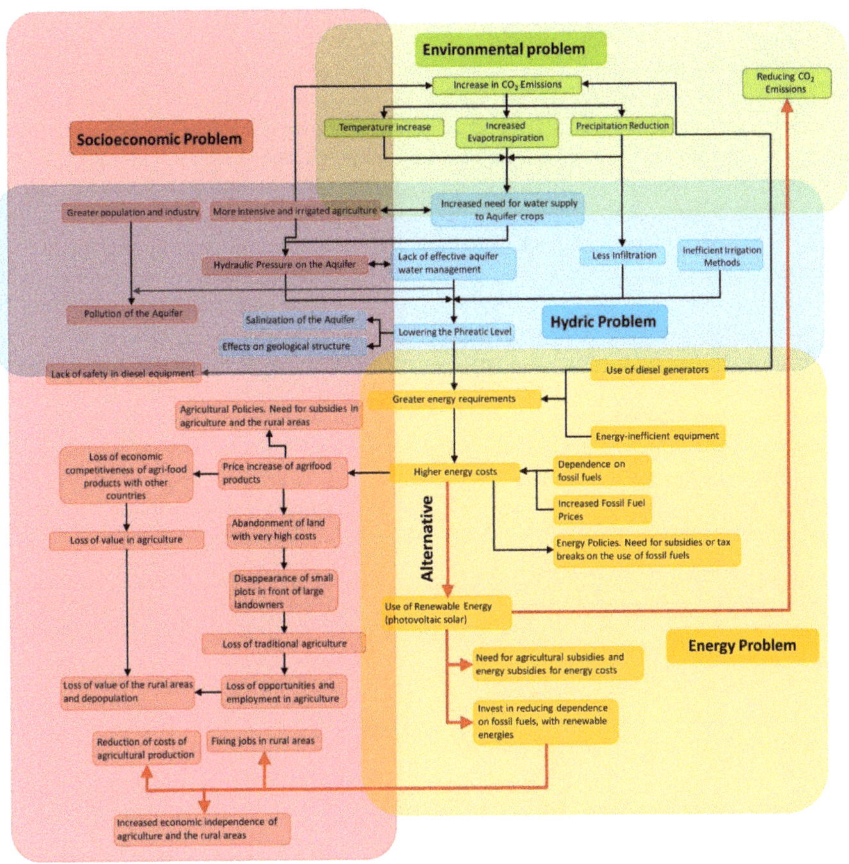

Figure 13. Integrating renewables into net-metering: advantages and multifocused approach.

Table 4. Economic benefits: Z1 zone (159 communities, 800 ha/community).

Crop Water Need (m³/ha)	Area (ha)	Economic Benefit Excluding Taxes (Euro)	Economic Benefit Including Taxes (Euro)
1500	440 (Aggregated Area)	52,800	26,800
	127,200 (Global Area)	8,395,134	4,547,364
3000	440 (Aggregated Area)	88,000	50,600
	127,200 (Global Area)	13,991,890	8,045,337

Table 5. Economic benefits: Z2 zone (28 communities, 800 ha/community).

Crop Water Need (m³/ha)	Area (ha)	Economic Benefit Excluding Taxes (Euro)	Economic Benefit Including Taxes (Euro)
1500	Community Area (616)	105,336	47,432
	Global Area (27,400)	2,955,596	1,330,883
3000	Community Area (616)	142,912	94,248
	Global Area (27,400)	4,009,932	2,644,481

5. Conclusions

The integration of PV solar installations connected to the grid into the agriculture sector is proposed and evaluated under net-metering and self-consumption scenarios. This solar resource allows us to decrease emissions and fossil fuel dependence and improve economic benefits from a surplus energy sale standpoint. This multifocused analysis is an exportable and scalable solution that can be applied in different locations depending on different parameters, such as crop water need, aquifer depth, and grouped crop areas. A Spanish aquifer highly overexploited over the decades is used to evaluate the proposed methodology. Different surplus energy sale scenarios are analyzed according to the typical crops in this location and the corresponding annual water requirements and common grouping areas. In this way, relevant annual benefits are estimated in grouped areas of 800 ha, accounting for 50,000 to 140,000 euros/year in a net-metering situation excluding taxes and fees; and 28,000 to 90,000 euros under current Spanish regulations. Regardless of the level of grouped areas, PV power plants interconnected with the grid for the use of surplus energy could generate nonnegligible global revenues: between 10 and 18 million euros/year with a legislation prone to net-metering and between 5 and 10 million euros/year under the current Spanish legislation framework. Therefore, global policies focused on water management and efficient agricultural objectives should be promoted for massive integration of such renewables into the agriculture sector. More specifically, energy policies in terms of net-metering and/or self-consumption schemes that provide regulatory stability to this energy model in agriculture are required by the sector.

Author Contributions: Conceptualization, A. M.-G.; Formal analysis, J.M.S.-L.; Funding acquisition, M.S.G.-C.; Investigation, A.R.-A.; Project administration, A.R.-A.; Validation, M.S.G.-C.; Writing—original draft, A.M.-G. and J.M.S.-L.; Writing—review and editing, A.M.-G.

Funding: This research was funded by the research project TIN2017-86647-P from the Spanish Ministry of Science and Innovation (including FEDER funds), and the Seneca Foundation 19882-GERM-15.

Conflicts of Interest: The authors declare no conflict of interest.

References

1. United Nations. *Session of the Conference of the Parties. 21st Yearly Session. Approval of the Paris Agreement*; United Nations Climate Change Conference; United Nations: New York, NY, USA, 2015.

2. Official Journal of the European Union. *Directive 2009/28/CE of the European Parliament and of the Council of 23 April 2009, On the Promotion of the Use of Energy from Renewable Sources*; 140/16; United Nations: New York, NY, USA, 2009.

3. Villamayor-Tomas, S. Chapter 2.1.3—The Water–Energy Nexus in Europe and Spain: An Institutional Analysis From the Perspective of the Spanish Irrigation Sector. In *Competition for Water Resources*; Ziolkowska, J.R., Peterson, J.M., Eds.; Elsevier: Berlin, Germany, 2017; pp. 105–122.

4. Glasnovic, Z.; Margeta, J. A model for optimal sizing of photovoltaic irrigation water pumping systems. *Sol. Energy* **2007**, *81*, 904–916. [CrossRef]

5. Abu-Aligh, M. Design of photovoltaic water pumping system and compare it with diesel powered pump. *Jordan J. Mech. Ind. Eng.* **2011**, *5*, 273–280.

6. Foster, R.; Cota, A. Solar water pumping advances and comparative economics. *Energy Procedia* **2014**, *57*, 1431–1436. [CrossRef]

7. Cuadros, F.; López-Rodríguez, F.; Marcos, A.; Coello, J. A procedure to size solar-powered irrigation (photoirrigation) schemes. *Sol. Energy* **2004**, *76*, 465–473. [CrossRef]

8. Odeh, I.; Yohanis, Y.; Norton, B. Influence of pumping head, insolation and PV array size on PV water pumping system performance. *Sol. Energy* **2006**, *80*, 51–64. [CrossRef]

9. Meah, K.; Ula, S.; Barrett, S. Solar photovoltaic water pumping—Opportunities and challenges. *Renew. Sustain. Energy Rev.* **2008**, *12*, 1162–1175.

10. Kelley, L.C.; Gilbertson, E.; Sheikh, A.; Eppinger, S.D.; Dubowsky, S. On the feasibility of solar-powered irrigation. *Renew. Sustain. Energy Rev.* **2010**, *14*, 2669–2682. [CrossRef]

11. Dursun, M.; Ozden, S. Application of Solar Powered Automatic Water Pumping in Turkey. *Int. J. Comput. Electr. Eng.* **2012**, *4*, 161. [CrossRef]

12. Rao, M.M.; Sahu, M.K.; Subudhi, P.K. Pv based water pumping system for agricultural sector. *Mater. Today Proc.* **2018**, *5*, 1008–1016.

13. Zahab, E.E.A.; Zaki, A.M.; El-sotouhy, M.M. Design and control of a standalone PV water pumping system. *J. Electr. Syst. Inf. Technol.* **2017**, *4*, 322–337. [CrossRef]

14. Singh, B.; Sharma, U.; Kumar, S. Standalone photovoltaic water pumping system using induction motor drive with reduced sensors. *IEEE Trans. Ind. Appl.* **2018**, *54*, 3645–3655. [CrossRef]

15. Binshad, T.; Vijayakumar, K.; Kaleeswari, M. PV based water pumping system for agricultural irrigation. *Front. Energy* **2016**, *10*, 319. [CrossRef]

16. González, A.; Riba, J.R.; Rius, A.; Puig, R. Optimal sizing of a hybrid grid-connected photovoltaic and wind power system. *Appl. Energy* **2015**, *154*, 752–762. [CrossRef]

17. Meah, K.; Fletcher, S.; Ula, S. Solar photovoltaic water pumping for remote locations. *Renew. Sustain. Energy Rev.* **2008**, *12*, 472–487. [CrossRef]

18. Langarita, R.; Chóliz, J.S.; Sarasa, C.; Duarte, R.; Jiménez, S. Electricity costs in irrigated agriculture: A case study for an irrigation scheme in Spain. *Renew. Sustain. Energy Rev.* **2017**, *68*, 1008–1019. [CrossRef]

19. Marano, V.; Rizzo, G.; Tiano, F.A. Application of dynamic programming to the optimal management of a hybrid power plant with wind turbines, photovoltaic panels and compressed air energy storage. *Appl. Energy* **2012**, *97*, 849–859. [CrossRef]

20. Maheshwari, Z.; Ramakumar, R. Smart Integrated Renewable Energy Systems (SIRES): A Novel Approach for Sustainable Development. *Energies* **2017**, *10*, 1145. [CrossRef]

21. Bacha, S.; Picault, D.; Burger, B.; Etxeberria-Otadui, I.; Martins, J. Photovoltaics in Microgrids: An Overview of Grid Integration and Energy Management Aspects. *IEEE Ind. Electron. Mag.* **2015**, *9*, 33–46. [CrossRef]

22. Guerrero, J.M.; Blaabjerg, F.; Zhelev, T.; Hemmes, K.; Monmasson, E.; Jemei, S.; Comech, M.P.; Granadino, R.; Frau, J.I. Distributed generation: Toward a new energy paradigm. *IEEE Ind. Electron. Mag.* **2010**, *1*, 52–64. [CrossRef]

23. Boehner, V.; Franz, P.; Hanson, J.; Gallart, R.; Martínez, S.; Sumper, A.; Girbau-Llistuella, F. Smart grids for rural conditions and e-mobility—Applying power routers, batteries and virtual power plants. In Proceedings of the International Council on Large Electric Systems CIGRE, Paris, France, 21–26 August 2016; pp. 1–9.

24. Alstone, P.; Gershenson, D.; Kammen, D.M. Decentralized energy systems for clean electricity access. *Nat. Clim. Chang.* **2015**, *5*, 305–314. [CrossRef]

25. Kou, L.; Sheng, W.; Wang, J.; Liang, Y.; Song, Q. Evaluation on the application mode of distributed generation. In Proceedings of the 2012 China International Conference on Electricity Distribution (CICED), Shanghai, China, 10–14 September 2012; pp. 1–5.

26. Girbau-Llistuella, F.; Sumper, A.; Díaz-González, F.; Sudrià-Andreu, A.; Gallart-Fernández, R. Local performance of the smart rural grid through the local energy management system. In Proceedings of the 7th International Conference on Modern Power Systems, Napoca, Romania, 6–9 June 2017; pp. 1–6.

27. Bassi, N. Solarizing groundwater irrigation in India: A growing debate. *Int. J. Water Resour. Dev.* **2018**, *34*, 132–145, [CrossRef]

28. Sajjad, I.A.; Manganelli, M.; Martirano, L.; Napoli, R.; Chicco, G.; Parise, G. Net-Metering Benefits for Residential Customers: The Economic Advantages of a Proposed User-Centric Model in Italy. *IEEE Ind. Appl. Mag.* **2018**, *24*, 39–49. [CrossRef]

29. Christoforidis, G.C.; Panapakidis, I.P.; Papadopoulos, T.A.; Papagiannis, G.K.; Koumparou, I.; Hadjipanayi, M.; Georghiou, G.E. A Model for the Assessment of Different Net-Metering Policies. *Energies* **2016**, *9*, 262. [CrossRef]

30. Romero-Rubio, C.; Andrés-Díaz, J.R. Spanish electrical system: Effect of energy reform in the development of distributed generation. In Proceedings of the 18th International Congress on Project Engineering, Alcañiz, Spain, 16–18 July 2014.

31. Ministry of Industry. *Royal Decree 900/2015, of October 9, Which Regulates the Administrative, Technical and Economic Conditions for the Supply of Electric Energy with Self-Consumption and Production with Self-Consumption*; Technical Report 243; State Official Newsletter BOE: Madrid, Spain, 2015. (In Spanish)

32. Arboleya, P.; Gonzalez-Moran, C.; Coto, M.; Garcia, J. Self-supply and net balance: The Spanish scenario. In Proceedings of the 2013 International Conference on New Concepts in Smart Cities: Fostering Public and Private Alliances (SmartMILE), Gijon, Spain, 11–13 December 2013; pp. 1–6.

33. Rubio-Aliaga, A.; Socorro, G.C.M.; Molina-Garcia, A.; Sánchez-Lozano, J. *Geographic Information System for Optimization and Integration of Photovoltaic Solar Energy in Agricultural Areas with Energy Deficiency and Water Scarcity*; Springer: Cham, Switzerland, 2017; pp. 181–197.

34. De Castro, M.; Martín-Vide, J.; Alonso, S. *Preliminary Evaluation of the Impacts in Spain due to Climate Change*; Ministry of Environment: Madrid, Spain, 2005. (In Spanish)

35. Chaturvedi, V.; Hejazi, M.I.; Edmonds, J.A.; Clarke, L.E.; Kyle, G.P.; Davies, E.; Wise, M.A.; Calvin, K.V. *Climate Policy Implications for Agricultural Water Demand*; Pacific Northwest National Laboratory Technical Report PNNL-22356; US Department of Energy: Richland, WA, USA, 2013.

36. El Moustaine, R.; Chahlaoui, A.; Rour, E. Relationships between the physico-chemical variables and groundwater biodiversity: A case study from meknes area, Morocco. *Int. J. Conserv. Sci.* **2014**, *5*, 203–214.

37. Capone, R.; Debs, P.; El Bilali, H.; Cardone, G.; Lamaddalena, N. Water footprint in the Mediterranean food chain: implications of food consumption patterns and food wastage. *Int. J. Nutr. Food Sci.* **2014**, *3*, 26–36. [CrossRef]

38. Closas, A.; Rap, E. Solar-based groundwater pumping for irrigation: Sustainability, policies, and limitations. *Energy Policy* **2017**, *104*, 33–37. [CrossRef]

39. Ruiz-Pulpón, A. Irrigation and sustainable management of water resources in the Guadiana basin: Territorial proposal prior to decision-making. *Geogr. Res.* **2006**, *40*, 183–200. (In Spanish)

40. Iglesias-Martínez, E. Economics and Sustainable Management of Groundwater: La Mancha Occidental Aquifer. Ph.D. Thesis, Technical School of Agricultural Engineers, Polytechnic University of Madrid (UPM), Madrid, Spain, 2001. (In Spanish)

41. Bosque-Maurel, J. Water as a scarce resource and its problems in Spain. *Geogr. Res.* **2008**, *69*, 453–493. (In Spanish)

42. Varela-Ortega, C.; Blanco-Gutiérrez, I.; Swartz, C.H.; Downing, T.E. Balancing groundwater conservation and rural livelihoods under water and climate uncertainties: An integrated hydro-economic modeling framework. *Glob. Environ. Chang.* **2011**, *21*, 604–619. Special Issue on The Politics and Policy of Carbon Capture and Storage. [CrossRef]

43. Faysse, N.; Hartani, T.; Frija, A.; Tazekrit, I.; Zairi, C.; Challouf, A. *Agricultural Use of Groundwater and Management Initiatives in the Maghreb: Challenges and Opportunities for Sustainable Aquifer Exploitation*; Technical Report; African Development Bank: Abidjan, Côte d'Ivoire, 2011.

44. Fernández-García, I.; Rodríguez-Díaz, J.; Camacho-Poyato, E.; Montesinos, P.; Berbel, J. Effects of modernization and medium term perspectives on water and energy use in irrigation districts. *Agric. Syst.* **2014**, *131*, 56–63. [CrossRef]

45. Bouwer, H. Integrated water management for the 21st century: Problems and solutions. *J. Irrig. Drain. Eng.* **2002**, *128*, 193–202. [CrossRef]

46. García-Cárdenas, R. Methodology for the Quantification and Control of Subsidence in Large Territorial Areas. Application to the Case of the Guadalentín Basin, Lorca (Murcia). Ph.D. Thesis, Department of Civil Engineering, San Antonio Catholic University of Murcia (UCAM), Murcia, Spain, 2017. (In Spanish)

47. Carpintero, O.; Naredo, J. On the evolution of energy balances in Spanish agriculture, 1950–2000. *J. Agric. Rural Hist.* **2006**, *16*, 531–554. (In Spanish)

48. Ghoneim, A. Design optimization of photovoltaic powered water pumping systems. *Energy Convers. Manag.* **2006**, *47*, 1449–1463. [CrossRef]

49. Chandel, S.; Naik, M.N.; Chandel, R. Review of solar photovoltaic water pumping system technology for irrigation and community drinking water supplies. *Renew. Sustain. Energy Rev.* **2015**, *49*, 1084–1099. [CrossRef]

50. Sumpsi, J. The crisis of modern agriculture. *Agric. Soc.* **1982**, *25*, 185–193. (In Spanish)

51. García-Álvarez, M.T.; Cabeza-García, L.; Soares, I. Assessment of energy policies to promote photovoltaic generation in the European Union. *Energy* **2018**, *151*, 864–874. [CrossRef]

52. Sauhats, A.; Zemite, L.; Petrichenko, L.; Moshkin, I.; Jasevics, A. Estimating the Economic Impacts of Net Metering Schemes for Residential PV Systems with Profiling of Power Demand, Generation, and Market Prices. *Energies* **2018**, *11*, 3222. [CrossRef]

53. Masson, G.; Briano, J.I.; Baez, M.J. *A Methodology for the Analysis of Pv Self-Consumption Policies*; Technical Report Task 1, Report IEA-PVPS T1-28; International Energy Agency: Paris, France, 2016.

54. Sobor, I. Photovoltaic pump system design for small irrigation. In Proceedings of the 2017 International Conference on Electromechanical and Power Systems (SIELMEN), Iasi, Romania, 11–13 October 2017; pp. 315–320.

55. Marchi, B.; Pasetti, M.; Zanoni, S.; Zavanella, L. The Italian reform of electricity tariffs for non household customers: The impact on distributed generation and energy storage. In Proceedings of the XXII Summer School Francesco Turco—Industrial Systems Engineering, At Palermo, Italy, 13–15 September 2017; pp. 1–7.

56. Berbel, J.; Expósito, A.; Gutiérrez-Martín, C.; Mateos, L. Effects of the Irrigation Modernization in Spain 2002–2015. *Water Resour. Manag.* **2019**, *33*, 1835–1849. [CrossRef]

57. Moreno, M.M.; Gutiérrez, J.L.; Cortina, L.M. Hydrogeological characteristics and groundwater evolution of the Western La Mancha unit: The influence of the wet period 2009–2011. *Boletín Geológico y Minero* **2012**, *123*, 91–108. (In Spanish)

58. Sanz, G.L. Irrigated agriculture in the Guadiana River high basin (Castilla-La Mancha, Spain): Environmental and socioeconomic impacts. *Agric. Water Manag.* **1999**, *40*, 171–181. [CrossRef]

59. Rubio-Aliaga, Á.; García-Cascales, M.; Sánchez-Lozano, J.; Molina-García, A. Multidimensional analysis of groundwater pumping for irrigation purposes: Economic, energy and environmental characterization for PV power plant integration. *Renew. Energy* **2019**, *138*, 174–186. [CrossRef]

60. Cui, L.; Huang, Y. Exploring the Schemes for Green Climate Fund Financing: International Lessons. *World Dev.* **2018**, *101*, 173–187. [CrossRef]

MDPI

St. Alban-Anlage 66

4052 Basel

Switzerland

Tel. +41 61 683 77 34

Fax +41 61 302 89 18

www.mdpi.com

Applied Sciences Editorial Office

E-mail: applsci@mdpi.com

www.mdpi.com/journal/applsci